Texts in Computer Science

Series Editors

David Gries, Department of Computer Science, Cornell University, Ithaca, NY, USA

Orit Hazzan ⓘ, Faculty of Education in Technology and Science, Technion—Israel Institute of Technology, Haifa, Israel

More information about this series at http://www.springer.com/series/3191

Gabriel Valiente

Algorithms on Trees and Graphs

With Python Code

Second Edition

 Springer

Gabriel Valiente
Department of Computer Science
Technical University of Catalonia
Barcelona, Spain

ISSN 1868-0941   ISSN 1868-095X (electronic)
Texts in Computer Science
ISBN 978-3-030-81884-5   ISBN 978-3-030-81885-2 (eBook)
https://doi.org/10.1007/978-3-030-81885-2

This Springer imprint is published by the registered company Springer Nature Switzerland AG
The registered company address is: Gewerbestrasse 11, 6330 Cham, Switzerland

To my child Aleksandr,
six years old,
who is eager to grow
and read it all

Preface to the Second Edition

The first edition of this book has been extensively used for graduate teaching and research all over the world in the last two decades. We have listed hundreds of citing publications in Appendix C, including books, scientific articles in journals and conference proceedings, M.Sc. and Ph.D. theses, and even United States patents.

In this new edition, we have substituted detailed pseudocode for both the literate programming description and the implementation of the algorithms using the LEDA library of efficient data structures and algorithms. Although the pseudocode is detailed enough to allow for a straightforward implementation of the algorithms in any modern programming language, we have added a proof-of-concept implementation in Python of all the algorithms in Appendix A. This is, therefore, a thoroughly revised and extended edition.

Regarding new material, we have added an adjacency map representation of trees and graphs, and both maximum cardinality and maximum weight bipartite matching as an additional application of graph traversal techniques. Further, we have revised the end-of-chapter problems and exercises and have included solutions to all the problems in Appendix B.

It has been a pleasure for the author to work out editorial matters together with Sriram Srinivas and, especially, Wayne Wheeler of Springer Nature, whose standing support and encouragement have made this new edition possible.

Last, but not least, any minor errors found so far have been corrected in this second edition, the bibliographic notes and references have been updated, and the index has been substantially enhanced. Even though the author and the publisher have taken much care in the preparation of this book, they make no representation, express or implied, with regard to the accuracy of the information contained herein and cannot accept any legal responsibility or liability for incidental or consequential damages arising out of the use of the information, algorithms, or program code contained in this book.

Gabriel Valiente

Barcelona, Spain
June 2021

Preface to the First Edition

Graph algorithms, a long-established subject in mathematics and computer science curricula, are also of much interest to disciplines such as computational molecular biology and computational chemistry. This book goes beyond the *classical* graph problems of shortest paths, spanning trees, flows in networks, and matchings in bipartite graphs, and addresses further algorithmic problems of practical application on trees and graphs. Much of the material presented on the book is only available in the specialized research literature.

The book is structured around the fundamental problem of isomorphism. Tree isomorphism is covered in much detail, together with the related problems of subtree isomorphism, maximum common subtree isomorphism, and tree comparison. Graph isomorphism is also covered in much detail, together with the related problems of subgraph isomorphism, maximal common subgraph isomorphism, and graph edit distance. A building block for solving some of these isomorphism problems are algorithms for finding maximal and maximum cliques.

Most intractable graph problems of practical application are not even approximable to within a constant bound, and several of the isomorphism problems addressed in this book are no exception. The book can thus be seen as a companion to recent texts on approximation algorithms [1, 16], but also as a complement to previous texts on combinatorial and graph algorithms [2–15, 17].

The book is conceived on the ground of first, introducing simple algorithms for these problems in order to develop some intuition before moving on to more complicated algorithms from the research literature and second, stimulating graduate research on tree and graph algorithms by providing together with the underlying theory, a solid basis for experimentation and further development.

Algorithms are presented on an intuitive basis, followed by a detailed exposition in a literate programming style. Correctness proofs are also given, together with a worst-case analysis of the algorithms. Further, full C++ implementation of all the algorithms using the LEDA library of efficient data structures and algorithms is given along the book. These implementations include result checking of implementation correctness using correctness certificates.

The choice of LEDA, which is becoming a de-facto standard for graduate courses on graph algorithms throughout the world is not casual, because it allows the student, lecturer, researcher, and practitioner to complement algorithmic graph theory with actual implementation and experimentation, building upon a thorough library of efficient implementations of modern data structures and fundamental algorithms.

An interactive demonstration including animations of all the algorithms using LEDA is given in an appendix. The interactive demonstration also includes visual checkers of implementation correctness.

The book is divided into four parts. Part I has an introductory nature and consists of two chapters. Chapter 1 includes a review of basic graph-theoretical notions and results used along the book, a brief primer of literate programming, and an exposition of the implementation correctness approach by result checking using correctness certificates. Chapter 2 is devoted exclusively to the fundamental algorithmic techniques used in the book: backtracking, branch-and-bound, divide-and-conquer, and dynamic programming. These techniques are illustrated by means of a running example: algorithms for the tree edit distance problem.

Part II also consists of two chapters. Chapter 3 addresses the most common methods for traversing general, rooted trees: depth-first prefix leftmost (preorder), depth-first prefix rightmost, depth-first postfix leftmost (postorder), depth-first postfix rightmost, breadth-first leftmost (top-down), breadth-first rightmost, and bottom-up traversal. Tree drawing is also discussed as an application of tree traversal methods. Chapter 4 addresses several isomorphism problems on ordered and unordered trees: tree isomorphism, subtree isomorphism, and maximum common subtree isomorphism. Computational molecular biology is also discussed as an application of the different isomorphism problems on trees.

Part III consists of three chapters. Chapter 5 addresses the most common methods for traversing graphs: depth-first and breadth-first traversal, which respectively generalize depth-first prefix leftmost (preorder) and breadth-first leftmost (top-down) tree traversal. Leftmost depth-first traversal of an undirected graph, a particular case of depth-first traversal, is also discussed. Isomorphism of ordered graphs is also discussed as an application of graph traversal methods. Chapter 6 addresses the related problems of finding cliques, independent sets, and vertex covers in trees and graphs. Multiple alignment of protein sequences in computational molecular biology is also discussed as an application of clique algorithms. Chapter 7 addresses several isomorphism problems on graphs: graph isomorphism, graph automorphism, subgraph isomorphism, and maximal common subgraph isomorphism. Chemical structure search is also discussed as an application of the different graph isomorphism problems.

Part IV consists of two appendices, followed by bibliographic references and an index. Appendix A gives an overview of LEDA, including a simple C++ representation of trees as LEDA graphs, and a C++ implementation of radix sort using LEDA. The interactive demonstration of graph algorithms presented along the book is put together in Appendix B. Finally, Appendix C contains a complete index to all program modules described in the book.

This book is suitable for use in upper undergraduate and graduate level courses on algorithmic graph theory. This book can also be used as a supplementary text in basic undergraduate and graduate level courses on algorithms and data structures, and in computational molecular biology and computational chemistry courses as well. Some basic knowledge of discrete mathematics, data structures, algorithms, and programming at the undergraduate level is assumed.

This book is based on lectures taught at the Technical University of Catalonia, Barcelona between 1996 and 2002, and the University of Latvia, Riga between 2000 and 2002. Numerous colleagues at the Technical University of Catalonia have influenced the approach to data structures and algorithms on trees and graphs expressed in this book. In particular, the author would like to thank José L. Balcázar, Rafel Casas, Jordi Cortadella, Josep Daz, Conrado Martnez, Xavier Messeguer, Roberto Nieuwenhuis, Fernando Orejas, Jordi Petit, Salvador Roura, and Maria Serna, to name just a few. It has been a pleasure to share teaching and research experiences with them over the last several years.

The author would also like to thank Ricardo Baeza-Yates, Francesc Rosselló, and Steven Skiena, for their standing support and encouragement, and Hans-Jörg Kreowski, for supporting basic and applied research on graph algorithms within the field of graph transformation. It has been a pleasure for the author to work out editorial matters together with Alfred Hofmann, Ingeborg Mayer, and Peter Straßer of Springer-Verlag. Special thanks are debt to the Technical University of Catalonia for funding the sabbatical year during which this book was written, and to the Institute of Mathematics and Computer Science, University of Latvia, in particular to Jānis Bārzdiņš and Rūsiņš Freivalds, for hosting the sabbatical visit.

Gabriel Valiente

Barcelona, Spain
July 2002

References

1. G. Ausiello, P. Crescenzi, G. Gambosi, V. Kahn, A. MarchettiSpaccamela, and M. Protasi. *Complexity and Approximation: Combinatorial Optimization Problems and their Approximability Properties.* Springer-Verlag, Berlin Heidelberg, 1999.
2. G. Chartrand and O. R. Oellermann. *Applied and Algorithmic Graph Theory.* McGraw-Hill, New York, 1993.
3. N. Christofides. *Graph Theory: An Algorithmic Approach.* Academic Press, New York, 1975.
4. S. Even. *GraphAlgorithms.* Computer Science Press, Rockville MD,1979.
5. A. Gibbons. *Algorithmic Graph Theory.* Cambridge University Press, Cambridge, 1985.
6. M. C. Golumbic. *Algorithmic Graph Theory and Perfect Graphs.* Academic Press, New York, 1980.
7. D. L. Kreher and D. R. Stinson. *Combinatorial Algorithms: Generation, Enumeration, and Search.* CRC Press, Boca Raton FL, 1999.
8. J. van Leeuwen. Graph algorithms. In *Handbook of Theoretical Computer Science*, volume A, chapter 10, pages 525-631. Elsevier, Amsterdam, 1990.
9. J. A. McHugh. *Algorithmic Graph Theory.* Prentice Hall, Englewood Cliffs NJ, 1990.
10. K. Mehlhorn. *Graph Algorithms and NP-Completeness, volume 2 of Data Structures and Algorithms.* Springer-Verlag, Berlin Heidelberg, 1984.
11. K. Mehlhorn and S. Naher. *The LEDA Platform of Combinatorial and Geometric Computing.* Cambridge University Press, Cambridge, 1999.

12. C. H. Papadimitriou and K. Steiglitz. *CombinatorialOptimization: Algorithms and Complexity*. Dover, Mineola, New York, 1998.
13. E. M. Reingold, 1. Nievergelt, and N. J. Deo. *Combinatorial Algorithms: Theory and Practice*. Prentice Hall, Englewood Cliffs NJ, 1977.
14. S. Skiena. *The Algorithm Design Manual*. Springer-Verlag, Berlin Heidelberg, 1998.
15. R. E. Tarjan. *Space-efficient Implementations of Graph Search Methods*. ACM Transactions on Mathematical Software. 1983 9(3), 326–339
16. V. V. Vazirani. *Approximation Algorithms*. Springer-Verlag, Berlin Heidelberg, 2001.
17. H. S. Wilf. *Combinatorial Algorithms: An Update*. SIAM, Philadelphia PA, 1989.

Contents

Part I
Introduction

1.1 Trees and Graphs

The notion of graph which is most useful in computer science is that of a directed graph or just a graph. A graph is a combinatorial structure consisting of a finite nonempty set of objects, called *vertices*, together with a finite (possibly empty) set of ordered pairs of vertices, called *directed edges* or *arcs*.

Definition 1.1 A **graph** $G = (V, E)$ consists of a finite nonempty set V of vertices and a finite set $E \subseteq V \times V$ of edges. The **order** of a graph $G = (V, E)$, denoted by n, is the number of vertices, $n = |V|$ and the **size**, denoted by m, is the number of edges, $m = |E|$. An edge $e = (v, w)$ is said to be **incident** with vertices v and w, where v is the **source** and w the **target** of edge e, and vertices v and w are said to be **adjacent**. Edges (u, v) and (v, w) are said to be **adjacent**, as are edges (u, v) and (w, v), and also edges (v, u) and (v, w).

Graphs are often drawn as a set of points in the plane and a set of arrows, each of which joins two (not necessarily different) points. In a drawing of a graph $G = (V, E)$, each vertex $v \in V$ is drawn as a point or a small circle and each edge $(v, w) \in E$ is drawn as an arrow from the point or circle of vertex v to the point or circle corresponding to vertex w.

Example 1.1 The graph $G = (V, E)$ of Fig. 1.1 has order 7 and size 12. The vertex set is $V = \{v_1, \ldots, v_7\}$ and the edge set is $E = \{e_1, \ldots, e_{12}\}$, where $e_1 = (v_1, v_2)$, $e_2 = (v_1, v_4)$, $e_3 = (v_2, v_5)$, $e_4 = (v_3, v_1)$, $e_5 = (v_4, v_2)$, $e_6 = (v_4, v_3)$, $e_7 = (v_4, v_6)$, $e_8 = (v_4, v_7)$, $e_9 = (v_5, v_4)$, $e_{10} = (v_5, v_7)$, $e_{11} = (v_6, v_3)$, and $e_{12} = (v_7, v_6)$.

© The Author(s), under exclusive license to Springer Nature Switzerland AG 2021
G. Valiente, *Algorithms on Trees and Graphs*, Texts in Computer Science,
https://doi.org/10.1007/978-3-030-81885-2_1

Fig. 1.1 A graph of order 7
and size 12

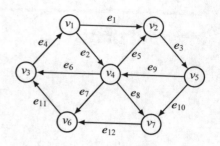

A vertex has two degrees in a graph, one given by the number of edges coming into the vertex and the other given by the number of edges in the graph going out of the vertex.

Definition 1.2 The **indegree** of a vertex v in a graph $G = (V, E)$ is the number of edges in G whose target is v, that is, $\text{indeg}(v) = |\{(u, v) \mid (u, v) \in E\}|$. The **outdegree** of a vertex v in a graph $G = (V, E)$ is the number of edges in G whose source is v, that is, $\text{outdeg}(v) = |\{(v, w) \mid (v, w) \in E\}|$. The **degree** of a vertex v in a graph $G = (V, E)$ is the sum of the indegree and the outdegree of the vertex, that is, $\text{deg}(v) = \text{indeg}(v) + \text{outdeg}(v)$.

Example 1.2 The degree of the vertices in the graph of Fig. 1.1 is the following:

Vertex	v_1	v_2	v_3	v_4	v_5	v_6	v_7	sum
Indegree	1	2	2	2	1	2	2	12
Outdegree	2	1	1	4	2	1	1	12

As can be seen in Example 1.2, the sum of the indegrees and the sum of the outdegrees of the vertices of the graph in Fig. 1.1 are both equal to 12, the number of edges of the graph. As a matter of fact, there is a basic relationship between the size of a graph and the degrees of its vertices, which will prove to be very useful in analyzing the computational complexity of algorithms on graphs.

Theorem 1.1 *Let $G = (V, E)$ be a graph with n vertices and m edges, and let $V = \{v_1, \ldots, v_n\}$. Then,*

$$\sum_{i=1}^{n} \text{indeg}(v_i) = \sum_{i=1}^{n} \text{outdeg}(v_i) = m .$$

Proof Let $G = (V, E)$ be a graph. Every edge $(v_i, v_j) \in E$ contributes one to $\text{indeg}(v_j)$ and one to $\text{outdeg}(v_i)$. $\square$

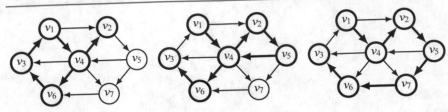

Fig. 1.2 A walk $[v_1, v_4, v_6, v_3, v_1, v_4, v_2]$, a trail $[v_1, v_4, v_2, v_5, v_4, v_6, v_3]$, and a path $[v_1, v_4, v_2, v_5, v_7, v_6, v_3]$ in the graph of Fig. 1.1

Walks, trails, and paths in a graph are alternating sequences of vertices and edges in the graph such that each edge in the sequence is preceded by its source vertex and followed by its target vertex. Trails are walks having no repeated edges, and paths are trails having no repeated vertices.

Definition 1.3 A **walk** from vertex v_i to vertex v_j in a graph is an alternating sequence

$$[v_i, e_{i+1}, v_{i+1}, e_{i+2}, \ldots, v_{j-1}, e_j, v_j]$$

of vertices and edges in the graph, such that $e_k = (v_{k-1}, v_k)$ for $k = i+1, \ldots, j$. A **trail** is a walk with no repeated edges, and a **path** is a trail with no repeated vertices (except, possibly, the initial and final vertices). The length of a walk, trail, or path is the number of edges in the sequence.

Since an edge in a graph is uniquely determined by its source and target vertices, a walk, trail, or path can be abbreviated by just enumerating either the vertices $[v_i, v_{i+1}, \ldots, v_{j-1}, v_j]$ or the edges $[e_{i+1}, e_{i+2}, \ldots, e_j]$ in the alternating sequence $[v_i, e_{i+1}, v_{i+1}, e_{i+2}, \ldots, v_{j-1}, e_j, v_j]$ of vertices and edges.

Example 1.3 The vertex sequence $[v_1, v_4, v_6, v_3, v_1, v_4, v_2]$ in Fig. 1.2 is a walk of length 6 which is not a trail, $[v_1, v_4, v_2, v_5, v_4, v_6, v_3]$ is a trail of length 6 which is not a path, and $[v_1, v_4, v_2, v_5, v_7, v_6, v_3]$ is a path of length 6.

Walks are closed if their initial and final vertices coincide.

Definition 1.4 A walk, trail, or path $[v_i, e_{i+1}, v_{i+1}, e_{i+2}, \ldots, v_{j-1}, e_j, v_j]$ is said to be **closed** if $v_i = v_j$. A **cycle** is a closed path of length at least one.

Example 1.4 The vertex sequence $[v_1, v_4, v_6, v_3, v_1]$ in Fig. 1.3 is a cycle of length 4.

The combinatorial structure of a graph encompasses two notions of the substructure. A subgraph of a graph is just a graph whose vertex and edge sets are contained in the vertex and edge sets of the given graph, respectively. The subgraph of a graph

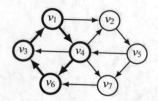

Fig. 1.3 A cycle $[v_1, v_4, v_6, v_3, v_1]$ in the graph of Fig. 1.1

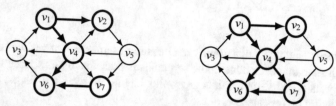

Fig. 1.4 A subgraph and an induced subgraph of the graph of Fig. 1.1

induced by a subset of its vertices has as edges the set of edges in the given graph whose source and target belong to the subset of vertices.

Definition 1.5 Let $G = (V, E)$ be a graph, and let $W \subseteq V$. A graph (W, S) is a **subgraph** of G if $S \subseteq E$. The subgraph of G **induced** by W is the graph $(W, E \cap W \times W)$.

Example 1.5 The subgraph with vertex set $\{v_1, v_2, v_4, v_6, v_7\}$ and edge set $\{(v_1, v_2), (v_1, v_4), (v_4, v_6), (v_7, v_6)\}$ shown in Fig. 1.4 is not an induced subgraph. The subgraph induced by $\{v_1, v_2, v_4, v_6, v_7\}$ has edge set $\{(v_1, v_2), (v_1, v_4), (v_4, v_2), (v_4, v_6), (v_4, v_7), (v_7, v_6)\}$.

Undirected Graphs

The notion of graph which is most often found in mathematics is that of an undirected graph. Unlike the directed edges or edges of a graph, edges of an undirected graph have no direction association with them and therefore, no distinction is made between the source and target vertices of an edge. In a mathematical sense, an undirected graph consists of a set of vertices and a finite set of undirected edges, where each edge has a set of one or two vertices associated with it. In the computer science view of undirected graphs, though, an undirected graph is the particular case of a directed graph in which for every edge (v, w) of the graph, the reversed edge (w, v) also belongs to the graph. Undirected graphs are also called *bidirected*.

Definition 1.6 A graph $G = (V, E)$ is **undirected** if $(v, w) \in E$ implies $(w, v) \in E$, for all $v, w \in V$.

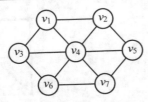
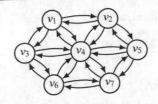

Fig. 1.5 The undirected graph underlying the graph of Fig. 1.1. The standard presentation is given in the drawing to the left, while the understanding of undirected edges as pairs of counter-parallel edges is emphasized in the drawing to the right

Undirected graphs are often drawn as a set of points in the plane and a set of line segments, each of which joins two (not necessarily different) points. In a drawing of an undirected graph $G = (V, E)$, each vertex $v \in V$ is drawn as a point or a small circle and each pair of counter-parallel edges (v, w), $(w, v) \in E$ is drawn as a line segment between the points or circles corresponding to vertices v and w.

Example 1.6 The undirected graph underlying the graph of Fig. 1.1 is shown in Fig. 1.5.

The terminology of directed graphs carries over to undirected graphs, although a few differences deserve mention. First of all, since an edge in an undirected graph is understood as a pair of counter-parallel edges in the corresponding bidirected graph, the number of edges coming into a given vertex and the number of edges going out of the vertex coincide and therefore, no distinction is made between the indegree and the outdegree of a vertex. That is, the *degree* of a vertex in an undirected graph is equal to the indegree and the outdegree of the vertex in the corresponding bidirected graph. For the same reason, the size of an undirected graph is half the size of the corresponding bidirected graph.

Definition 1.7 The **degree** of a vertex v in an undirected graph $G = (V, E)$ is the number of edges in G that are incident with v. The **degree sequence** of G is the sequence of n non-negative integers obtained by arranging the vertex degrees in non-decreasing order.

Example 1.7 There are three vertices of degree 2 in the graph of Fig. 1.6, two vertices of degree 3, two vertices of degree 4, and two vertices of degree 5. The degree sequence of the graph is $[2, 2, 2, 3, 3, 4, 4, 5, 5]$.

A walk, trail, or path in an undirected graph is a walk, trail, or path, respectively, in the corresponding bidirected graph. Another important graph-theoretical notion is that of connected and disconnected graphs. An undirected graph is *connected* if every pair of its vertices is joined by some walk, and an undirected graph that is not connected is said to be *disconnected*. On the other hand, a graph is connected if for all vertices v and w in the graph, there is a walk from v to w, and it is *strongly*

Fig. 1.6 Undirected graph
with degree sequence
$[2, 2, 2, 3, 3, 4, 4, 5, 5]$

Fig. 1.7 A graph with five
strong components whose
underlying undirected graph
is connected

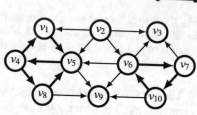

connected if there are walks from v to w and from w back to v, for all vertices v and w in the graph.

Definition 1.8 An undirected graph $G = (V, E)$ is **connected** if for every pair of vertices $v, w \in E$, there is a walk between v and w. A graph is **connected** if the underlying undirected graph is connected. Graph G is **strongly connected** if for every pair of vertices $v, w \in E$, there are walks from v to w and from w to v.

Connectivity is an equivalence relation on the vertex set of a graph, which induces a partition of the graph into subgraphs induced by connected vertices, called *connected components* or just *components*.

Definition 1.9 A **component** of an undirected graph G is a connected subgraph of G which is not properly contained in any other connected subgraph of G. A **strong component** of a graph G is a strongly connected subgraph of G which is not properly contained in any other strongly connected subgraph of G.

Example 1.8 The graph of Fig. 1.7 has five strong components, induced, respectively, by the vertex sets $\{v_1, v_4, v_5, v_8\}$, $\{v_2\}$, $\{v_3\}$, $\{v_6, v_7, v_{10}\}$, and $\{v_9\}$. The underlying undirected graph is, however, connected.

Some common families of undirected graphs are the trees, the complete graphs, the path graphs, the cycle graphs, the wheel graphs, the bipartite graphs, and the regular graphs. A *tree* is a connected graph having no cycles.

Definition 1.10 An undirected graph $G = (V, E)$ is said to be an **undirected tree** if it is connected and has no cycles.

Example 1.9 The undirected trees with one, two, three, four, five, and six vertices are shown in Fig. 1.8.

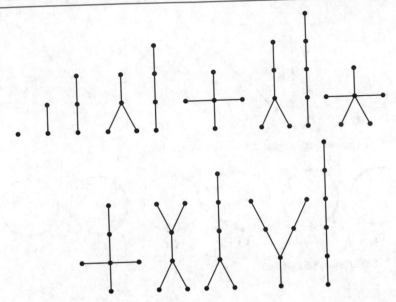

Fig. 1.8 The first fourteen undirected trees

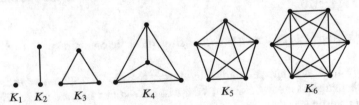

Fig. 1.9 The first six complete graphs

In a *complete* graph, every pair of distinct vertices is joined by an edge.

Definition 1.11 An undirected graph $G = (V, E)$ is said to be a **complete graph** if for all $v, w \in V$ with $v \neq w$, $(v, w) \in E$. The complete graph on n vertices is denoted by K_n.

Example 1.10 The complete graphs on one, two, three, four, five, and six vertices are shown in Fig. 1.9.

A *path* graph can be drawn such that all vertices lie on a straight line.

Definition 1.12 A connected undirected graph $G = (V, E)$ with $n = m + 1$ is said to be a **path graph** if it can be drawn such that all vertices lie on a straight line. The path graph on n vertices is denoted by P_n.

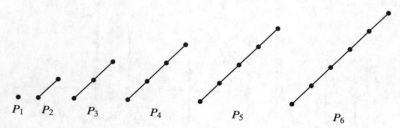

Fig. 1.10 The first six path graphs

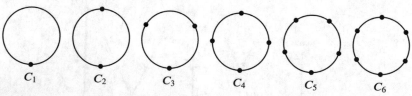

Fig. 1.11 The first six cycle graphs

Example 1.11 The path graphs on one, two, three, four, five, and six vertices are shown in Fig. 1.10.

A *cycle* graph can be drawn such that all vertices lie on a circle.

Definition 1.13 A connected undirected graph $G = (V, E)$ with $n = m$ is said to be a **cycle graph** if it can be drawn such that all vertices lie on a circle. The cycle graph on n vertices is denoted by C_n.

Example 1.12 The cycle graphs on one, two, three, four, five, and six vertices are shown in Fig. 1.11.

A *wheel* graph has a distinguished (inner) vertex that is connected to all other (outer) vertices, and it can be drawn such that all outer vertices lie on a circle centered at the inner vertex.

Definition 1.14 A connected undirected graph $G = (V, E)$ is said to be a **wheel graph** if it can be drawn such that all but a single vertex lie on a circle centered at the single vertex, which is connected to all other vertices. The wheel graph with n outer vertices is denoted by W_{n+1}.

Example 1.13 The wheel graphs with one, two, three, four, five, and six outer vertices are shown in Fig. 1.12.

Another common family of undirected graphs is the *bipartite* graphs. The vertex set of a bipartite graph can be partitioned into two subsets in such a way that every

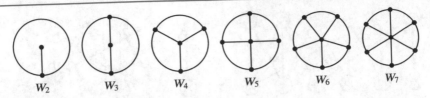

Fig. 1.12 The first six wheel graphs

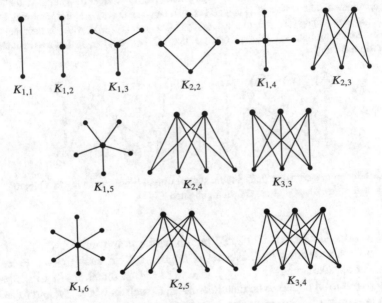

Fig. 1.13 The first twelve complete bipartite graphs

edge of the graph joins a vertex of one subset with a vertex of the other subset. In a *complete bipartite* graph, every vertex of one subset is joined by some edge with every vertex of the other subset.

Definition 1.15 An undirected graph $G = (V, E)$ is said to be a **bipartite graph** if V can be partitioned into two disjoint subsets X and Y such that for all $(x, y) \in E$, either $x \in X$ and $y \in Y$, or $x \in Y$ and $y \in X$, and it is said to be a **complete bipartite graph** if, furthermore, $(x, y), (y, x) \in E$ for all $x \in X$ and $y \in Y$. The complete bipartite graph on $p + q$ vertices, where subset X has p vertices and subset Y has q vertices, is denoted by $K_{p,q}$.

Example 1.14 The complete bipartite graphs with two, three, four, five, six, and seven vertices are shown in Fig. 1.13. The bipartite sets are distinguished by drawing vertices either in red or in black. Then, every edge joins a red vertex with a black vertex.

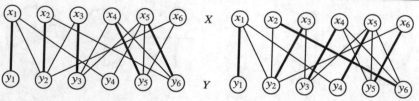

Fig. 1.14 Maximal (left) and maximum (right) matching in a bipartite graph. Matchings are shown highlighted. The matching $\{(x_1, y_1), (x_2, y_2), (x_3, y_3), (x_4, y_5)\}$ is not maximal, because it is included in matching $\{(x_1, y_1), (x_2, y_2), (x_3, y_3), (x_4, y_5), (x_5, y_6)\}$, which is maximal but not maximum. Matching $\{(x_1, y_1), (x_2, y_6), (x_3, y_3), (x_4, y_5), (x_5, y_4), (x_6, y_2)\}$ is maximal and also a maximum matching

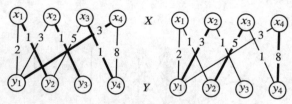

Fig. 1.15 Maximum cardinality (left) and maximum weight (right) matching in a bipartite graph with weighted edges. Matchings are shown highlighted

Definition 1.16 Let $G = (X \cup Y, E)$ be a bipartite graph with $E \subseteq X \times Y$ and $X \cap Y = \emptyset$. A **matching** in G is a set of edges $M \subseteq E$ such that $x_1 \neq x_2$ and $y_1 \neq y_2$ for all distinct edges $(x_1, y_1), (x_2, y_2) \in M$. A matching M in a bipartite graph G is **maximal** if there is no matching M' in G such that $M \subseteq M'$ and $M \neq M'$, and it is **maximum** if there is no matching in G with more edges than M. Further, a matching M in a bipartite graph G is **perfect** if $|M| = \min(|X|, |Y|)$.

Example 1.15 There are 577 non-trivial matchings in the bipartite graph of Fig. 1.14, 69 of which are maximal. The smallest one has four edges, and there are 22 of them, such as $\{(x_1, y_1), (x_2, y_2), (x_4, y_3), (x_5, y_5)\}$. There are also 44 maximal matchings with five edges, such as $\{(x_1, y_1), (x_2, y_2), (x_3, y_3), (x_4, y_5), (x_5, y_6)\}$, and three maximal matchings with six edges, which are also maximum matchings: $\{(x_1, y_1), (x_2, y_2), (x_3, y_3), (x_4, y_6), (x_5, y_4), (x_6, y_5)\}, \{(x_1, y_1), (x_2, y_6), (x_3, y_2), (x_4, y_3), (x_5, y_4), (x_6, y_5)\}$, which is shown to the right of the figure, and $\{(x_1, y_1), (x_2, y_6), (x_3, y_3), (x_4, y_5), (x_5, y_4), (x_6, y_2)\}$.

A maximum matching in a bipartite graph is also called a *maximum cardinality bipartite matching*, in order to distinguish it from a *maximum weight bipartite matching* in a bipartite graph with weighted edges, where the weight of a matching is the total weight of the edges in the matching.

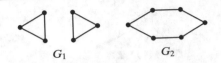

Fig. 1.16 Regular graphs G_1 and G_2 of same order and size

Example 1.16 In the bipartite graph of Fig. 1.15, $\{(x_1, y_2), (x_2, y_3), (x_3, y_4),$ $(x_4, y_1)\}$ is a maximum cardinality matching, with total edge weight 6, and $\{(x_2, y_1),$ $(x_3, y_2), (x_4, y_4)\}$ is a maximum weight matching of weight 16.

A natural generalization of bipartite graphs is the *k-partite* graphs. The vertex set of a *k*-partite graph can be partitioned into *k* subsets in such a way that every edge of the graph joins a vertex of one subset with a vertex of another subset.

Definition 1.17 An undirected graph $G = (V, E)$ is said to be a *k*-**partite graph** if V can be partitioned into $k \geqslant 2$ subsets $V_1, V_2, \ldots, V_k$ such that for all $(v, w) \in E$ with $v \in V_i$ and $w \in V_j$, it holds that $i \neq j$.

Complete graphs and cycle graphs are trivial examples of *regular* graphs. In a regular graph, all vertices have the same degree.

Definition 1.18 An undirected graph $G = (V, E)$ is said to be a **regular graph** if $\deg(v) = \deg(w)$ for all $v, w \in V$.

Example 1.17 The regular graphs G_1 and G_2 shown in Fig. 1.16 share the same degree sequence, $[2, 2, 2, 2, 2, 2]$.

Labeled Graphs

Most applications of graphs in computer science involve *labeled* graphs, where vertices and edges have additional attributes such as color or weight. For a labeled graph $G = (V, E)$, the label of a vertex $v \in V$ is denoted by *label*[v], and the label of an edge $(v, w) = e \in E$ is denoted by *label*[e] or just by *label*[v, w].

Example 1.18 An undirected graph of the geographical adjacencies between some of the first European states to adopt the European currency is shown in Fig. 1.17. Vertices represent European states and are labeled with a standard abbreviation for the name of the state. There is an edge between two vertices if the corresponding states share a border. Edges are labeled with the distance in kilometers between the capital cities of the states they join.

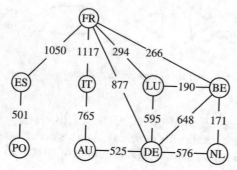

Fig. 1.17 A labeled undirected graph of geographical adjacencies between some European states

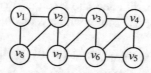

Fig. 1.18 An ordered undirected graph. The relative order of the vertices adjacent with a given vertex reflects the counter-clockwise ordering of the outgoing edges of the vertex in the drawing

Ordered Graphs

An ordered graph is a graph in which the relative order of the adjacent vertices is fixed for each vertex. Ordered graphs arise when a graph is drawn or *embedded* on a certain surface, for instance, in the Euclidean plane.

Example 1.19 The relative order of the vertices adjacent with each of the vertices of the ordered undirected graph shown in Fig. 1.18 reflects the counter-clockwise ordering of the outgoing edges of the vertex in the drawing. For instance, the the ordered sequence of vertices $[v_1, v_8, v_7, v_3]$ adjacent with vertex v_2 reflects the counter-clockwise ordering $\{v_2, v_1\}, \{v_2, v_8\}, \{v_2, v_7\}, \{v_2, v_3\}$ of the edges incident with vertex v_2 in the drawing of the graph.

Trees

The families of undirected graphs introduced above have a directed graph counterpart. In particular, while the notion of tree most often found in discrete mathematics is that of an undirected tree, the notion of the tree which is most useful in computer science is that of the rooted directed tree or just tree. A tree is the particular case of a graph in which there is a distinguished vertex, called the root of the tree, such that there is a unique walk from the root to any vertex of the tree. The vertices of a tree are called *nodes*.

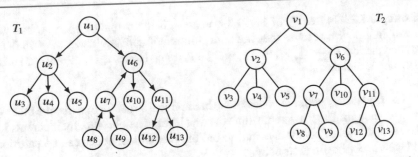

Fig. 1.19 Graph T_1 is not a rooted tree, although it is connected and has no cycles. Graph T_2 is a tree rooted at node v_1

Definition 1.19 A connected graph $G = (V, E)$ is said to be a **tree** if the underlying undirected graph has no cycles and there is a distinguished node $r \in V$, called the **root** of the tree and denoted by $root[T]$ such that for all nodes $v \in V$, there is a path in G from the root r to node v.

Example 1.20 Two connected graphs are shown in Fig. 1.19. Graph T_1 is not a tree rooted at node u_1, although the underlying undirected graph has no cycles. As a matter of fact, there is no path in T_1 from node u_1 to any of nodes u_7, u_8, and u_9. Graph T_2, on the other hand, is indeed a tree rooted at node v_1.

The existence of a unique path in a tree from the root to any other node imposes a hierarchical structure in the tree. The root lies at the top, and nodes can be partitioned in hierarchical levels according to their distance from the root of the tree.

Definition 1.20 Let $T = (V, E)$ be a tree. The **depth** of node $v \in V$, denoted by $depth[v]$, is the length of the unique path from the root node $root[T]$ to node v, for all nodes $v \in V$. The depth of T is the maximum among the depth of all nodes $v \in V$.

The hierarchical structure embodied in a tree leads to further distinguishing among the nodes of the tree: *parent* nodes are joined by edges to their *children* nodes, nodes sharing the same parent are called *sibling*, and nodes having no children are called *leaves*.

Definition 1.21 Let $T = (V, E)$ be a tree. Node $v \in V$ is said to be the **parent** of node $w \in V$, denoted by $parent[w]$, if $(v, w) \in E$ and, in such a case, node w is said to be a **child** of node v. The **children** of node $v \in V$ are the set of nodes $W \subseteq V$ such that $(v, w) \in E$ for all $w \in W$. A node having no children is said to be a **leaf** node. Non-root nodes $v, w \in V$ are said to be **sibling** nodes if $parent[v] = parent[w]$. The number of children of node $v \in V$ is denoted by $children[v]$. The tree is said to be a **binary tree** if $|children[v]| \leqslant 2$ for every node $v \in V$, and a **complete binary tree** if, furthermore, it has $\lfloor n/2 \rfloor$ non-leaf nodes, where $n = |V|$, and at most one node $v \in V$ with $|children[v]| = 1$, whose only child is a leaf node.

Example 1.21 In tree T_2 of Fig. 1.19, node v_2 is the parent of node v_5, and v_8, v_9 are sibling nodes, because they are both children of node v_7. The root of T_2 is node v_1, and the leaves are v_3, v_4, v_5, v_8, v_9, v_{10}, v_{12}, v_{13}. Nodes v_2, v_6, v_7, v_{11} are neither root nor leaf nodes.

The reader may prefer to consider the hierarchical structure of a tree the other way around, though. The leaves constitute the first level of height zero, their parents form the second level of height one, and so on. In other words, nodes can be partitioned in levels according to their height.

Definition 1.22 Let $T = (V, E)$ be a tree. The **height** of node $v \in V$, denoted by $height[v]$, is the length of a longest path from node v to any node in the subtree of T rooted at node v, for all nodes $v \in V$. The height of T is the maximum among the height of all nodes $v \in V$.

It follows from Definition 1.20 that in a tree $T = (V, E)$, $depth[v] \geqslant 0$ for all nodes $v \in V$, and that $depth[v] = 0$ if and only if $v = root[T]$. Further, it follows from Definition 1.22 that $height[v] \geqslant 0$ for all nodes $v \in V$, and that $height[v] = 0$ if and only if v is a leaf node.

Example 1.22 In the tree shown twice in Fig. 1.20, node v_1 is the root and has depth zero and height three, which is also the height of the tree. Node v_2 has depth and height one; nodes v_3, v_4, v_5, v_{10} have depth two and height zero; node v_6 has depth one and height two; nodes v_7, v_{11} have depth two and height one; and nodes v_8, v_9, v_{12}, v_{13} have depth three, which is also the depth of the tree, and height zero, which is the height of all the leaves in the tree.

Trees are, indeed, the least connected graphs. The number of edges in a tree is one less than the number of nodes in the tree.

Theorem 1.2 Let $T = (V, E)$ be a tree. Then $m = n - 1$.

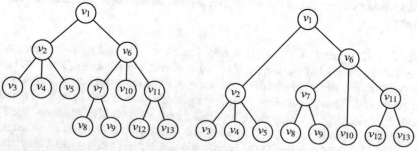

Fig. 1.20 The tree of Fig. 1.19 is shown twice, emphasizing the depth (left) and the height (right) of all nodes in the tree

Proof By induction on the number of nodes n in the tree. For $n = 1$, a tree with one node has no edges, that is, $m = 0$. Assume, then, that every tree with $n \geqslant 1$ nodes has $n - 1$ edges.

Let $T = (V, E)$ be a tree with $n + 1$ nodes, and let $v \in V$ be any leaf node of T. The subgraph of T induced by $V \setminus \{v\}$ is a tree with n nodes and has, by induction hypothesis, $n - 1$ edges. Now, since there is only one edge $(w, v) \in E$ (namely, with w the parent in T of node v), it follows that the subgraph of T induced by $V \setminus \{v\}$ has one edge less than T. Therefore, T has $n + 1$ nodes and n edges. □

As with graphs, there is also a basic relationship between the number of nodes in a tree and the number of children of the nodes of the tree, which will prove to be very useful in analyzing the computational complexity of algorithms on trees.

Lemma 1.1 *Let $T = (V, E)$ be a tree on n nodes, and let $V = \{v_1, \ldots, v_n\}$. Then,*

$$\sum_{i=1}^{n} children[v_i] = n - 1 .$$

Proof Let $T = (V, E)$ be a tree. Since every non-root node $w \in V$ is the target of a different edge $(v, w) \in E$, it holds by Theorem 1.1 that $\sum_{i=1}^{n} children[v_i] = \sum_{i=1}^{n} outdeg(v_i) = m$ and then, by Theorem 1.2, $\sum_{i=1}^{n} children[v_i] = n - 1$. □

Now, given that trees are a particular case of graphs, it is natural to consider trees as subgraphs of a given graph and, as a matter of fact, those trees that span all the vertices of a graph arise in several graph algorithms. A *spanning tree* of a graph is a subgraph that contains all the vertices of the graph and is a tree.

Definition 1.23 Let $G = (V, E)$ be a graph. A subgraph (W, S) of G is a **spanning tree** of graph G if $W = V$ and (W, S) is a tree.

Example 1.23 The graph of Fig. 1.21 has $n = 4$ vertices and $m = 6$ edges, and thus $\binom{6}{3} = 20$ subgraphs with 4 vertices and 3 edges. However, only 6 of these subgraphs are trees. These spanning trees of the graph are shown highlighted.

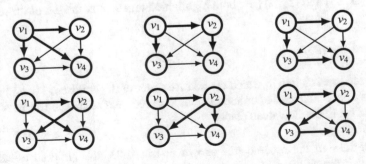

Fig. 1.21 Spanning trees of a graph

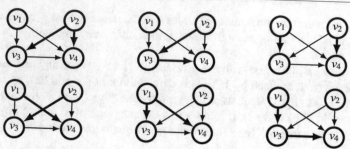

Fig. 1.22 Spanning forests of a graph

Fig. 1.23 An ordered tree.
The relative order of the
children of a given node
reflects the
counter-clockwise ordering
of the outgoing edges of the
node in the drawing

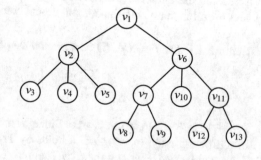

Not every graph has a spanning tree, though, but every graph has a *spanning forest*, that is, an ordered set of pairwise-disjoint subgraphs that are trees and which, together, span all the vertices of the graph.

Definition 1.24 Let $G = (V, E)$ be a graph. A sequence $[(W_1, S_1), \ldots, (W_k, S_k)]$ of $k \geq 1$ subgraphs of G is a **spanning forest** of graph G if $W_1 \cup \cdots \cup W_k = V$; $W_i \cap W_j = \emptyset$ for all $1 \leq i, j \leq k$ with $i \neq j$; and (W_i, S_i) is a tree, for all $1 \leq i \leq k$.

Example 1.24 The graph of Fig. 1.22 has no spanning tree, but there are 12 forests that span all the vertices of the graph. These spanning forests of the graph are shown highlighted, up to a permutation of the sequence of trees in a forest. Consider, for instance, trees $T_1 = (\{v_1\}, \emptyset)$ and $T_2 = (\{v_2, v_3, v_4\}, \{(v_2, v_3), (v_2, v_4)\})$. Spanning forests $[T_1, T_2]$ and $[T_2, T_1]$ are both highlighted together, at the left of the top row.

Ordered Trees

An ordered tree is a tree in which the relative order of the children is fixed for each node. As a particular case of ordered graphs, ordered trees arise when a tree is drawn or *embedded* in the Euclidean plane.

Example 1.25 In the ordered tree shown in Fig. 1.23, the children of node v_1 are $[v_2, v_6]$; the children of node v_2 are $[v_3, v_4, v_5]$; the children of node v_6 are

$[v_7, v_{10}, v_{11}]$; the children of node v_7 are $[v_8, v_9]$; and the children of node v_{11} are $[v_{12}, v_{13}]$. The remaining nodes have no children. The relative order of the children of each of the nodes reflects the counter-clockwise ordering of the outgoing edges of the node in the drawing. For instance, the ordered sequence $[v_7, v_{10}, v_{11}]$ of the children of node v_6 reflects the counter-clockwise ordering (v_6, v_7), (v_6, v_{10}), (v_6, v_{11}) of the edges going out of node v_6 in the drawing of the ordered tree.

The relative order of children nodes leads to further distinguishing among the nodes of an ordered tree: children nodes are ordered from the *first* up to the *last*, non-first children nodes have a *previous sibling*, and non-last children nodes have a *next sibling*. For sibling nodes v and w, let $v < w$ denote that v precedes w in the ordered tree.

Definition 1.25 Let $T = (V, E)$ be an ordered tree, and let $(v, w) \in E$. Node $w \in V$ is said to be the **first child** of node $v \in V$, denoted by *first*$[v]$, if there is no node $z \in V$ such that $(v, z) \in E$ and $z < w$. Node $w \in V$ is said to be the **last child** of node $v \in V$, denoted by *last*$[v]$, if there is no node $z \in V$ such that $(v, z) \in E$ and $w < z$. Node $z \in V$ is said to be the **next sibling** of node $w \in V$, denoted by *next*$[w]$, if $(v, z) \in E$, $w < z$, and there is no node $x \in V$ such that $(v, x) \in E$ and $w < x < z$. Node $w \in V$ is said to be the **previous sibling** of node $z \in V$, denoted by *previous*$[z]$, if *next*$[w] = z$.

Example 1.26 In the ordered tree shown in Fig. 1.23, *first*$[v_6] = v_7$, *last*$[v_6] = v_{11}$, and *next*$[v_7] = v_{10} = previous[v_{11}]$.

1.2 Basic Data Structures

Some of the basic data structures needed for the description of algorithms on trees and graphs are illustrated below using a fragment of *pseudocode*. Pseudocode conventions follow modern programming guidelines, such as avoiding global side effects (with the only exception of object attributes, which are hidden behind dot-notation) and the unconditional transfer of control by way of goto or gosub statements.

- Assignment of value a to variable x is denoted by $x \leftarrow a$.
- Comparison of equality between the values of two variables x and y is denoted by $x = y$; comparison of strict inequality is denoted by either $x \neq y$, $x < y$, or $x > y$; and comparison of non-strict inequality is denoted by either $x \leqslant y$ or $x \geqslant y$.
- Logical *true* and *false* are denoted by true and false, respectively.
- Logical negation, conjunction, and disjunction are denoted by not, and, and or, respectively.
- Non-existence is denoted by *nil*.

- Mathematical notation is preferred over programming notation. For example, the cardinality of a set S is denoted by $|S|$, membership of an element x in a set S is denoted by $x \in S$, insertion of an element x into a set S is denoted by $S \leftarrow S \cup \{x\}$, and deletion of an element x from a set S is denoted by $S \leftarrow S \setminus \{x\}$.
- Control structures use the following reserved words: **all**, **break**, **do**, **else**, **for**, **if**, **repeat**, **return**, **then**, **to**, **until**, and **while**.
- Blocks of statements are shown by means of indention.

if ... **then**	**for all** ... **do**	**while** ... **do**	**repeat**
...	...	...	...
else	...	**if** ... **then**	...
...	**return** ...	...	**until** ...

The collection of abstract operations on arrays, matrices, lists, stacks, queues, priority queues, sets, and dictionaries is presented next by way of examples.

Arrays

An **array** is a one-dimensional array indexed by non-negative integers. The $[i]$ operation returns the i-th element of the array, assuming there is such an element.

```
let A[1..n] be a new array
for i = 1 to n do
    A[i] ← false
```

Matrices

A **matrix** is a two-dimensional array indexed by non-negative integers. The $[i, j]$ operation returns the element in the i-th row and the j-th column of the matrix, assuming there is such an element.

```
let M[1..m][1..n] be a new matrix
for i = 1 to m do
    for j = 1 to n do
        A[i, j] ← 0
```

Lists

A **list** is just a sequence of elements. The **front** operation returns the first element and the **back** operation returns the last element in the list, assuming the list is not empty. The **prev** operation returns the element before a given element in the list, assuming the given element is not at the front of the list. The **next** operation returns the element after a given element in the list, assuming the given element is not at the back of the list. The **append** operation inserts an element at the rear of the list. The **concatenate** operation deletes the elements of another list and inserts them at the rear of the list.

let L be an empty list $\hspace{4cm}$ $L.append(x)$
append x to L
let L' be an empty list
append x to L'
concatenate L' to L $\hspace{4cm}$ $L.concatenate(L')$
let x be the element at the front of L $\hspace{3cm}$ $x \leftarrow L.front()$
while $x \neq nil$ **do**
$\quad$ output x
$\quad x \leftarrow L.next(x)$
let x be the element at the back of L $\hspace{3cm}$ $x \leftarrow L.back()$
while $x \neq nil$ **do**
$\quad$ output x
$\quad x \leftarrow L.prev(x)$

Stacks

A **stack** is a sequence of elements that are inserted and deleted at the same end (the top) of the sequence. The **top** operation returns the top element in the stack, assuming the stack is not empty. The **pop** operation deletes and returns the top element in the stack, also assuming the stack is not empty. The **push** operation inserts an element at the top of the stack.

let S be an empty stack $\hspace{4cm}$ $S.push(x)$
push x onto S
let x be the element at the top of S $\hspace{3cm}$ $x \leftarrow S.top()$
output x
while S is not empty **do** $\hspace{4cm}$ $\neg S.empty()$
$\quad$ pop from S the top element x $\hspace{3cm}$ $x \leftarrow S.pop()$
$\quad$ output x

Queues

A **queue** is a sequence of elements that are inserted at one end (the rear) and deleted at the other end (the front) of the sequence. The **front** operation returns the front element in the queue, assuming the queue is not empty. The **dequeue** operation deletes and returns the front element in the queue, assuming the queue is not empty. The **enqueue** operation inserts an element at the rear end of the queue.

let Q be an empty queue
enqueue x into Q $Q.enqueue(x)$
let x be the element at the front of Q $x \leftarrow Q.front()$
output x
repeat
   dequeue from Q the front element x $x \leftarrow Q.dequeue()$
   output x
until Q is empty

                                      $Q.empty()$

Priority Queues

A **priority queue** is a queue of elements with both an information and a priority associated with each element, where there is a linear order defined on the priorities. The **front** operation returns an element with the minimum priority, assuming the priority queue is not empty. The **dequeue** operation deletes and returns an element with the minimum priority in the queue, assuming the priority queue is not empty. The **enqueue** operation inserts an element with a given priority in the priority queue.

let Q be an empty priority queue
enqueue (x, y) into Q $Q.enqueue(x, y)$
let (x, y) be an element x
   with the minimum priority y in Q $(x, y) \leftarrow Q.front()$
output (x, y)
repeat
   dequeue from Q an element x
     with the minimum priority y
   output (x, y) $(x, y) \leftarrow Q.dequeue()$
until Q is empty

                                        $Q.empty()$

Sets

A **set** is just a set of elements. The **insert** operation inserts an element in the set. The **delete** operation deletes an element from the set. The **member** operation returns true if an element belongs to the set and false otherwise.

> let S be an empty set
> $S = S \cup \{x\}$
> **for all** $x \in S$ **do**
>   output x
>   delete x from S

$S.insert(x)$
$S.member(x)$

$S.delete(x)$

Dictionaries

A **dictionary** is an associative container, consisting of a set of elements with both an information and a unique key associated with each element, where there is a linear order defined on the keys and the information associated with an element is retrieved on the basis of its key. The **member** operation returns true if there is an element with a given key in the dictionary, and false otherwise. The **lookup** operation returns the element with a given key in the dictionary, or *nil* if there is no such element. The **insert** operation inserts and returns an element with a given key and given information in the dictionary, replacing the element (if any) with the given key. The **delete** operation deletes the element with a given key from the dictionary if there is such an element.

> let D be an empty dictionary
> $D[x] \leftarrow y$
> **for all** $x \in D$ **do**
> $y \leftarrow D[x]$
>   output (x, y)
>   delete x from D

$D.insert(x, y)$
$D.member(x)$
$y \leftarrow D.lookup(x)$

$D.delete(x)$

1.3 Representation of Trees and Graphs

There are several different ways in which graphs and, in particular, trees can be represented in a computer, and the choice of a data structure often depends on the efficiency with which the operations that access and update the data need to be supported. As a matter of fact, there is often a tradeoff between the space complexity of a data structure and the time complexity of the operations.

Representation of Graphs

A graph representation consists of a collection of abstract operations, a concrete representation of graphs by appropriate data structures, and an implementation of the abstract operations using the concrete data structures.

The choice of abstract operations to be included in a graph representation depends on the particular application area. For instance, the representation of graphs in the LEDA library of efficient data structures and algorithms [43] supports about 120 different operations, roughly half of which address the needs of computational geometry algorithms, and the representation of graphs in the BGL library of graph algorithms [55] supports about 50 different operations. The following collection of 32 abstract operations, though, covers the needs of most of the algorithms on graphs presented in this book.

- $G.vertices()$ and $G.edges()$ give, respectively, a list of the vertices and a list of the edges of graph G in the order fixed by the representation of G.
- $G.incoming(v)$ and $G.outgoing(v)$ give a list of the edges of graph G coming into and going out of vertex v, respectively, in the order fixed by the representation of G.
- $G.number\_of\_vertices()$ and $G.number\_of\_edges()$ give, respectively, the order n and the size m of graph G.
- $G.indeg(v)$ and $G.outdeg(v)$ give, respectively, the number of edges coming into and going out of vertex v in graph G.
- $G.adjacent(v, w)$ is true if there is an edge in graph G going out of vertex v and coming into vertex w, and false otherwise.
- $G.source(e)$ and $G.target(e)$ give, respectively, the source and the target vertex of edge e in graph G.
- $G.opposite(v, e)$ gives $G.target(e)$ if vertex v is the source of edge e in graph G, and $G.source(e)$ otherwise.
- $G.first\_vertex()$ and $G.last\_vertex()$ give, respectively, the first and the last vertex in the representation of graph G. Further, $G.pred\_vertex(v)$ and $G.succ\_vertex(v)$ give, respectively, the predecessor and the successor of vertex v in the representation of graph G. These operations support iteration over the vertices of the graph.
- $G.first\_edge()$ and $G.last\_edge()$ give, respectively, the first and the last edge in the representation of graph G. Further, $G.pred\_edge(e)$ and $G.succ\_edge(e)$ give, respectively, the predecessor and the successor of edge e in the representation of graph G. These operations support iteration over the edges of the graph.
- $G.first\_in\_edge(v)$ and $G.last\_in\_edge(v)$ give, respectively, the first and the last edge in the representation of graph G coming into vertex v. Further, $G.in\_pred(e)$ and $G.in\_succ(e)$ give, respectively, the previous and the next edge after e in the representation of graph G coming into vertex $G.target(e)$. These operations support iteration over the vertices of the graph adjacent with a given vertex.

- $G.first\_adj\_edge(v)$ and $G.last\_adj\_edge(v)$ give, respectively, the first and the last edge in the representation of graph G going out of vertex v. Further, $G.adj\_pred(e)$ and $G.adj\_succ(e)$ give, respectively, the previous and the next edge after e in the representation of graph G going out of vertex $G.source(e)$. These operations also support iteration over the vertices of the graph adjacent with a given vertex.
- $G.new\_vertex()$ inserts a new vertex in graph G, and $G.new\_edge(v, w)$ inserts a new edge in graph G going out of vertex v and coming into vertex w. Further, $G.del\_vertex(v)$ deletes vertex v from graph G, together with all those edges going out of or coming into vertex v, and $G.del\_edge(e)$ deletes edge e from graph G. These operations support dynamic graphs.

Notice that some of these operations are actually defined in terms of a smaller set of 10 abstract operations on graphs,

- $G.vertices()$
- $G.edges()$
- $G.incoming(v)$
- $G.outgoing(v)$
- $G.source(e)$

- $G.target(e)$
- $G.new\_vertex()$
- $G.new\_edge(v, w)$
- $G.del\_vertex(v)$
- $G.del\_edge(e)$

using the abstract operations on basic data structures presented in Sect. 1.2, as follows:

- $G.number\_of\_vertices()$ is given by $G.vertices().size()$.
- $G.number\_of\_edges()$ is given by $G.edges().size()$.
- $G.indeg(v)$ is given by $G.incoming(v).size()$.
- $G.outdeg(v)$ is given by $G.outgoing(v).size()$.
- $G.adjacent(v, w)$ is true if there is an edge $e \in G.outgoing(v)$ such that $G.target(e) = w$, and it is false otherwise.
- $G.opposite(v, e)$ is given by $G.target(e)$ if $G.source(e) = v$, and it is given by $G.source(e)$ otherwise, that is, if $G.target(e) = v$.
- $G.first\_vertex()$ is given by $G.vertices().front()$.
- $G.last\_vertex()$ is given by $G.vertices().back()$.
- $G.pred\_vertex(v)$ is given by $G.vertices().prev(v)$.
- $G.succ\_vertex(v)$ is given by $G.vertices().next(v)$.
- $G.first\_edge()$ is given by $G.edges().front()$.
- $G.last\_edge()$ is given by $G.edges().back()$.
- $G.pred\_edge(e)$ is given by $G.edges().prev(e)$.
- $G.succ\_edge(e)$ is given by $G.edges().next(e)$.
- $G.first\_in\_edge(v)$ is given by $G.incoming(v).front()$.
- $G.last\_in\_edge(v)$ is given by $G.incoming(v).back()$.
- $G.in\_pred(e)$ is given by $G.incoming(v).prev(e)$.
- $G.in\_succ(e)$ is given by $G.incoming(v).next(e)$.
- $G.first\_adj\_edge(v)$ is given by $G.outgoing(v).front()$.

- $G.last\_adj\_edge(v)$ is given by $G.outgoing(v).back()$.
- $G.adj\_pred(e)$ is given by $G.outgoing(v).prev(e)$.
- $G.adj\_succ(e)$ is given by $G.outgoing(v).next(e)$.

Together with the abstract operations on the underlying basic data structures, these operations support iteration over the vertices and edges of a graph. For instance, in the following procedure, variable v is assigned each of the vertices of graph G in turn,

> **for all** vertices v of G **do**
> ...
> $\hfill v \in G.vertices()$

and in the following procedure, all those vertices of graph G which are adjacent with vertex v are assigned in turn to variable w.

> **for all** vertices w adjacent with vertex v in G **do**
> ...
> $\hfill e \in G.outgoing(v)$
> $\hfill w = G.target(e)$

Together, they provide for a simple traversal of a graph, in which vertices and edges are visited in the order fixed by the representation of the graph.

> **for all** vertices v of G **do**
>   **for all** vertices w adjacent with vertex v in G **do**
> ...
> $\hfill v \in G.vertices()$
> $\hfill e \in G.outgoing(v)$
> $\hfill w = G.target(e)$

Adjacency Matrix

The data structures most often used for representing graphs are adjacency matrices and adjacency lists. The adjacency matrix representation of a graph is a Boolean matrix, with an entry for each ordered pair of vertices in the graph, where the entry corresponding to vertices v and w has the value *true* if there is an edge in the graph going out of vertex v and coming into vertex w, and it has the value *false* otherwise.

Definition 1.26 Let $G = (V, E)$ be a graph with n vertices. The **adjacency matrix** representation of G is an $n \times n$ Boolean matrix with an entry for each ordered pair of vertices of G, where the entry corresponding to vertices v and w is equal to *true* if $(v, w) \in E$ and is equal to *false* otherwise, for all vertices $v, w \in V$.

Example 1.27 The adjacency matrix representation of the graph of Fig. 1.1 is shown in Fig. 1.24. Entries equal to *true* are denoted by 1, and *false* entries are denoted by 0.

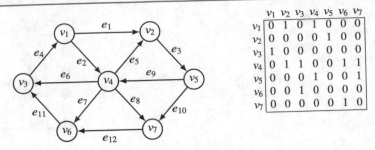

Fig. 1.24 Adjacency matrix representation of the graph of Fig. 1.1

Notice that with most data structures used for representing graphs, there is an order on the vertices fixed by the representation. In the case of the adjacency matrix representation A of a graph $G = (V, E)$, vertex v precedes vertex w if and only if the row of A corresponding to v lies above the row of A corresponding to w or, in an equivalent formulation, the column of A corresponding to v lies to the left of the column of A corresponding to w, for all vertices $v, w \in V$.

Now, since the edges are implicit in the adjacency matrix representation of a graph, those operations having an edge as argument or giving an edge as result cannot be implemented using an adjacency matrix representation. These operations are: $G.edges()$, $G.incoming(v)$, $G.outgoing(v)$, $G.source(e)$, $G.target(e)$, and $G.del\_edge(e)$. Furthermore, $G.new\_vertex()$ and $G.del\_vertex(v)$ cannot be implemented unless the adjacency matrix can be dynamically resized, and $G.new\_edge(v, w)$ can be implemented by setting to *true* the entry of A at the row corresponding to vertex v and the column corresponding to vertex w, and takes $O(1)$ time, but the operation cannot give the new edge as result.

Let $G = (V, E)$ be a graph with n vertices, let us assume the vertices are numbered $1, 2, \ldots, n$ in some arbitrary manner and their number is stored in the attribute *index*, and let A be the adjacency matrix representation of G. The only remaining operation, $G.vertices()$, is just the sequence of all vertices $v \in V$ ordered by the *index* attribute.

The adjacency matrix representation of a graph $G = (V, E)$ with n vertices takes $\Theta(n^2)$ space. Therefore, no graph algorithm can be implemented using an adjacency matrix representation to run in $O(n)$ time. The main advantage of the adjacency matrix representation is the support of the adjacency test in $O(1)$ time.

Adjacency List

The adjacency list representation of a graph, on the other hand, is an array of lists, one for each vertex in the graph, where the list corresponding to vertex v contains the target vertices of the edges coming out of vertex v.

Fig. 1.25 Adjacency list representation of the graph of Fig. 1.1. Lists of adjacent vertices are shown as sequences

vertex	adjacent vertices
v_1	$[v_2, v_4]$
v_2	$[v_5]$
v_3	$[v_1]$
v_4	$[v_2, v_3, v_6, v_7]$
v_5	$[v_4, v_7]$
v_6	$[v_3]$
v_7	$[v_6]$

Definition 1.27 Let $G = (V, E)$ be a graph with n vertices and m edges. The **adjacency list** representation of G consists of a list of n elements (the vertices of the graph) and a list of n lists with a total of m elements (the target vertices for the edges of the graph). The list corresponding to vertex v contains all vertices $w \in V$ with $(v, w) \in E$, for all vertices $v \in V$.

Notice that adjacency lists are not necessarily arranged in any particular order although there is, as a matter of fact, an order on the vertices and edges fixed by the adjacency list representation of the graph. Given the adjacency list representation of a graph $G = (V, E)$, vertex precedence is given by the order of the corresponding entries in the list of lists, and edge (v, w) precedes edge (v, z) if and only if vertex w precedes vertex z in the list corresponding to vertex v, for all edges (v, w), $(v, z) \in E$.

Example 1.28 The adjacency list representation of the graph of Fig. 1.1 is shown in simplified form in Fig. 1.25, where lists of adjacent vertices are shown as sequences.

As in the case of adjacency matrices, edges are implicit in the adjacency list representation of a graph, and those operations having an edge as argument or giving an edge as result cannot be implemented using an adjacency list representation. These operations are: $G.edges()$, $G.incoming(v)$, $G.outgoing(v)$, $G.source(e)$, $G.target(e)$, and $G.del\_edge(e)$. Furthermore, $G.new\_vertex()$ and $G.del\_vertex(v)$ cannot be implemented unless the list of lists can be dynamically resized, and $G.new\_edge(v, w)$ can be implemented by appending vertex w to the list corresponding to vertex v, and takes $O(1)$ time, but the operation cannot give the new edge as result. The only remaining operation, $G.adjacent(v, w)$, can be implemented by scanning the list corresponding to vertex v and takes $O(\text{outdeg}(v))$ time.

The adjacency list representation of a graph $G = (V, E)$ with n vertices and m edges takes $\Theta(n + m)$ space. Beside the low space requirement, the main advantage of the adjacency list representation is the support of iteration operations in $O(1)$ time. Their main disadvantage, though, lies in the fact that an adjacency test $G.adjacent(v, w)$ is not supported in $O(1)$ time but in $O(\text{outdeg}(v))$ time.

Notice that the adjacency matrix representation of a graph can be easily obtained from the adjacency list representation, assuming the vertices are numbered $1, 2, \ldots, n$ in some arbitrary manner and their number is stored in attribute *index*. The following procedure makes A the adjacency matrix of graph G and runs in $O(n^2+m) = O(n^2)$ time.

```
let A[1..n][1..n] be a new matrix
for i = 1 to n do
    for j = 1 to n do
        A[i, j] ← false
for all edges e of G do
    A[G.source(e).index, G.target(e).index] ← true
```

Then, implementing the adjacency test using the adjacency matrix representation of a graph is straightforward. The following function $adjacent(A, v, w)$ returns *true* if $(v, w) \in E$ and false otherwise and, given the adjacency matrix representation A of the graph, runs in $O(\text{outdeg}(v))$ time.

```
function adjacent(A, v, w)
    return A[v.index, w.index]
```

An alternative implementation of the adjacency test consists of scanning adjacent vertices in the adjacency list representation. Let $G = (V, E)$ be a graph, and let $v, w \in V$. The following function $adjacent(v, w)$ returns *true* if $(v, w) \in E$, and false otherwise, and runs in $O(\text{outdeg}(v))$ time.

```
function adjacent(v, w)
    for all vertices x adjacent with vertex v in G do
        if x = w then
            return true
    return false
```

Extended Adjacency List

The adjacency list representation can be extended by making edges explicit. The extended adjacency list representation of a graph consists of a list of vertices and a list of edges. Associated with each vertex are two lists of incoming and outgoing edges. Associated with each edge are its source and target vertices.

Definition 1.28 Let $G = (V, E)$ be a graph with n vertices and m edges. The **extended adjacency list** representation of G consists of a list of n elements (the vertices of the graph), a list of m elements (the edges of the graph), and two lists of n lists of a total of m elements (the edges of the graph). The *incoming* list corresponding

Fig. 1.26 Extended adjacency list representation for the graph of Fig. 1.1. Lists of incoming and outgoing edges are shown as sequences

vertex	incoming edges	outgoing edges	edge	source	target
v_1	$[e_4]$	$[e_1, e_2]$	e_1	v_1	v_2
v_2	$[e_1, e_5]$	$[e_3]$	e_2	v_1	v_4
v_3	$[e_6, e_{11}]$	$[e_4]$	e_3	v_2	v_5
v_4	$[e_2, e_9]$	$[e_5, e_6, e_7, e_8]$	e_4	v_3	v_1
v_5	$[e_3]$	$[e_9, e_{10}]$	e_5	v_4	v_2
v_6	$[e_7, e_{12}]$	$[e_{11}]$	e_6	v_4	v_3
v_7	$[e_8, e_{10}]$	$[e_{12}]$	e_7	v_4	v_6
			e_8	v_4	v_7
			e_9	v_5	v_4
			e_{10}	v_5	v_7
			e_{11}	v_6	v_3
			e_{12}	v_7	v_6

to vertex v contains all edges $(u, v) \in E$ coming into vertex v, for all vertices $v \in V$. The *outgoing* list corresponding to vertex v contains all edges $(v, w) \in E$ going out of vertex v, for all vertices $v \in V$. The source vertex v and the target vertex w are associated with each edge $(v, w) \in E$.

Example 1.29 The extended adjacency list representation of the graph of Fig. 1.1 is shown in a simplified form in Fig. 1.26, where lists of incoming and outgoing edges are shown as sequences.

Notice that edges are explicit in the extended adjacency list graph representation. Therefore, all of the previous operations can be implemented using the extended adjacency list representation and take $O(1)$ time, with the exception of $G.del\_node(v)$, which takes $O(\deg(v))$ time. The operations can be implemented as follows:

- $G.vertices()$ and $G.edges()$ are, respectively, the list of vertices and the list of edges of graph G.
- $G.incoming(v)$ and $G.outgoing(v)$ are, respectively, the list of edges coming into vertex v and the list of edges going out of vertex v.
- $G.source(e)$ and $G.target(e)$ are, respectively, the source and the target vertex associated with edge e.
- $G.new\_vertex()$ is implemented by appending a new vertex v to the list of vertices of graph G and returning vertex v.
- $G.new\_edge(v, w)$ is implemented by appending a new edge e to the list of edges of graph G, setting to v the source vertex associated with edge e, setting to w the target vertex associated with edge e, appending e to the list of edges going out of vertex v and to the list of edges coming into vertex w, and returning edge e.
- $G.del\_vertex(v)$ is implemented by performing $G.del\_edge(e)$ for each edge e in the list of edges coming into vertex v and for each edge e in the list of edges going out of vertex v and then deleting vertex v from the list of vertices of graph G.

- $G.del\_edge(e)$ is implemented by deleting edge e from the list of edges of graph G, from the list of edges coming into vertex $G.target(e)$, and from the list of edges going out of vertex $G.source(e)$.

The extended adjacency list representation of a graph $G = (V, E)$ with n vertices and m edges takes $\Theta(n + m)$ space.

Adjacency Map

The adjacency list representation can also be extended by making edges explicit in a different way. The lists of incoming and outgoing edges can be replaced with dictionaries of source vertices to incoming edges and target vertices to outgoing edges. This allows for a more efficient adjacency test, although adding a logarithmic factor to the cost to all of the operations (when dictionaries are implemented using balanced trees) or turning the worst-case cost for all of the operations into the expected cost (when dictionaries are implemented using hashing).

The adjacency map representation of a graph consists of a dictionary D of vertices to a pair of dictionaries of vertices to edges: a first dictionary I of source vertices to incoming edges, and a second dictionary O of target vertices to outgoing edges.

Definition 1.29 Let $G = (V, E)$ be a graph with n vertices and m edges. The **adjacency map** representation of G consists of a dictionary of n elements (the vertices of the graph) to a pair of dictionaries of m elements (the source and target vertices for the edges of the graph, respectively). The *incoming* dictionary corresponding to vertex v contains the mappings $(u, (u, v))$ for all edges $(u, v) \in E$ coming into vertex v, for all vertices $v \in V$. The *outgoing* dictionary corresponding to vertex v contains the mappings $(v, (v, w))$ for all edges $(v, w) \in E$ going out of vertex v, for all vertices $v \in V$.

Example 1.30 The adjacency map representation of the graph of Fig. 1.1 is shown in simplified form in Fig. 1.27, where dictionaries are shown as sequences of vertex-edge mappings.

vertex	incoming edges	outgoing edges
v_1	$[v_3 \rightarrow e_4]$	$[v_2 \rightarrow e_1, v_4 \rightarrow e_2]$
v_2	$[v_1 \rightarrow e_1, v_4 \rightarrow e_5]$	$[v_5 \rightarrow e_3]$
v_3	$[v_4 \rightarrow e_6, v_6 \rightarrow e_{11}]$	$[v_1 \rightarrow e_4]$
v_4	$[v_1 \rightarrow e_2, v_5 \rightarrow e_9]$	$[v_2 \rightarrow e_5, v_3 \rightarrow e_6, v_6 \rightarrow e_7, v_7 \rightarrow e_8]$
v_5	$[v_2 \rightarrow e_3]$	$[v_4 \rightarrow e_9, v_7 \rightarrow e_{10}]$
v_6	$[v_4 \rightarrow e_7, v_7 \rightarrow e_{12}]$	$[v_3 \rightarrow e_{11}]$
v_7	$[v_4 \rightarrow e_8, v_5 \rightarrow e_{10}]$	$[v_6 \rightarrow e_{12}]$

Fig. 1.27 Adjacency map representation for the graph of Fig. 1.1. Dictionaries are shown as sequences of vertex-edge mappings

Notice that both vertices and edges are explicit in the adjacency map representation of a graph. The vertices are the keys of the dictionary. The edges are the values in the dictionary of outgoing edges, and also the values in the dictionary of incoming edges, for all vertices of the graph. Therefore, all of the previous abstract operations can be implemented using the adjacency map representation and take expected $O(1)$ time, with the exception of $G.del\_vertex(v)$, which takes $O(\deg(v))$ expected time. The operations can be implemented as follows:

- $G.vertices()$ are the keys in dictionary D.
- $G.edges()$ are the values $D[v].O[w]$ for all keys v in dictionary D and for all keys w in dictionary $D[v].O$.
- $G.incoming(v)$ are the values $D[v].I[u]$ for all keys u in dictionary $D[v].I$.
- $G.outgoing(v)$ are the values $D[v].O[w]$ for all keys w in dictionary $D[v].O$.
- $G.source(e)$ and $G.target(e)$ are, respectively, the source and the target vertex associated with edge e.
- $G.new\_vertex()$ is implemented by inserting an entry in dictionary D, with key a new vertex v and value a pair of empty dictionaries $D[v].I$ and $D[v].O$, and returning vertex v.
- $G.new\_edge(v, w)$ is implemented by setting to v the source vertex associated with a new edge e, setting to w the target vertex associated with edge e, inserting an entry in dictionary $D[v].O$ with key w and value e, inserting an entry in dictionary $D[w].I$ with key v and value e, and returning edge e.
- $G.del\_vertex(v)$ is implemented by performing $G.del\_edge(e)$ for each entry with key u and value e in dictionary $D[v].I$ and for each entry with key w and value e in dictionary $D[v].O$ and then deleting the entry with key v from dictionary D.
- $G.del\_edge(e)$ is implemented by deleting the entry with key w from dictionary $D[v].O$ and deleting the entry with key v from dictionary $D[w].I$, where $v = G.source(e)$ and $w = G.target(e)$.

Recall that the adjacency test $G.adjacent(v, w)$ can be defined in terms of the small set of abstract operations on graphs, using the abstract operations on basic data structures, by scanning the list of outgoing edges $G.outgoing(v)$ for an edge e such that $G.target(e) = w$. Now, besides the low space requirement, the main advantage of the adjacency map representation is the support of adjacency test in expected $O(1)$ time, when dictionaries are implemented using hashing, as follows:

- $G.adjacent(v, w)$ is true if $(w, e) \in D[v].O$ (where $e = D[v].O[w]$), and false otherwise.

The adjacency map representation of a graph $G = (V, E)$ with n vertices and m edges takes $\Theta(n + m)$ space.

Representation of Trees

As in the case of graphs, a tree representation consists of a collection of abstract operations, a concrete representation of trees by appropriate data structures, and implementation of the abstract operations using the concrete data structures.

The following collection of 13 abstract operations covers the needs of most of the algorithms on trees presented in this book.

- $T.number\_of\_nodes()$ gives the number of nodes $|V|$ of tree $T = (V, E)$.
- $T.root()$ gives $root[T]$.
- $T.is\_root(v)$ is true if node $v = root[T]$, and false otherwise.
- $T.number\_of\_children(v)$ gives $children[v]$, that is, the number of nodes $w \in V$ such that $(v, w) \in E$.
- $T.parent(v)$ gives $parent[v]$, that is, the parent in T of node v.
- $T.children(v)$ gives a list of the children of node v in T.
- $T.is\_leaf(v)$ is true if node v is a leaf of T, and false otherwise.
- $T.first\_child(v)$ and $T.last\_child(v)$ give, respectively, the first and the last of the children of node v, according to the order on the children of v fixed by the representation of T, or nil if v is a leaf node of T.
- $T.previous\_sibling(v)$ and $T.next\_sibling(v)$ give, respectively, the previous child of $T.parent(v)$ before vertex v and the next child of $T.parent(v)$ after vertex v, according to the order on the children of $T.parent(v)$ fixed by the representation of T, or nil if v is a first child or a last child, respectively.
- $T.is\_first\_child(v)$ and $T.is\_last\_child(v)$ are true if node v is, respectively, the first and the last of the children of node $T.parent(v)$, according to the order on the children of $T.parent(v)$ fixed by the representation of T, and false otherwise.

The previous operations support iteration over the children of a node in a tree, much in the same way that the graph operations support iteration over the vertices and edges of a graph. In the following procedure, the children of node v in tree T are assigned in turn to variable w.

> let w be the first child of node v in T $w = T.first\_child(v)$
> **while** $w \neq nil$ **do**
>   process node w
>   let w be the next sibling of node w in T $w = T.next\_sibling(w)$

The following procedure is an alternative form of iteration over the children w of a node v in a tree T.

> **if** v is not a leaf node of T **then** $\neg T.is\_leaf(v)$

> let w be the first child of node v in T $w = T.first\_child(v)$
> process node w
> **while** w is not the last child of node v in T **do** $w \neq T.last\_child(v)$
>   let w be the next sibling of node w in T $w = T.next\_sibling(w)$
>   process node w

The following procedure is still another, more compact form of iteration over the children w of a node v in a tree T.

> **for all** children w of node v in T **do** $w \in T.children(v)$
>   process node w

Array of Parents

The data structures most often used for representing trees are the array-of-parents representation and the first-child, next-sibling representation. The array-of-parents representation of a tree is an array of nodes, with an entry for each node in the tree, where the entry corresponding to node v has the value $parent[v]$ if v is not the root of the tree, and it has the value nil otherwise.

Definition 1.30 Let $T = (V, E)$ be a tree with n nodes. The **array-of-parents** representation of T is an array P of n nodes, indexed by the nodes of the tree, where $P[v] = parent[v]$ if $v \neq root[T]$, and $P[v] = nil$ otherwise, for all nodes $v \in V$.

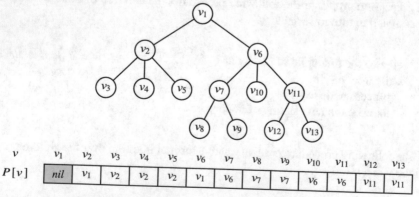

v	v_1	v_2	v_3	v_4	v_5	v_6	v_7	v_8	v_9	v_{10}	v_{11}	v_{12}	v_{13}
$P[v]$	nil	v_1	v_2	v_2	v_2	v_1	v_6	v_7	v_7	v_6	v_6	v_{11}	v_{11}

Fig. 1.28 Array-of-parents representation of a tree

Example 1.31 The array-of-parents representation P of the tree $T = (V, E)$ of Fig. 1.23 is shown in Fig. 1.28. The parent of node v is given by $P[v]$, for all nodes $v \in V$.

The array of parent nodes is not necessarily arranged in any particular order although there is, as a matter of fact, an order on the nodes fixed by the representation. Given the array-of-parents representation of a tree, node precedence is given by the order of the nodes as array indices, and precedence between sibling nodes is also given by the order of array indices.

Let $T = (V, E)$ be a tree with n nodes, and let P be the array-of-parents representation of T. Operations $T.root()$, $T.number\_of\_children(v)$, $T.children(v)$, $T.is\_leaf(v)$, $T.first\_child(v)$, $T.last\_child(v)$, $T.previous\_sibling(v)$, $T.next\_sibling(v)$, $T.is\_first\_child(v)$, and $T.is\_last\_child(v)$ require scanning the whole array and thus take $O(n)$ time. The remaining operations can be implemented to take $O(1)$ time as follows:

- $T.number\_of\_nodes()$ is the size of array P.
- $T.is\_root(v)$ is true if $P[v] = nil$, and false otherwise.
- $T.parent(v)$ is equal to $P[v]$.

First-Child, Next-Sibling

The first-child, next-sibling representation, on the other hand, consists of two arrays of nodes, each of them with an entry for each node in the tree, where the entries corresponding to node v have, respectively, the value $first[v]$, or nil if v is a leaf node, and $next[v]$, or nil if v is a last sibling.

Definition 1.31 Let $T = (V, E)$ be a tree with n nodes. The **first-child next-sibling** representation of T is a pair (F, N) of arrays of n nodes, indexed by the nodes of the tree, where $F[v] = first[v]$ if v is not a leaf node, $F[v] = nil$ otherwise, $N[v] = next[v]$ if v is not a last child node, and $N[v] = nil$ otherwise, for all nodes $v \in V$.

Example 1.32 The array-of-parents representation of the tree $T = (V, E)$ of Fig. 1.23 is shown in Fig. 1.29. The first child and the next sibling of node v are given by $F[v]$ and $N[v]$, respectively, for all nodes $v \in V$.

As with the array-of-parents representation, the arrays of first children and next siblings are not necessarily arranged in any particular order, as long as they are arranged in the same order. Anyway, there is an order on the nodes fixed by the representation. Given the first-child, next-sibling representation of a tree, node precedence is given by the order of the nodes as array indices, and precedence between sibling nodes is also given by the order of array indices.

Let $T = (V, E)$ be a tree with n nodes, and let (F, N) be the first-child, next-sibling representation of T. Operations $T.root()$, $T.is\_root(v)$, $T.parent(v)$, and $T.children(v)$ require scanning both of arrays F and N, and $T.previous\_sibling(v)$ and $T.is\_first\_child(v)$ require scanning array N, and thus take $O(n)$ time. Further, $T.number\_of\_children(v)$ and $T.last\_child(v)$ require following up to $children[v]$ next-sibling links, where the former also requires following a first-child link, and take $O(children[v])$ time. The remaining operations can be implemented to take $O(1)$ time as follows:

- $T.number\_of\_nodes()$ is the size of array F, or the size of array N.
- $T.is\_leaf(v)$ is true if $F[v] = nil$, and false otherwise.
- $T.first\_child(v)$ is given by $F[v]$.
- $T.next\_sibling(v)$ is given by $N[v]$.
- $T.is\_last\_child(v)$ is true if $N[v] = nil$, and false otherwise.

Graph-Based Representation of Trees

Trees can also be represented using the small set of abstract operations on graphs, which can be implemented using either the extended adjacency list representation or the adjacency map representation. The following operations are defined in terms of the graph representation, using the abstract operations on basic data structures presented in Sect. 1.2, as follows:

- $T.number\_of\_nodes()$ is given by $T.vertices().size()$.
- $T.root()$ requires following the path of parent nodes starting off with any node of the tree and then storing the root of the tree in the attribute *root*.

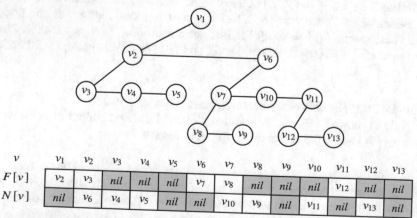

v	v_1	v_2	v_3	v_4	v_5	v_6	v_7	v_8	v_9	v_{10}	v_{11}	v_{12}	v_{13}
$F[v]$	v_2	v_3	nil	nil	nil	v_7	v_8	nil	nil	nil	v_{12}	nil	nil
$N[v]$	nil	v_6	v_4	v_5	nil	nil	v_{10}	v_9	nil	v_{11}	nil	v_{13}	nil

Fig. 1.29 First-child, next-sibling representation of a tree. Vertical arrows denote first-child links, and horizontal arrows denote next-sibling links

```
function root(T)
    if T.root ≠ nil then
        return T.root
    if T is empty then
        return nil
    v ← T.vertices().front()
    while T.incoming(v) is not empty do
        v ← T.parent(v)
    T.root ← v
    return v
```

- $T.is\_root(v)$ is given by $T.incoming(v).empty()$.
- $T.number\_of\_children(v)$ is given by $T.outgoing(v).size()$.
- $T.parent(v)$ is given by $T.incoming(v).front().source()$.
- $T.children(v)$ is given by the list $[T.target(e) : e \in T.outgoing(v)]$.
- $T.is\_leaf(v)$ is given by $T.outgoing(v).empty()$.
- $T.first\_child(v)$ is given by $T.outgoing(v).front().target()$.
- $T.last\_child(v)$ is given by $T.outgoing(v).back().target()$.
- $T.previous\_sibling(v)$ requires accessing the previous edge in the list of edges coming out of $parent[v]$.

```
function previous_sibling(T, v)
    if T.is_root(v) then
        return nil
    e ← T.incoming(v).front()
    u ← T.source(e)
    if T.outgoing(u).prev(e) = nil then
        return nil
    else
        return T.outgoing(u).prev(e).target()
```

- $T.next\_sibling(v)$ requires accessing the next edge in the list of edges coming out of $parent[v]$.

```
function previous_sibling(T, v)
    if T.is_root(v) then
        return nil
    e ← T.incoming(v).front()
    u ← T.source(e)
    if T.outgoing(u).next(e) = nil then
        return nil
    else
        return T.outgoing(u).next(e).target()
```

- $T.is\_first\_child(v)$ is true if $T.previous\_sibling(v)$ is equal to nil, and false otherwise.
- $T.is\_last\_child(v)$ is true if $T.next\_sibling(v)$ is equal to nil, and false otherwise.

All of the operations take $O(1)$ time with the only exception of $T.root()$, which takes time linear in the depth or height of the tree. These are worst-case time bounds when using extended adjacency lists, and expected time bounds when using adjacency maps. The representation uses $O(n)$ space.

Summary

Basic graph-theoretical notions underlying algorithms on trees and graphs, together with appropriate data structures for their representation, are reviewed in this chapter. References to more compact, implicit representations are given in the bibliographic notes below.

Bibliographic Notes

A more detailed exposition of the graph-theoretical notions reviewed in this chapter can be found in [4,6–10,23,25,33,47,52,61]. Ordered or embedded graphs are the subject of topological graph theory. See, for instance, [22,46,63], and also [43, Chap. 8]. General references to algorithms and data structures include [1,5,13,36–38, 54,62], and further references to combinatorial and graph algorithms include [11,12, 15,19,20,39–43,48,53,56,59,64]. Pseudocode conventions follow, to a large extent, [13] and are based on the ALGOL notation, which was adopted by the Association for Computer Machinery for the publication of algorithms in textbooks and academic papers and, especially, in a regular column of the *Communications of the ACM* magazine between 1960 and 1975, which became part of the Collected Algorithms of the ACM [29,30].

Adjacency matrices are most often used in combinatorics and graph theory. Adjacency lists are the preferred graph representation in computer science, though, because a large number of graph algorithms can be implemented to run in time linear in the number of vertices and edges in the graph using an adjacency list representation, while no graph algorithm can be implemented to run in linear time using an adjacency matrix representation. The use of adjacency lists was advocated in [28,58] and traversing incident edges or edges, instead of adjacent vertices, was proposed in [14]. The adjacency map representation is advocated in [21] and was adopted in the NetworkX package for complex network analysis in Python [24]. More compact, implicit representations of static graphs are studied in [2,18,26,27,34,49,57,60]. See also [3,16]. Implicit representations of static trees include the Prüfer sequences [50], which are used for counting and generating trees. See also [17,44,45,51] and the bibliographic notes for Chap. 4.

Fig. 1.30 Graph for
problem 1.3

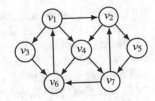

Problems

1.1 Determine the size of the complete graph K_n on n vertices and the complete bipartite graph $K_{p,q}$ on $p + q$ vertices.

1.2 Determine the values of n for which the circle graph C_n on n vertices is bipartite, and also the values of n for which the complete graph K_n is bipartite.

1.3 Give all the spanning trees of the graph in Fig. 1.30, and also the number of spanning trees of the underlying undirected graph.

1.4 Extend the adjacency matrix graph representation by replacing those operations having an edge as argument or giving an edge or a list of edges as result, by corresponding operations having as argument or giving as result the source and target vertices of the edge or edges: $G.del\_edge(v, w)$, $G.edges()$, $G.incoming(v)$, $G.outgoing(v)$, $G.source(v, w)$, and $G.target(v, w)$.

1.5 Extend the first-child, next-sibling tree representation, in order to support the collection of basic operations but $T.root()$, $T.number\_of\_children(v)$, and $T.children(v)$ in $O(1)$ time.

1.6 Show how to double check that the graph-based representation of a tree is indeed a tree, in time linear in the size of the tree.

Exercises

Exercise 1.1 The standard representation of an undirected graph in the format adopted for the DIMACS Implementation Challenges [31,32] consists of a problem definition line of the form "p edge n m", where n and m are, respectively, the number of vertices and the number of edges, followed by m edge descriptor lines of the form "e i j", each of them giving an edge as a pair of vertex numbers in the range 1 to n. Comment lines of the form "c ..." are also allowed. Implement procedures to read a DIMACS graph and to write a graph in DIMACS format.

Exercise 1.2 The external representation of a graph in the Stanford GraphBase (SGB) format [35] consists essentially of a first line of the form "* GraphBase `graph(util`$_{\text{t}}$`ypes ... ,`n`V ,`m`A)` ", where n and m are, respectively, the number of vertices and the number of edges; a second line containing an identification string; a "`* Vertices`" line; n vertex descriptor lines of the form "*label* , Ai , 0 , 0", where i is the number of the first edge in the range 0 to $m - 1$ going out of the vertex and *label* is a string label; an "`* Arcs`" line; m edge descriptor lines of the form "Vj , Ai , *label* , 0", where j is the number of the target vertex in the range 0 to $n - 1$, i is the number of the next edge in the range 0 to $m - 1$ going out of the same source vertex, and *label* is an integer label; and a last "`* Checksum ...`" line. Further, in the description of a vertex with no outgoing edge, or an edge with no successor going out of the same source vertex, "Ai" becomes "0". Implement procedures to read a SGB graph and to write a graph in SGB format.

Exercise 1.3 Implement algorithms to generate the path graph P_n, the circle graph C_n, and the wheel graph W_n on n vertices, using the collection of 32 abstract operations from Sect. 1.3.

Exercise 1.4 Implement an algorithm to generate the complete graph K_n on n vertices and the complete bipartite graph $K_{p,q}$ with $p + q$ vertices, using the collection of 32 abstract operations from Sect. 1.3.

Exercise 1.5 Implement the extended adjacency matrix graph representation given in Problem 1.4, wrapped in a Python class, using Python lists together with the internal numbering of the vertices.

Exercise 1.6 Enumerate all perfect matchings in the complete bipartite graph $K_{p,q}$ on $p + q$ vertices.

Exercise 1.7 Implement an algorithm to generate the complete binary tree with n nodes, using the collection of 13 abstract operations from Sect. 1.3.

Exercise 1.8 Implement an algorithm to generate random trees with n nodes, using the collection of 13 abstract operations from Sect. 1.3. Give the time and space complexity of the algorithm.

Exercise 1.9 Give an implementation of operation $T.previous\_sibling(v)$ using the array-of-parents tree representation.

Exercise 1.10 Implement the extended first-child, next-sibling tree representation of Problem 1.5, wrapped in a Python class, using Python lists together with the internal numbering of the nodes.

References

1. Aho AV, Hopcroft JE, Ullman JD (1974) The Design and Analysis of Computer Algorithms. Addison-Wesley, Reading MA
2. Arikati SR, Maheshwari A, Zaroliagis CD (1997) Efficient computation of implicit representations of sparse graphs. Discrete Applied Mathematics 78(1–3):1–16
3. Balcázar JL, Lozano A (1989) The complexity of graph problems for succintly represented graphs. In: Nagl M (ed) Proc. 15th Int. Workshop Graph-Theoretic Concepts in Computer Science, vol 411. Lecture Notes in Computer Science. Springer, Berlin Heidelberg, pp 277–285
4. Bang-Jensen J, Gutin G (2001) Digraphs: Theory. Algorithms and Applications. Springer, Berlin Heidelberg
5. Bentley JL (2000) Programming Pearls, 2nd edn. ACM Press, New York NY
6. Berge C (1985) Graphs and Hypergraphs, 2nd edn. North-Holland, Amsterdam
7. Bollobás B (1979) Graph Theory: An Introductory Course, Graduate Texts in Mathematics, vol 63. Springer, Berlin Heidelberg
8. Bollobás B (1998) Modern Graph Theory, Graduate Texts in Mathematics, vol 184. Springer, Berlin Heidelberg
9. Busacker RG, Saaty TL (1965) Finite Graphs and Networks: An Introduction with Applications. McGraw-Hill, New York NY
10. Chartrand G, Lesniak L (1996) Graphs and Digraphs, 3rd edn. Chapman and Hall, London, England
11. Chartrand G, Oellermann OR (1993) Applied and Algorithmic Graph Theory. McGraw-Hill, New York NY
12. Christofides N (1975) Graph Theory: An Algorithmic Approach. Academic Press, New York NY
13. Cormen TH, Leiserson CE, Rivest RL, Stein C (2009) Introduction to Algorithms, 3rd edn. MIT Press, Cambridge MA
14. Ebert J (1987) A versatile data structure for edge-oriented graph algorithms. Communications of the ACM 30(6):513–519
15. Even S (1979) Graph Algorithms. Computer Science Press, Rockville MD
16. Feigenbaum J, Kannan S, Vardi MY, Viswanathan M (1998). In: Morvan M, Meinel C, Krob D (eds) Complexity of problems on graphs represented as ordered binary decision diagrams, vol 1373. Springer, Berlin Heidelberg, pp 216–226
17. Furnas GW (1984) The generation of random, binary unordered trees. Journal of Classification 1(1):187–233
18. Galperin H, Wigderson A (1983) Succint representations of graphs. Information and Control 56(3):183–198
19. Gibbons A (1985) Algorithmic Graph Theory. Cambridge University Press, Cambridge, England
20. Golumbic MC (1980) Algorithmic Graph Theory and Perfect Graphs. Academic Press, New York NY
21. Goodrich MT, Tamassia R, Goldwasser MH (2013) Data Structures and Algorithms in Python. John Wiley & Sons Inc, Hoboken NJ
22. Gross J, Tucker TW (1987) Topological Graph Theory. John Wiley & Sons, New York NY
23. Gross J, Yellen J (1999) Graph Theory and its Applications. CRC Press, Boca Raton FL
24. Hagberg AA, Schult DA, Swart PJ (2008). In: Varoquaux G, Vaught T, Millman J (eds) Exploring network structure, dynamics, and function using NetworkX. SciPy.org, Pasadena CA, pp 11–16

25. Harary F (1969) Graph Theory. Addison-Wesley, Reading MA
26. He X, Kao MY, Lu HI (1999) Linear-time succinct encodings of planar graphs via canonical orderings. SIAM Journal on Discrete Mathematics 12(3):317–325
27. He X, Kao MY, Lu HI (2000) A fast general methodology for information-theoretically optimal encodings of graphs. SIAM Journal on Computing 30(3):838–846
28. Hopcroft JE, Tarjan RE (1973) An $O(n \log n)$ algorithm for isomorphism of triconnected planar graphs. Journal of Computer and System Sciences 7(3):323–331
29. Hopkins T (2002) Renovating the collected algorithms from ACM. ACM Transactions on Mathematical Software 28(1):59–74
30. Hopkins T (2009) The collected algorithms of the ACM. WIREs Computational Statistics 1(3):316–324
31. Johnson DS, McGeoch CC (eds) (1993) Network Flows and Matching: First DIMACS Implementation Challenge, DIMACS: Series in Discrete Mathematics and Theoretical Computer Science, vol 12. American Mathematical Society, Providence RI
32. Johnson DS, Trick MA (eds) (1996) Cliques, Coloring, and Satisfiability: Second DIMACS Implementation Challenge, DIMACS: Series in Discrete Mathematics and Theoretical Computer Science, vol 26. American Mathematical Society, Providence RI
33. Jungnickel D (1999) Graphs. Networks and Algorithms. Springer, Berlin Heidelberg
34. Kannan S, Moor M, Rudich S (1992) Implicit representation of graphs. Discrete Applied Mathematics 5(4):596–603
35. Knuth DE (1993) The Stanford GraphBase: A Platform for Combinatorial Computing. ACM Press, New York NY
36. Knuth DE (1997) Fundamental Algorithms, The Art of Computer Programming, vol 1, 3rd edn. Addison-Wesley, Reading MA
37. Knuth DE (1998) Seminumerical Algorithms, The Art of Computer Programming, vol 2, 3rd edn. Addison-Wesley, Reading MA
38. Knuth, D.E.: Sorting and Searching, (1998) The Art of Computer Programming, vol 3, 3rd edn. Addison-Wesley, Reading MA
39. Kreher DL, Stinson DR (1999) Combinatorial Algorithms: Generation, Enumeration, and Search. CRC Press, Boca Raton FL
40. van Leeuwen, J.: Graph algorithms. In: J. van Leeuwen (ed.) Handbook of Theoretical Computer Science, vol. A, chap. 10, pp. 525–631. Elsevier, Amsterdam (1990)
41. McHugh JA (1990) Algorithmic Graph Theory. Prentice Hall, Englewood Cliffs NJ
42. Mehlhorn K (1984) Graph Algorithms and NP-Completeness, Data Structures and Algorithms, vol 2. Springer, Berlin Heidelberg
43. Mehlhorn K, Näher S (1999) The LEDA Platform of Combinatorial and Geometric Computing. Cambridge University Press, Cambridge, England
44. Mäkinen E (1999) Generating random binary trees: A survey. Information Sciences 115(1–4):123–136
45. Mlinarić D, Mornar V, Milašinović B (2020) Generating trees for comparison. Computers 9(2):35
46. Nishizeki T, Chiba N (1988) Planar Graphs: Theory and Algorithms, North-Holland Mathematical Studies, vol 140. North-Holland, Amsterdam
47. Ore O (1962) Theory of Graphs, Colloquium Publications, vol 38. American Mathematical Society, Providence RI
48. Papadimitriou CH, Steiglitz K (1998) Combinatorial Optimization: Algorithms and Complexity. Dover, Mineola NY
49. Papadimitriou CH, Yannakakis M (1987) A note on succint representations of graphs. Information and Control 71(3):181–185
50. Prüfer H (1918) Neuer Beweis eines Satzes über Permutationen. Archiv der Mathematik und Physik 27(1):142–144

51. Quiroz AJ (1989) Fast random generation of binary, t-ary and other types of trees. Journal of Classification 6(1):223–231
52. Read RC, Wilson RJ (1998) An Atlas of Graphs. Oxford University Press, Oxford, England
53. Reingold EM, Nievergelt J, Deo NJ (1977) Combinatorial Algorithms: Theory and Practice. Prentice Hall, Englewood Cliffs NJ
54. Sedgewick R (1992) Algorithms in C++. Addison-Wesley, Reading MA
55. Siek JG, Lee LQ, Lumsdaine A (2001) The Boost Graph Library: User Guide and Reference Manual. Addison-Wesley, Reading MA
56. Skiena SS (1998) The Algorithm Design Manual, 1st edn. Springer, Berlin Heidelberg
57. Talamo M, Vocca P (2001) Representing graphs implicitly using almost optimal space. Discrete Applied Mathematics 108(1–2):193–210
58. Tarjan RE (1972) Depth-first search and linear graph algorithms. SIAM Journal on Computing 1(2):146–160
59. Tarjan RE (1983) Space-efficient implementations of graph search methods. ACM Transactions on Mathematical Software 9(3):326–339
60. Torán G (1984) Succinct representations of graphs. Discrete Applied Mathematics 8(1):289–294
61. Tutte WT (1998) Graph Theory As I Have Known It, Oxford Lecture Series in Mathematics and its Applications, vol 11. Oxford University Press, Oxford, England
62. Weiss MA (1999) Data Structures and Algorithm Analysis in C++, 2nd edn. Addison-Wesley, Reading MA
63. White AT (2001) Graphs of Groups on Surfaces: Interactions and Models, North-Holland Mathematical Studies, vol 188. North-Holland, Amsterdam
64. Wilf HS (1989) Combinatorial Algorithms: An Update. SIAM, Philadelphia PA

Algorithmic Techniques

2.1 The Tree Edit Distance Problem

The problem of transforming or *editing* trees is a primary means of tree comparison, with practical applications in combinatorial pattern matching, pattern recognition, chemical structure search, computational molecular biology, and other areas of engineering and life sciences. In the most general sense, trees can be transformed by the application of elementary edit operations, which have a certain cost or weight associated with them, and the edit distance between two trees is the cost of a least-cost sequence of elementary edit operations, or the length of the shortest sequence of elementary edit operations, that allows to transform one of the trees into the other.

Rooted ordered trees are considered in this chapter, with the following elementary edit operations: deletion of a leaf node from a tree, insertion of a new leaf node into a tree, and substitution or replacement of a node in a tree with a node in another tree.

Definition 2.1 Let $T_1 = (V_1, E_1)$ and $T_2 = (V_2, E_2)$ be ordered trees. An **elementary edit operation** on T_1 and T_2 is either the **deletion** from T_1 of a leaf node $v \in V_1$, denoted by $v \mapsto \lambda$ or (v, λ); the **substitution** or replacement of a node $v \in V_1$ with a node $w \in V_2$, denoted by $v \mapsto w$ or (v, w); or the **insertion** into T_2 of a node $w \notin V_2$ as a new leaf, denoted by $\lambda \mapsto w$ or (λ, w). Deletion of a non-root node $v \in V_1$ implies the deletion of the edge $(parent[v], v) \in E_1$, and insertion of a non-root node $w \notin V_2$ as a child of a leaf node $parent[w] \in V_2$ implies insertion of an edge $(parent[w], w) \notin E_2$.

Deletion and insertion operations are thus made on leaves only. The deletion of a non-leaf node requires first the deletion of the whole subtree rooted at the node, and the same applies to the insertion of non-leaves.

Now, a tree can be transformed into another tree by application of a sequence of elementary edit operations. Recall that a relation $R \subseteq A \times B$ is a set of ordered

© The Author(s), under exclusive license to Springer Nature Switzerland AG 2021
G. Valiente, *Algorithms on Trees and Graphs*, Texts in Computer Science,
https://doi.org/10.1007/978-3-030-81885-2_2

pairs $\{(a, b) \mid a \in A, b \in B\}$. An *ordered relation* $R \subseteq A \times B$ is an ordered set, or sequence, of ordered pairs $[(a, b) \mid a \in A, b \in B]$.

Definition 2.2 Let $T_1 = (V_1, E_1)$ and $T_2 = (V_2, E_2)$ be ordered trees. A **transformation** of T_1 into T_2 is an ordered relation $E \subseteq (V_1 \cup \{\lambda\}) \times (V_2 \cup \{\lambda\})$ such that

- $\{v \in V_1 \mid (v, w) \in E, w \in V_2 \cup \{\lambda\}\} = V_1$.
- $\{w \in V_2 \mid (v, w) \in E, v \in V_1 \cup \{\lambda\}\} = V_2$.
- $(v_1, w), (v_2, w) \in E$ implies $v_1 = v_2$, for all nodes $v_1, v_2 \in V_1 \cup \{\lambda\}$ and $w \in V_2$.
- $(v, w_1), (v, w_2) \in E$ implies $w_1 = w_2$, for all nodes $v \in V_1$ and $w_1, w_2 \in V_2 \cup \{\lambda\}$.

Remark 2.1 Notice that in a transformation $E \subseteq (V_1 \cup \{\lambda\}) \times (V_2 \cup \{\lambda\})$ of an ordered tree $T_1 = (V_1, E_1)$ into an ordered tree $T_2 = (V_2, E_2)$, the relation $E \cap V_1 \times V_2$ is a bijection. In fact, condition $(v_1, w), (v_2, w) \in E$ implies $v_1 = v_2$ establishes injectivity, and $(v, w_1), (v, w_2) \in E$ implies $w_1 = w_2$ establishes the surjectivity of $E \cap V_1 \times V_2$.

Example 2.1 A transformation of an ordered tree T_1 into another ordered tree T_2 is illustrated in Fig. 2.1. The transformation consists of deleting leaves v_3, v_4 from T_1, substituting or replacing nodes v_1, v_2, v_5 of T_1 with nodes w_1, w_2, w_3 of T_2, respectively, and inserting leaves w_4, w_5, w_6, w_7 into T_2. Substitution of corresponding nodes is left implicit in the figure. The transformation is denoted by $[v_1 \mapsto w_1, v_2 \mapsto w_2, v_3 \mapsto \lambda, v_4 \mapsto \lambda, v_5 \mapsto w_3, \lambda \mapsto w_4, \lambda \mapsto w_5, \lambda \mapsto w_6, \lambda \mapsto w_7]$ and also $[(v_1, w_1), (v_2, w_2), (v_3, \lambda), (v_4, \lambda), (v_5, w_3), (\lambda, w_4), (\lambda, w_5), (\lambda, w_6), (\lambda, w_7)]$.

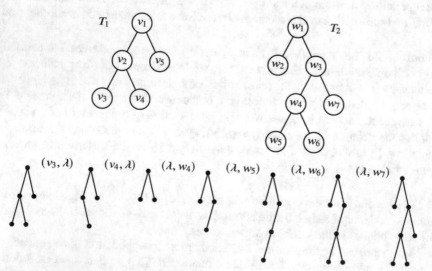

Fig. 2.1 A transformation between two ordered trees. Nodes are numbered according to the order in which they are visited during a preorder traversal

Not every sequence of elementary edit operations corresponds to a valid transformation between two ordered trees, though. On the one hand, deletions and insertions must appear in a bottom-up order—for instance, according to a postorder traversal of the trees—in order to ensure that deletions and insertions are indeed made on leaves. In the postorder traversal of an ordered tree, which will be introduced in Sect. 3.2, a postorder traversal of the subtree rooted in turn at each of the children of the root is performed first, and the root of the tree is visited last, where the subtree rooted at the first child is traversed first, followed by the subtree rooted at the next sibling, etc.

On the other hand, for a tree transformation to be valid, both parent and sibling order must be preserved by the transformation, in order to ensure that the result of the transformation is indeed an ordered tree. That is, in a valid transformation of an ordered tree T_1 into another ordered tree T_2, the parent of a non-root node of T_1 which is substituted or replaced with a non-root node of T_2 must be substituted or replaced with the parent of the node of T_2. Further, whenever sibling nodes of T_1 are substituted or replaced with sibling nodes of T_2, the substitution must preserve their relative order.

The latter requirement for a transformation between two ordered trees to be valid is formalized by the notion of *mapping*, also known as *trace*.

Definition 2.3 Let $T_1 = (V_1, E_1)$ and $T_2 = (V_2, E_2)$ be ordered trees, let $W_1 \subseteq V_1$, and let $W_2 \subseteq V_2$. A **mapping** M of T_1 to T_2 is a bijection $M \subseteq W_1 \times W_2$ such that

- $(root[T_1], root[T_2]) \in M$ if $M \neq \emptyset$.
- $(v, w) \in M$ only if $(parent[v], parent[w]) \in M$, for all non-root nodes $v \in W_1$ and $w \in W_2$.
- v_1 is a left sibling of v_2 if and only if w_1 is a left sibling of w_2, for all nodes $v_1, v_2 \in W_1$ and $w_1, w_2 \in W_2$ with $(v_1, w_1), (v_2, w_2) \in M$.

Example 2.2 The mapping M corresponding to the transformation of $T_1 = (V_1, E_1)$ to $T_2 = (V_2, E_2)$ given in Example 2.1 is illustrated in Fig. 2.2. As a matter of fact, $M \subseteq \{v_1, v_2, v_5\} \times \{w_1, w_2, w_3\} \subseteq V_1 \times V_2$ is indeed a bijection such that node v_1, the parent of node v_2, is mapped to node w_1, the parent of node w_2, because node v_2 is mapped to node w_2 and, furthermore, node v_2, a left sibling

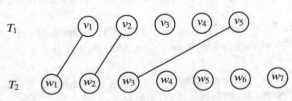

Fig. 2.2 The mapping underlying the tree transformation of Fig. 2.1. Nodes are numbered according to the order in which they are visited during a preorder traversal

of node v_5, is mapped to node w_2, a left sibling of node w_3, and node v_5 is mapped to node w_3.

Lemma 2.1 *Let M be a mapping of an ordered tree $T_1 = (V_1, E_1)$ to another ordered tree $T_2 = (V_2, E_2)$. Then, depth$[v] = depth[w]$ for all $(v, w) \in M$.*

Proof Let $T_1 = (V_1, E_1)$ and $T_2 = (V_2, E_2)$ be ordered trees, and let $M \subseteq W_1 \times W_2$ be a mapping of T_1 to T_2, where $W_1 \subseteq V_1$, and $W_2 \subseteq V_2$. Suppose $depth[v] \neq depth[w]$ for some nodes $v \in V_1$ and $w \in V_2$ with $(v, w) \in M$ and assume, without loss of generality, that $depth[v] < depth[w]$.

By induction on the depth of node v. If $depth[v] = 0$, then $v = root[T_1]$, and it must be $w \neq root[T_2]$, because $depth[w] > 0 = depth[v]$ and, since M is a bijection, $(root[T_1], root[T_2]) \notin M \neq \emptyset$, contradicting the assumption that M is a mapping. Then, $depth[v] = depth[w]$.

Assume now that $depth[x] = depth[y]$ for all $(x, y) \in M$ with $depth[x] \leqslant n$, where $n < depth[T_1]$. If $depth[v] = n + 1$, it must be $(x, y) \in M$ for some children $x \in V_1$ and $y \in V_2$ with $parent[x] = v$ and $parent[y] = w$, because M is a mapping, and $depth[x] = depth[y] = n$ by induction hypothesis. Then, $depth[v] = depth[x] + 1 = depth[y] + 1 = depth[w]$. Therefore, $depth[v] = depth[w]$ for all $(v, w) \in M$. $\square$

A transformation is thus valid if node deletions and insertions are made on leaves only, and node substitutions constitute a mapping.

Definition 2.4 Let $T_1 = (V_1, E_1)$ and $T_2 = (V_2, E_2)$ be ordered trees. A transformation $E \subseteq (V_1 \cup \{\lambda\}) \times (V_2 \cup \{\lambda\})$ of T_1 into T_2 is said to be a **valid transformation** if

- (v_j, λ) occurs before (v_i, λ) in E, for all $(v_i, \lambda), (v_j, \lambda) \in E \cap V_1 \times \{\lambda\}$ such that node v_j is a descendant of node v_i in T_1.
- (λ, w_i) occurs before (λ, w_j) in E, for all $(\lambda, w_i), (\lambda, w_j) \in E \cap \{\lambda\} \times V_2$ such that node w_j is a descendant of node w_i in T_2.
- $E \cap V_1 \times V_2$ is a mapping of T_1 to T_2.

Now, the elementary edit operations of deletion, substitution, and insertion are, as a matter of fact, sufficient for the transformation of any tree into any other tree.

Lemma 2.2 *Let $T_1 = (V_1, E_1)$ and $T_2 = (V_2, E_2)$ be ordered trees. Then, there is a sequence of elementary edit operations transforming T_1 into T_2.*

Proof Let $[v_i, \ldots, v_j]$ be the set of nodes V_1 in the order in which they are first visited during a postorder traversal of T_1, and let $[w_k, \ldots, w_\ell]$ be the set of nodes V_2 in the order in which they are first visited during a postorder traversal of T_2. Then, the sequence of elementary edit operations $[v_i \mapsto \lambda, \ldots, v_j \mapsto \lambda, \lambda \mapsto w_k, \ldots, \lambda \mapsto w_\ell]$ allows to transform T_1 into T_2. $\square$

It follows from the proof of Lemma 2.2 that the elementary edit operation of substitution is not necessary for the transformation of a tree into another tree. However, substitutions are convenient because they allow to find shortest or, in general, least-cost transformations between trees.

Definition 2.5 Let $T_1 = (V_1, E_1)$ and $T_2 = (V_2, E_2)$ be ordered trees. The **cost** of an elementary edit operation on T_1 and T_2 is given by a function $\gamma : V_1 \cup V_2 \cup \{\lambda\} \times V_1 \cup V_2 \cup \{\lambda\} \to \mathbb{R}$ such that

- $\gamma(v, w) \geqslant 0$,
- $\gamma(v, w) = 0$ if and only if $v \neq \lambda$ and $w \neq \lambda$,
- $\gamma(v, w) = \gamma(w, v)$, and
- $\gamma(v, w) \leqslant \gamma(v, z) + \gamma(z, w)$

for all $v, w, z \in V_1 \cup V_2 \cup \{\lambda\}$.

The first two conditions in Definition 2.5 are known as *nonnegative definiteness*, the third condition is the *symmetry* of the cost function, and the fourth condition is known as *triangularity* or *triangular inequality*. Together, these conditions on the cost of elementary edit operations turn ordered trees into a metric space.

Definition 2.6 Let $T_1 = (V_1, E_1)$ and $T_2 = (V_2, E_2)$ be ordered trees. The **cost** of a transformation $E \subseteq (V_1 \cup \{\lambda\}) \times (V_2 \cup \{\lambda\})$ of T_1 into T_2 is given by $\gamma(E) = \sum_{(v,w) \in E} \gamma(v, w)$.

Now, the edit distance between two ordered trees is the cost of a least-cost valid transformation between then trees.

Definition 2.7 The **edit distance** between ordered trees T_1 and T_2 is $\delta(T_1, T_2) = \min\{\gamma(E) \mid E \text{ is a valid transformation of } T_1 \text{ to } T_2\}$.

Example 2.3 Assume that the cost of elementary edit operations is such that $\gamma(v, w) = 1$ if either $v = \lambda$ or $w = \lambda$, and $\gamma(v, w) = 0$ otherwise. That is, node deletions and insertions have cost equal to one, and node substitutions have zero cost. Then, the cost of the transformation of T_1 into T_2 given in Example 2.1 is equal to 6. A least-cost valid transformation of T_1 into T_2 is given by $[(v_1, w_1), (\lambda, w_2), (v_2, w_3), (v_3, w_4), (\lambda, w_5), (\lambda, w_6), (v_4, w_7), (v_5, \lambda)]$, and the edit distance between T_1 and T_2 is thus equal to 4.

An alternative formulation of the tree edit problem consists of building a kind of grid graph on the nodes of the ordered trees, called the *edit graph* of the trees, and then reducing the problem of finding a valid transformation of one tree into the other to that of finding a path in the edit graph from one corner down to the opposite corner.

Recall that in the preorder traversal of an ordered tree, which will be discussed in more detail in Sect. 3.1, the root of the tree is visited first, followed by a preorder traversal of the subtree rooted in turn at each of the children of the root, where the subtree rooted at the first child is traversed first, followed by the subtree rooted at the next sibling, etc.

Definition 2.8 Let $T_1 = (V_1, E_1)$ and $T_2 = (V_2, E_2)$ be ordered trees. The **edit graph** of T_1 and T_2 has a vertex of the form vw for each pair of nodes $v \in \{v_0\} \cup V_1$ and $w \in \{w_0\} \cup V_2$, where $v_0 \notin V_1$ and $w_0 \notin V_2$ are dummy nodes. Further,

- $(v_i w_j, v_{i+1} w_j) \in E$ if and only if $depth[v_{i+1}] \geqslant depth[w_{j+1}]$,
- $(v_i w_j, v_{i+1} w_{j+1}) \in E$ if and only if $depth[v_{i+1}] = depth[w_{j+1}]$, and
- $(v_i w_j, v_i w_{j+1}) \in E$ if and only if $depth[v_{i+1}] \leqslant depth[w_{j+1}]$

for $0 \leqslant i < n_1$ and $0 \leqslant j < n_2$, where nodes are numbered according to the order in which they are visited during a preorder traversal of the trees. Moreover, $(v_i w_{n_2}, v_{i+1} w_{n_2}) \in E$ for $0 \leqslant i < n_1$, and $(v_{n_1} w_j, v_{n_1} w_{j+1}) \in E$ for $0 \leqslant j < n_2$.

In the edit graph of ordered trees T_1 and T_2, a *vertical* edge of the form $(v_i w_j, v_{i+1} w_j)$ represents the deletion of node v_{i+1} from T_1, where the condition $depth[v_{i+1}] \geqslant depth[w_{j+1}]$ ensures that node w_{j+1} does not belong to the subtree of T_2 rooted at node w_j, that is, node w_{j+1} does not have to be inserted into T_2 before node v_{i+1} can be deleted from T_1. Further, a *diagonal* edge of the form $(v_i w_j, v_{i+1} w_{j+1})$ represents the substitution or replacement of node v_{i+1} of T_1 with node w_{j+1} of T_2, where the condition $depth[v_{i+1}] = depth[w_{j+1}]$ follows from Lemma 2.1. A *horizontal* edge, on the other hand, of the form $(v_i w_j, v_i w_{j+1})$ represents the insertion of node w_{j+1} into T_2, where the condition $depth[v_{i+1}] \leqslant depth[w_{j+1}]$ ensures that node v_{i+1} does not belong to the subtree of T_1 rooted at node v_i, that is, node v_{i+1} does not have to be deleted from T_1 before node w_{j+1} can be inserted into T_2.

Missing horizontal and diagonal edges ensure thus that once a path in the edit graph traverses a vertical edge, corresponding to the deletion of a certain node, that path can only be extended by traversing further vertical edges, corresponding to the deletion of all the nodes in the subtree rooted at that node. Conversely, missing vertical and diagonal edges ensure that once a path in the edit graph traverses a horizontal edge, corresponding to the insertion of a certain node, that path can only be extended by traversing further horizontal edge, corresponding to the insertion of all the nodes in the subtree rooted at that node.

Example 2.4 The edit graph for the trees of Fig. 2.1 in shown in Fig. 2.3. In the drawing of a tree edit graph, nodes are placed in a rectangular grid in the order in which they are visited during a preorder traversal of the trees, deletions correspond to vertical edges, substitutions correspond to diagonal edges, and insertions correspond to horizontal edges. Further, edges are indeed arcs, directed from left to right and from top to bottom. The highlighted path in Fig. 2.3 corresponds to the transformation of

Fig. 2.3 Edit graph for the
trees of Fig. 2.1. The path in
the graph corresponding to
the valid transformation of
T_1 to T_2 given in
Example 2.1 is shown
highlighted

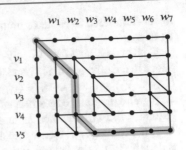

T_1 to T_2 given in Example 2.1, consisting of the substitution or replacement of v_1 with w_1, the substitution or replacement of v_2 with w_2, the deletion of v_3 and v_4 from T_1, the substitution or replacement of v_5 with w_3, and the insertion of w_4, w_5, w_6, and w_7 into T_2.

The correspondence of valid transformations between ordered trees T_1 and T_2 and paths in the edit graph of T_1 and T_2 from the top-left to the bottom-right corner is given by the fact that node substitutions (diagonal edges) along a path in the edit graph of T_1 and T_2 from the top-left to the bottom-right corner constitute a mapping of T_1 to T_2, which can thus be extended to a valid transformation of T_1 into T_2.

Lemma 2.3 *Let P be a path in the edit graph of ordered trees $T_1 = (V_1, E_1)$ and $T_2 = (V_2, E_2)$ from the top-left to the bottom-right corner, and let $M = \{(v_{i+1}, w_{j+1}) \in V_1 \times V_2 \mid (v_i w_j, v_{i+1} w_{j+1}) \in P\}$. Then, M is a mapping of T_1 to T_2. Conversely, let $M \subseteq V_1 \times V_2$ be a mapping of T_1 to T_2. Then, there is a path P in the edit graph of T_1 and T_2 from the top-left to the bottom-right corner such that $\{(v_{i+1}, w_{j+1}) \in V_1 \times V_2 \mid (v_i w_j, v_{i+1} w_{j+1}) \in P\} = M$.*

Proof Let $T_1 = (V_1, E_1)$ and $T_2 = (V_2, E_2)$ be ordered trees, let P be a path in the edit graph of T_1 and T_2 from the top-left to the bottom-right corner, and let $M = \{(v_{i+1}, w_{j+1}) \in V_1 \times V_2 \mid (v_i w_j, v_{i+1} w_{j+1}) \in P\}$. If $(v_1, w_1) \notin M$, then $(v_0 v_1, w_0 w_1) \notin P$ and then, P has no diagonal edges and thus $M = \emptyset$, that is, M is a trivial mapping of T_1 to T_2.

Otherwise, $(v_1, w_1) \in M$ and, since nodes of T_1 and T_2 are numbered according to the order in which they are visited during a preorder traversal of the trees, $root[T_1] = v_1$ and $root[T_2] = w_1$. Then, $(root[T_1], root[T_2]) \in M$.

Let now $v_{i+1} \in V_1$ and $w_{j+1} \in V_2$ be nodes with $(v_{i+1}, w_{j+1}) \in M$, where $i, j \neq 0$, let $v_{k+1} = parent[v_{i+1}]$, let $w_{\ell+1} = parent[w_{j+1}]$, and suppose $(v_{k+1}, w_{\ell+1}) \notin M$. Since $(v_k w_\ell, v_{k+1} w_{\ell+1}) \notin P$, there is no path contained in P from node $v_k w_\ell$ to node $v_i w_j$, because $k < i$, $\ell < j$, $depth[v_k] < depth[v_i]$, and $depth[w_\ell] < depth[w_j]$, contradicting thus the hypothesis that $(v_{i+1}, w_{j+1}) \in M$, that is, $(v_i w_j, v_{i+1} w_{j+1}) \in P$. Then, $(parent[v_{i+1}], parent[w_{j+1}]) \in M$.

Finally, let $v_{i+1}, v_{k+1} \in V_1$ and $w_{j+1}, w_{\ell+1} \in V_2$, $i, j, k, \ell \neq 0$, be nodes with $(v_{i+1}, w_{j+1}), (v_{k+1}, w_{\ell+1}) \in M$. Then, v_{i+1} is a left sibling of v_{k+1} if and only if w_{j+1} is a left sibling of $w_{\ell+1}$, because nodes in the edit graph are numbered according to a preorder traversal of the trees, and edges in the edit graph go from lower-numbered to equal or higher-numbered nodes. Therefore, M is a mapping of T_1 to T_2.

Conversely, let $M \subseteq V_1 \times V_2$ be a mapping of T_1 to T_2, and let X be the set of edges in the edit graph of T_1 and T_2 induced by M, that is, $X = \{(v_i w_j, v_{i+1} w_{j+1}) \in (\{v_0\} \cup V_1) \times (\{w_0\} \cup V_2) \mid (v_{i+1}, w_{j+1}) \in M\}$. Since M is an injection, there are no parallel diagonal edges in X of the form $(v_i w_j, v_{i+1} w_{j+1})$ and $(v_k w_j, v_{k+1} w_{j+1})$ with $i \neq k$. Further, since M is a mapping, there are no parallel diagonal edges in X of the form $(v_i w_j, v_{i+1} w_{j+1})$ and $(v_i w_\ell, v_{i+1} w_{\ell+1})$ with $j \neq \ell$. Therefore, there is a path $P \supseteq X$ in the edit graph of T_1 and T_2 from the top-left to the bottom-right corner such that $\{(v_{i+1}, w_{j+1}) \in V_1 \times V_2 \mid (v_i w_j, v_{i+1} w_{j+1}) \in P\} = M$. $\square$

Remark 2.2 Notice that a valid transformation of T_1 into T_2 can be obtained from a path in the edit graph of T_1 and T_2 from the top-left to the bottom-right corner, by just reordering the edit operations between each pair of successive node substitutions according to the preorder traversal of the trees. That is, in such a way that deletion and insertion of children nodes do not precede deletion and insertion of their parent or their left sibling nodes.

Now, computing the edit distance between two ordered trees can be reduced to the problem of finding the shortest path in the edit graph of the trees, with edges in the edit graph labeled or weighted by the cost of the respective tree edit operation.

Lemma 2.4 *Let G be the edit graph of ordered trees $T_1 = (V_1, E_1)$ and $T_2 = (V_2, E_2)$, with edges of the form $(v_i w_j, v_k w_\ell) \in E$ weighted by $\gamma(v_k, \lambda)$ if $k = i + 1$ and $\ell = j$, weighted by $\gamma(v_k, w_\ell)$ if $k = i + 1$ and $\ell = j + 1$, and weighted by $\gamma(\lambda, w_\ell)$ if $k = i$ and $\ell = j + 1$. Let also P be a shortest path in G from the top-left to the bottom-right corner, and let $E = \{(v_{i+1}, \lambda) \in V_1 \times \{\lambda\} \mid (v_i w_j, v_{i+1} w_j) \in P\} \cup \{(v_{i+1}, w_{j+1}) \in V_1 \times V_2 \mid (v_i w_j, v_{i+1} w_{j+1}) \in P\} \cup \{(\lambda, w_{j+1}) \in \{\lambda\} \times V_2 \mid (v_i w_j, v_i w_{j+1}) \in P\}$. Then, $\delta(T_1, T_2) = \sum_{(v,w) \in E} \gamma(v, w)$.*

Proof Let G be the weighted edit graph of ordered trees $T_1 = (V_1, E_1)$ and $T_2 = (V_2, E_2)$, let P be a shortest path in G from the top-left to the bottom-right corner, and let $E = \{(v_{i+1}, \lambda) \in V_1 \times \{\lambda\} \mid (v_i w_j, v_{i+1} w_j) \in P\} \cup \{(v_{i+1}, w_{j+1}) \in V_1 \times V_2 \mid (v_i w_j, v_{i+1} w_{j+1}) \in P\} \cup \{(\lambda, w_{j+1}) \in \{\lambda\} \times V_2 \mid (v_i w_j, v_i w_{j+1}) \in P\}$. Then, $E \cap V_1 \times V_2$ is a mapping of T_1 to T_2, by Lemma 2.3. Further, E can be reordered into a valid transformation of T_1 into T_2, by Remark 2.2 and then, $\delta(T_1, T_2) = \sum_{(v,w) \in E} \gamma(v, w)$ by Definition 2.7. $\square$

Example 2.5 A shortest path in the edit graph for the trees of Fig. 2.1, corresponding to the least-cost transformation of T_1 to T_2 given in Example 2.3, is shown highlighted in Fig. 2.4.

The following function builds the edit graph G of ordered trees T_1 and T_2. The vertices of the edit graph are labeled by the preorder number of the nodes of T_1

Fig. 2.4 Shortest path in the edit graph of two ordered trees

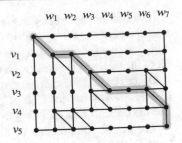

and T_2 they correspond to, and the edges are labeled by a string indicating the tree edit operation they correspond to: "del" for deletion of a node from T_2, "sub" for substitution or replacement of a node of T_1 with a node of T_2, and "ins" for insertion of a node into T_2.

```
function tree_edit_graph(T_1, T_2)
    let G be an empty graph
    let n_1 and n_2 be the number of nodes of T_1 and T_2
    preorder_tree_traversal(T_1)
    preorder_tree_traversal(T_2)
    preorder_tree_depth(T_1)
    preorder_tree_depth(T_2)
    let d_1[1..n_1] and d_2[1..n_2] be new arrays (of integers)
    for all nodes v of T_1 do
        d_1[v.order] ← v.depth
    for all nodes w of T_2 do
        d_2[w.order] ← w.depth
    let A[0..n_1][0..n_2] be a new matrix (of nodes)
    for i = 0 to n_1 do
        for j = 0 to n_2 do
            add a new vertex A[i, j] to G labeled [i, j]
    for i = 0 to n_1 − 1 do
        add a new edge to G from A[i, n_2] to A[i + 1, n_2] labeled "del"
    for j = 0 to n_2 − 1 do
        add a new edge to G from A[n_1, j] to A[n_1, j + 1] labeled "ins"
    for i = 0 to n_1 − 1 do
        for j = 0 to n_2 − 1 do
            if d_1[i + 1] ⩾ d_2[j + 1] then
                add a new edge to G from A[i, j] to A[i + 1, j] labeled "del"
            if d_1[i + 1] = d_2[j + 1] then
                add a new edge from A[i, j] to A[i + 1, j + 1] labeled "sub"
            if d_1[i + 1] ⩽ d_2[j + 1] then
                add a new edge to G from A[i, j] to A[i, j + 1] labeled "ins"
    return G
```

The cost of elementary tree edit operations can be set to $\gamma(v, w) = 1$, for all edit operations of the form $v \mapsto w$ with $v \in V_1 \cup \{\lambda\}$ and $w \in V_2 \cup \{\lambda\}$, by defining the weight or cost of the respective edges in the edit graph of the trees. However, since all of the edges in the edit graph correspond to these elementary tree edit operations, it suffices to find the shortest path in the edit graph without edge weights, by means of breadth-first traversal. See Sect. 5.2 for details about the breadth-first traversal and the breadth-first spanning tree of a graph.

The actual computation of the shortest path P in the edit graph G of ordered trees T_1 and T_2 is done by the following function. The shortest path P is recovered from the predecessor edges *pred* computed by the acyclic shortest path function (see Sect. 5.3), starting off with the last vertex of G (the bottom-right corner of the edit graph) and following up the thread of predecessor edges until reaching the first vertex of G (the top-left corner of the edit graph), which has no predecessor edge. The shortest path P is represented by a list of edges (of graph G).

> **function** *tree_edit*(T_1, T_2, P)
> $G \leftarrow tree\_edit\_graph(T_1, T_2)$
>   let s and t be the first vertex and the last vertex of G
> $(W, S) \leftarrow breadth\_first\_spanning\_subtree(G, s)$
>   **return** *acyclic_shortest_path*(G, s, t, W, S)

Lemma 2.5 *The tree edit algorithm based on shortest paths in the edit graph for finding a least-cost transformation of an ordered tree T_1 to an ordered tree T_2 with, respectively, n_1 and n_2 nodes runs in $O(n_1 n_2)$ time using $O(n_1 n_2)$ additional space.*

Proof Let $T_1 = (V_1, E_1)$ and $T_2 = (V_2, E_2)$ be ordered trees with, respectively, n_1 and n_2 nodes. The edit graph of T_1 and T_2 has $(n_1 + 1)(n_2 + 1)$ vertices and at most $3n_1 n_2$ edges, and building the edit graph takes thus $O(n_1 n_2)$ time and uses $O(n_1 n_2)$ additional space. Further, the breadth-first spanning tree procedure and the acyclic shortest path procedure also take $O(n_1 n_2)$ time. Therefore, the tree edit algorithm runs in $O(n_1 n_2)$ time using $O(n_1 n_2)$ additional space. $\qquad\square$

Throughout this chapter, the representation of the trees is assumed to be arranged in such a way that the order of the nodes fixed by the representation of each of the trees coincides with the order in which they are visited during a preorder traversal of the tree, as required by the formulation of the tree edit problem. Further, for any trees T_1 and T_2, the preorder number of all nodes in the trees is assumed to be stored in the attributes $v.order$ and $w.order$, for all nodes v of T_1 and w of T_2.

2.2 Backtracking

The solution to combinatorial problems on trees and graphs often requires an exhaustive search of the set of all possible solutions. As a matter of fact, most combinatorial problems on trees and graphs can be stated in the general framework of assigning values of a finite domain to each of a series of variables, where the value for each variable must be taken from a corresponding finite domain. An ordered set of values satisfying some specified constraints represents a solution to the problem, and an exhaustive search for a solution must consider the elements of the cartesian product of variable domains as potential solutions.

Consider, for instance, the tree edit problem. Let $T_1 = (V_1, E_1)$ and $T_2 = (V_2, E_2)$ be ordered trees. A potential solution to the problem of finding a transformation of T_1 into T_2 is given by a bijection $M \subseteq W_1 \times W_2$, where $W_1 \subseteq V_1$ and $W_2 \subseteq V_2$, which can also be seen as an assignment of a node $w \in V_2 \cup \{\lambda\}$ to each node $v \in V_1$. The dummy node λ represents the deletion of a node from T_1, and the insertion of a node into T_2 remains implicit, that is, all nodes $w \in V_2 \setminus W_2$ are inserted into T_2.

However, not every assignment of a node $w \in V_2 \cup \{\lambda\}$ to each node $v \in V_1$ corresponds to a valid transformation of T_1 into T_2. The bijection $M \subseteq W_1 \times W_2 \subseteq V_1 \times V_2$ must be a mapping of T_1 to T_2, according to Definition 2.4. Notice that in such a case, a valid transformation of T_1 into T_2 can be obtained by first deleting from T_1 all those nodes of T_1 which were assigned to the dummy node λ, in the order given by a preorder traversal of T_1; substituting or replacing all nodes of T_1 which were assigned to a non-dummy node of T_2 with the node they were assigned to; and inserting into T_2 all those nodes of T_2 which were not assigned to any node of T_1, in the order given by a preorder traversal of T_2.

Example 2.6 For the transformation of the ordered tree $T_1 = (V_1, E_1)$ to the ordered tree $T_2 = (V_2, E_2)$ of Fig. 2.1, there are $8^5 = 32,768$ possible assignments of the $8 = 7 + 1$ nodes of $V_2 \cup \{\lambda\}$ to the 5 nodes of V_1. However, only $9,276$ of them correspond to a bijection of V_1 to V_2. Further, 98 of these bijections map nodes of same depth, 20 of them also preserve the parent-child relation, and only 12 of them, shown in Fig. 2.5, also preserve sibling order and represent thus a valid transformation of T_1 to T_2. These valid transformations are further illustrated in Fig. 2.6.

Backtracking is a general technique for organizing the exhaustive search for a solution to a combinatorial problem. Roughly, the technique consists of repeatedly extending a partial solution to the problem—represented as an assignment of values of a finite domain, satisfying certain constraints, to an ordered finite set of variables—to a complete solution to the problem, by extending the representation of the partial solution of one variable at a time and shrinking the representation of the partial solution (backtracking) whenever a partial solution cannot be further extended.

The backtracking technique can be applied to those problems that exhibit the *domino principle*, meaning that if a constraint is not satisfied by a given partial solution, the constraint will not be satisfied by any extension of the partial solution at all. The domino principle allows to stop extending a partial solution as soon as

M	v_1	v_2	v_3	v_4	v_5
$\{\,\}$	λ	λ	λ	λ	λ
$\{(v_1, w_1)\}$	w_1	λ	λ	λ	λ
$\{(v_1, w_1), (v_5, w_2)\}$	w_1	λ	λ	λ	w_2
$\{(v_1, w_1), (v_5, w_3)\}$	w_1	λ	λ	λ	w_3
$\{(v_1, w_1), (v_2, w_2)\}$	w_1	w_2	λ	λ	λ
$\{(v_1, w_1), (v_2, w_2), (v_5, w_3)\}$	w_1	w_2	λ	λ	w_3
$\{(v_1, w_1), (v_2, w_3)\}$	w_1	w_3	λ	λ	λ
$\{(v_1, w_1), (v_2, w_3), (v_4, w_4)\}$	w_1	w_3	λ	w_4	λ
$\{(v_1, w_1), (v_2, w_3), (v_4, w_7)\}$	w_1	w_3	λ	w_7	λ
$\{(v_1, w_1), (v_2, w_3), (v_3, w_4)\}$	w_1	w_3	w_4	λ	λ
$\{(v_1, w_1), (v_2, w_3), (v_3, w_4), (v_4, w_7)\}$	w_1	w_3	w_4	w_7	λ
$\{(v_1, w_1), (v_2, w_3), (v_3, w_7)\}$	w_1	w_3	w_7	λ	λ

Fig. 2.5 Valid transformations between the ordered trees of Fig. 2.1

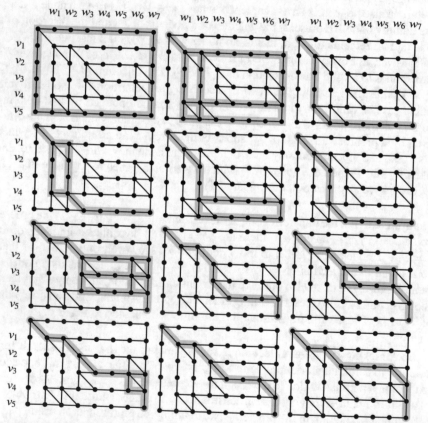

Fig. 2.6 Transformations between the ordered trees of Fig. 2.1. All paths in the edit graphs that correspond to a valid transformation of T_1 to T_2 are shown highlighted

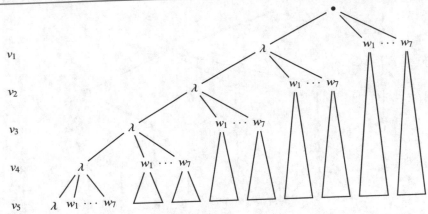

Fig. 2.7 Schematic view of the backtracking tree for the tree edit problem. Each of values $\lambda, w_1, \ldots, w_7$ can, in principle, be assigned to variables $v_1, \ldots, v_5$

it can be established that the extension will not lead to a solution to the problem, because the partial solution already violates some constraint.

The procedure of extending a partial solution toward a complete solution to the problem and shrinking a partial solution that cannot be further extended can be better understood in terms of a *backtracking tree* of possible assignments of values to the variables that represent the problem. The root of the backtracking tree is a dummy node, the nodes at level one represent the possible values which the first variable can be assigned to, the nodes at level two represent the possible values which the second variable can be assigned to, given the value which the first variable was assigned to, the nodes at level three represent the possible values which the third variable can be assigned to, given the values which the first and second variables were assigned to, and so on. The backtracking tree for an instance of the tree edit problem is illustrated in Fig. 2.7.

The domino principle allows *pruning* the backtracking tree at those nodes representing partial solutions that violate some constraint of the problem.

Finding a solution to a tree edit problem by backtracking means finding a node in the backtracking tree for which all the constraints of the problem are satisfied, and finding all solutions to the tree edit problem means finding all such nodes in the backtracking tree. These solution nodes will be leaves in the backtracking tree for a tree edit problem, because of the problem formulation, but this is not always the case.

Example 2.7 The backtracking tree for the transformation between the ordered trees of Fig. 2.1 is shown in Fig. 2.8. A path from the root to a leaf in the backtracking tree corresponds to a valid transformation of T_1 to T_2. Pruned subtrees are not shown, for clarity.

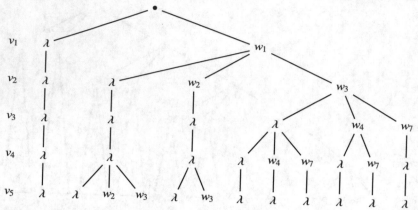

Fig. 2.8 Backtracking tree for the tree edit problem

Notice that the backtracking tree for a combinatorial problem is not explicit but remains implicit in the formulation of the problem. Further, the size of the backtracking tree is often exponential in the number of variables and possible values. Therefore, it is crucial for a backtracking procedure to build only that portion of the backtracking tree that is needed at each stage, during the search for one or more solutions to the problem.

Now, the following backtracking algorithm for enumerating node assignments that correspond to valid tree transformations extends an (initially empty) mapping M in all possible ways, in order to enumerate all assignments $M \subseteq V_1 \times V_2 \cup \{\lambda\}$ of nodes of an ordered tree $T_2 = (V_2, E_2)$, including a dummy node λ, to the nodes of an ordered tree $T_1 = (V_1, E_1)$. The node assignments found are collected in a list L of dictionaries of nodes (of tree T_1) to nodes (of tree T_2).

Further, the preorder number *order* and depth *depth* of the nodes in the trees will be used for testing the constraints at each stage, and the candidate nodes which each node can be mapped to are represented by a dictionary C of nodes (of tree T_1) to lists of nodes (of tree T_2).

```
function backtracking_tree_edit(T_1, T_2)
    preorder_tree_traversal(T_1)
    preorder_tree_traversal(T_2)
    let M be an empty dictionary (of nodes to nodes)
    let L be an empty list (of dictionaries of nodes to nodes)
    C ← set_up_candidate_nodes(T_1, T_2)
    let v be the first node of T_1 in preorder
    (M, L) ← extend_tree_edit(T_1, T_2, M, L, C, v)
    return L
```

The candidate nodes which a node of T_1 can be mapped to are just those nodes of T_2 of same depth as the node of T_1, together with the dummy node λ representing the deletion of the node from T_1. That is, $C[v] = \{w \in V_2 \mid depth[v] = depth[w]\} \cup \{\lambda\}$.

> **function** *set_up_candidate_nodes*(T_1, T_2)
>   *preorder_tree_depth*(T_1)
>   *preorder_tree_depth*(T_2)
>   add a new node $T_2.dummy$ to T_2
>   let C be an empty dictionary (of nodes to lists of nodes)
>   **for all** nodes v of T_1 **do**
>   let $C[v]$ be an empty list (of nodes)
>   append $T_2.dummy$ to $C[v]$
>   **for all** nodes w of T_2 **do**
>   **if** $w \neq T_2.dummy$ and $v.depth = w.depth$ **then**
>   append w to $C[v]$
>   **return** C

The actual extension of a partial solution is done by the following recursive function, which extends a node assignment $M \subseteq V_1 \times V_2 \cup \{\lambda\}$ by mapping node $v \in V_1$ to each node $w \in C[v] \subseteq V_2 \cup \{\lambda\}$ in turn, that is, to each remaining candidate node, making a recursive call upon the successor (in preorder) of node $v \in V_1$ whenever $M \cup \{(v, w)\}$ is a node assignment corresponding to a valid transformation of the subtree of T_1 induced by the nodes up to, and including node v into the subtree of T_2 induced by those nodes of T_2 which were already mapped to nodes of T_1.

All such node assignments M which are defined for all nodes of T_1, that is, all node assignments that correspond to a valid transformation of T_1 to T_2, are collected in a list L of node assignments. Notice that the actual backtracking is hidden in the mechanism used by the programming language to implement the recursion.

> **function** *extend_tree_edit*(T_1, T_2, M, L, C, v)
>   **for all** w in $C[v]$ **do**
> $M[v] \leftarrow w$
>   **if** v is the last node of T_1 in preorder **then**
>   append a copy of M to L
>   **else**
>   let N be a copy of C
> $N \leftarrow$ *refine_candidate_nodes*(T_1, T_2, N, v, w)
>   let v' be the next node after v in T_1 in preorder
> $(M, L) \leftarrow$ *extend_tree_edit*(T_1, T_2, M, L, N, v')
>   **return** (M, L)

After having assigned a node $v \in V_1$ to a node $w \in V_2 \cup \{\lambda\}$, the candidate nodes which the remaining nodes of T_1 can still be mapped to are found by removing from a copy N of C all those candidate nodes which violate some constraint.

First, if node $v \in V_1$ was just assigned to a non-dummy node $w \in V_2$, assigning another node of T_1 to node w would violate the constraint that the assignment be a bijection. All such occurrences of node $w \in V_2$ as a candidate which the remaining nodes of T_1 could be assigned to are removed from N.

Next, those candidate nodes $y \in V_2$ for each child x of node $v \in V_1$, which are not children of node $w \in V_2$, are removed from N.

Finally, those candidate nodes which violate a sibling order constraint are also removed from N, ensuring thus that no right sibling x of node $v \in V_1$ is mapped later to a left sibling y of node $w \in V_2$. Notice that for sibling nodes $v, x \in V_1$, node x is a right sibling of node v if and only if $preorder[v] < preorder[x]$. Further, for sibling nodes $w, y \in V_2$, node y is a left sibling of node w if and only if $preorder[y] < preorder[w]$.

> **function** $refine\_candidate\_nodes(T_1, T_2, C, v, w)$
>   **if** $w \neq T_2.dummy$ **then**
>   **for all** nodes x of T_1 **do**
>   **for all** y in $C[x]$ **do**
>   **if** $y = w$ **then**
>   delete y from $C[x]$
>   **for all** children x of node v **do**
>   **for all** y in $C[x]$ **do**
>   **if** $y \neq T_2.dummy$ and the parent of node y in T_2 is not w **then**
>   delete y from $C[x]$
>   **if** v is not the root of T_1 and $w \neq T_2.dummy$ **then**
>   **for all** children x of the parent of node v in T_1 **do**
>   **if** $v.order < x.order$ **then**
>   **for all** y in $C[x]$ **do**
>   **if** $y \neq T_2.dummy$ and $w.order > y.order$ **then**
>   delete y from $C[x]$
>   **return** C

The depth of all nodes in the trees is computed by the following procedure, during an iterative preorder traversal of the trees. See Sect. 3.1 for a detailed discussion of the procedure.

```
procedure preorder_tree_depth(T)
    let deepest be the first node of T
    for all nodes v of T in preorder do
        if v is the root of T then
            v.depth ← 0
        else
            v.depth ← parent[v].depth + 1
            if v.depth > deepest.depth then
                deepest ← v
    T.depth ← deepest.depth
```

2.3 Branch-and-Bound

The backtracking technique can be used for finding either one solution or all solutions to a combinatorial problem. When a cost can be associated with each partial solution, a least-cost solution can be found in a more efficient way by a simple variation of backtracking, called *branch-and-bound*. The technique consists of remembering the lowest-cost solution found at each stage of the backtracking search for a solution and using the cost of the lowest-cost solution found so far as a lower bound on the cost of a least-cost solution to the problem, in order to discard partial solutions as soon as it can be established that they will not improve on the lowest-cost solution found so far.

The branch-and-bound technique can be applied to those problems that exhibit an extension of the *domino principle*, meaning not only that if a constraint is not satisfied by a given partial solution, the constraint will not be satisfied by any extension of the partial solution at all—as in the case of backtracking—but also that the cost of any extension of a partial solution will be greater than or equal to the cost of the partial solution itself. Such an extended domino principle allows to stop extending a partial solution as soon as it can be established that the extension will not lead to a solution to the problem, because the partial solution already violates some constraint, and also as soon as it can be established that the extension will not lead to a least-cost solution to the problem, because the cost of the partial solution already reaches or exceeds the cost of the lowest-cost solution found so far.

As in the case of backtracking, the procedure of extending a partial solution toward a least-cost solution to the problem and shrinking a partial solution that cannot be further extended to a solution of lesser cost can be better understood in terms of a *branch-and-bound tree* of possible assignments of values to the variables that represent the problem. The root of the branch-and-bound tree is a dummy node of zero cost, the nodes at level one represent the possible values which the first variable can be assigned to, the nodes at level two represent the possible values which the second variable can be assigned to, given the value which the first variable was assigned to, the nodes at level three represent the possible values which the third variable can be assigned to, given the values which the first and second variables were assigned to, and so on. Further, $cost[v] \geqslant cost[w]$ for all leaves $v \in V_1$ and all successor nodes w of node v in a preorder traversal of the branch-and-bound tree. That is, subtrees in the backtracking tree rooted at nodes of cost greater than the cost of a previous leaf node are pruned off the branch-and-bound tree.

Consider, for instance, the tree edit problem. Let $T_1 = (V_1, E_1)$ and $T_2 = (V_2, E_2)$ be ordered trees, and recall that a solution to the problem of finding a valid transformation of T_1 into T_2 is given by an assignment $M \subseteq V_1 \times V_2 \cup \{\lambda\}$ of nodes of T_2 to the nodes of T_1, where the dummy node λ represents the deletion of a node from T_1, and all nodes in $\{w \in V_2 \mid \nexists v \in V_1, (v, w) \in M\}$ are inserted into T_2.

Let also the cost of a (partial) solution $M \subseteq V_1 \times V_2 \cup \{\lambda\}$ be given by the number of nodes deleted from T_1, that is, $cost[M] = |\{(v, w) \in M \mid w = \lambda\}|$. Then, finding a least-cost solution to a tree edit problem by branch-and-bound means finding a node of least cost in the branch-and-bound tree for which all the constraints of the problem are satisfied. Again, because of the problem formulation, a least-cost solution node will be a leaf in the branch-and-bound tree for a tree edit problem.

Example 2.8 The branch-and-bound tree for the transformation between the ordered trees of Fig. 2.1 is shown in Fig. 2.9. Leaves of cost lesser than the cost of all predecessor leaves in preorder are distinguished by showing their cost in a framed box, and the remaining leaves, of cost equal to the cost of their previous leaf in preorder, have their cost shown in a dashed box. The least-cost solution found is $\{(v_1, w_1), (v_2, w_3), (v_3, w_4), (v_4, w_7), (v_5, \lambda)\}$ and has cost 1.

Now, the following branch-and-bound algorithm for finding a node assignment that corresponds to a valid tree transformation of least cost extends an (initially empty) mapping M in all possible ways, in order to find a least-cost assignment $M \subseteq V_1 \times V_2 \cup \{\lambda\}$ of nodes of an ordered tree $T_2 = (V_2, E_2)$, including a dummy node λ, to the nodes of an ordered tree $T_1 = (V_1, E_1)$. Both the node assignment M tried at each stage and the least-cost node assignment A found are represented by a dictionary of nodes (of tree T_1) to nodes (of tree T_2).

Further, as in the case of backtracking, the preorder number $order$ and depth $depth$ of the nodes in the trees will be used for testing the constraints at each stage, and the candidate nodes which each node can be mapped to are represented by a dictionary C of nodes (of tree T_1) to lists of nodes (of tree T_2).

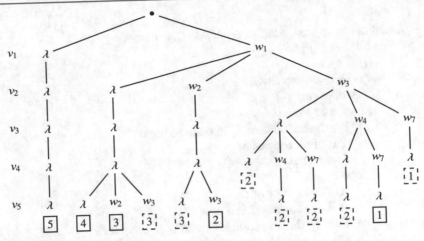

Fig. 2.9 Branch-and-bound tree for the tree edit problem

> **function** *branch_and_bound_tree_edit*(T_1, T_2)
> *preorder_tree_traversal*(T_1)
> *preorder_tree_traversal*(T_2)
> let M and A be empty dictionaries (of nodes to nodes)
> $C \leftarrow$ *set_up_candidate_nodes*(T_1, T_2)
> let v be the first node of T_1 in preorder
> $cost \leftarrow 0$
> let *low* be the number of nodes of T_1
> $(M, A, cost, low) \leftarrow$ *extend_tree_edit*($T_1, T_2, M, A, C, v, cost, low$)
> **return** A

The actual extension of a partial solution is done by the following recursive function, which extends a node assignment $M \subseteq V_1 \times V_2 \cup \{\lambda\}$ by mapping node $v \in V_1$ to each node $w \in C[v] \subseteq V_2 \cup \{\lambda\}$ in turn, that is, to each remaining candidate node, making a recursive call upon the successor (in preorder) of node $v \in V_1$ whenever $M \cup \{(v, w)\}$ is a node assignment corresponding to a valid transformation of the subtree of T_1 induced by the nodes up to, and including node v into the subtree of T_2 induced by those nodes of T_2 which were already mapped to nodes of T_1 and, furthermore, the cost *cost* of $M \cup \{(v, w)\}$ is lesser than the cost *low* of the lowest-cost node assignment A found so far.

```
function extend_tree_edit(T₁, T₂, M, A, C, v, cost, low)
    for all w in C[v] do
        M[v] ← w
        if w = T₂.dummy then
            cost ← cost + 1
        if cost < low then
            if v is the last node of T₁ in preorder then
                let A be a copy of M
                low ← cost
            else
                let N be a copy of C
                N ← refine_candidate_nodes(T₁, T₂, N, v, w)
                let v' be the next node after v in T₁ in preorder
                (M, A, cost, low) ←
                    extend_tree_edit(T₁, T₂, M, A, N, v', cost, low)
        if w = T₂.dummy then
            cost ← cost − 1
    return (M, A, cost, low)
```

2.4 Divide-and-Conquer

The backtracking technique can be used for finding either one solution or all solutions to a combinatorial problem, and the branch-and-bound technique can be used for finding, in a more efficient way, a least-cost solution. The *divide-and-conquer* technique can also be used to find a solution and, in particular, a least-cost solution to a combinatorial problem.

Divide-and-conquer is the general technique of solving a problem by dividing it into smaller, easier-to-solve subproblems; recursively solving these smaller subproblems; and then combining their solutions into a solution to the original problem. A subproblem whose size is small enough is just solved in a straightforward way. Further, the subproblem size must be a fraction of the problem size for the divide-and-conquer algorithm to be most efficient, as follows from the solution to the recurrence relations that describe the time and space complexity of divide-and-conquer algorithms. Dividing a problem into subproblems of about the same size as the original problem leads to less efficient, sometimes called *subtract-and-surrender* algorithms.

The divide-and-conquer technique can be applied to those problems that exhibit the *independence principle*, meaning that a problem instance can be divided into a series of smaller problem instances that are independent of each other.

Consider, for instance, the tree edit problem. Let $T_1 = (V_1, E_1)$ and $T_2 = (V_2, E_2)$ be ordered trees, and recall that a solution to the problem of finding a valid transformation of T_1 into T_2 of least cost is given by an assignment $M \subseteq V_1 \times V_2 \cup \{\lambda\}$

of nodes of T_2 to the nodes of T_1 of least cost, where the dummy node λ represents the deletion of a node from T_1, all nodes in $\{w \in V_2 \mid \nexists v \in V_1, (v, w) \in M\}$ are inserted into T_2, and the cost of a solution $M \subseteq V_1 \times V_2 \cup \{\lambda\}$ is given by the number of nodes deleted from T_1, that is, $cost[M] = |\{(v, w) \in M \mid w = \lambda\}|$.

As a matter of fact, it is more convenient to state the least-cost tree transformation problem as a greatest-benefit problem, where the *benefit* of a least-cost solution $M \subseteq V_1 \times V_2 \cup \{\lambda\}$ is given by the number of nodes of T_1 replaced with nodes of T_2, that is, $benefit[M] = |\{(v, w) \in M \mid w \neq \lambda\}|$. In the rest of this chapter, it will not be distinguished between least-cost and greatest-benefit tree transformations.

Then, finding a least-cost solution to a tree edit problem by divide-and-conquer means dividing T_1 and T_2 into subtrees, recursively finding a least-cost transformation of each of the subtrees of T_1 to the corresponding subtree of T_2, and then combining these transformations of subtrees into a least-cost transformation of T_1 to T_2. The transformation of a trivial subtree of T_1 of the form $(\{v\}, \emptyset)$, that is, a subtree consisting of a single node, to a trivial subtree of T_2 of the form $(\{w\}, \emptyset)$ is just $M = \{(v, w)\}$, that is, the replacement of node $v \in V_1$ with node $w \in V_2$.

The division of T_1 and T_2 into subtrees can be made on the basis of a node $v \in V_1$ and a node $w \in V_2$ of same depth, as follows. Let A_1 be the subtree of T_1 containing all nodes from the root down to the predecessor of node v in preorder, let B_1 be the subtree of T_1 rooted at node v, let A_2 be the subtree of T_2 containing all nodes from the root down to the predecessor of node w in preorder, and let B_2 be the subtree of T_2 rooted at node w. Then, in a least-cost transformation of T_1 to T_2, either node v is deleted from T_1 (and A_1 is transformed to T_2), node w is inserted into T_2 (and T_1 is transformed to A_2), or node v is replaced with node w (and A_1 is transformed to A_2, and B_1 to B_2). Notice that the case of node v being replaced with a node of T_2 different from node w is implied by the insertion of node w into T_2, and the case of a node of T_1 different from node v being replaced with node w is implied by the deletion of node v from T_1.

Example 2.9 A division of ordered trees $T_1 = (V_1, E_1)$ and $T_2 = (V_2, E_2)$ of Fig. 2.1 is illustrated in Fig. 2.10. A least-cost transformation of T_1 to T_2 can be obtained by dividing tree T_1 upon node $v_5 \in V_1$ in subtrees A_1 and B_1, dividing tree T_2 upon node $w_3 \in V_2$ in subtrees A_2 and B_2, and then taking the transformation of lesser cost among a least-cost transformation of A_1 to T_2, a

Fig. 2.10 A division of the ordered trees $T_1 = (V_1, E_1)$ and $T_2 = (V_2, E_2)$ of Fig. 2.1 upon nodes $v_5 \in V_1$ and $w_3 \in V_2$. Nodes are numbered according to the order in which they are visited during a preorder traversal of the trees

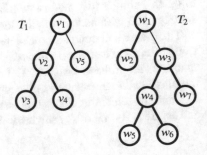

least-cost transformation of T_1 to A_2, and a least-cost transformation of A_1 to A_2 together with a least-cost transformation of B_1 to B_2. The least-cost transformation $M = \{(v_1, w_1), (v_2, w_3), (v_3, w_4), (v_4, w_7)\}$ of A_1 to T_2 gives a least-cost transformation $M \cup \{(v_5, \lambda)\}$ of T_1 to T_2, of cost 1, corresponding to the deletion of node v_5 from T_1 (and the insertion of nodes w_2, w_5, w_6 into T_2).

Now, given the recursive nature of the divide-and-conquer algorithm for the tree edit problem, it is convenient to delimit the subtrees involved in each recursive call to the algorithm by giving the first and last node in preorder of each subtree, together with the whole trees, instead of making explicit copies of the subtrees. In this sense, a recursive call to the divide-and-conquer algorithm will involve the subtree of T_1 with nodes from v_1 to v_2 and the subtree of T_2 with nodes from w_1 to w_2. Several cases need to be distinguished.

- The particular case of $v_1 = v_2$ and $w_1 = w_2$ will correspond to the transformation $M = \{(v_1, w_1)\}$ of the trivial subtree $(\{v_1\}, \emptyset)$ of T_1 to the trivial subtree $(\{w_2\}, \emptyset)$ of T_2, with benefit 1.
- The particular case of $v_1 \neq v_2$ and $w_1 = w_2$ will correspond to the transformation of the subtree A_1 of T_1 (the subtree of T_1 with nodes from v_1 down to the predecessor in preorder of node v_2) to tree T_2, that is, the deletion of the subtree of T_1 rooted at node v_2.
- The particular case of $v_1 = v_2$ and $w_1 \neq w_2$ will correspond to the transformation of tree T_1 to the subtree A_2 of T_2 (the subtree of T_2 with nodes from w_1 down to the predecessor in preorder of node w_2), that is, the insertion of the subtree of T_2 rooted at node w_2.
- The general case of $v_1 \neq v_2$ and $w_1 \neq w_2$ will correspond to either the transformation of the subtree A_1 of T_1 to tree T_2, the transformation of tree T_1 to the subtree A_2 of T_2, or the transformation of the subtree A_1 of T_1 to the subtree A_2 of T_2 together with the transformation of the subtree B_1 of T_1 (the subtree of T_1 rooted at node v_2) to the subtree B_2 of T_2 (the subtree of T_2 rooted at node w_2).

Now, the following choice of nodes $v \in V_1$ and $w \in V_2$ upon which to make the division of the subtree of T_1 with nodes from v_1 down to v_2, to the subtree of T_2 with nodes from w_1 down to w_2, will be considered in the rest of this section. Node v will be the last child of node v_1 that belongs to the subtree of T_1 with nodes from v_1 down to v_2. In a similar way, node w will be the last child of node w_1 that belongs to the subtree of T_2 with nodes from w_1 down to w_2.

The divide-and-conquer procedure for the tree edit problem can be better understood in terms of a *divide-and-conquer tree* of all recursive calls made for the solution to a tree edit problem instance. The root of the divide-and-conquer tree stands for the initial call to the recursive procedure (upon the whole tree T_1 and the whole tree T_2), and the remaining nodes represent recursive calls (upon subtrees A_1, B_1 of T_1 and subtrees A_2, B_2 of T_2) and have either none, one, or four children depending on the relative size of the subtrees. That is, the node corresponding to a recursive call upon the subtree of T_1 with nodes from v_1 down to v_2 and the subtree of T_2 with nodes from w_1 down to w_2 will have no children if $A_1 = \emptyset$ and $B_1 = \emptyset$ (that is, if $v_1 = w_1$ and $w_1 = w_2$), will have one child if either $A_1 \neq \emptyset$ and $B_1 = \emptyset$ or $A_1 = \emptyset$ and $B_1 \neq \emptyset$, and will have four children if both $A_1 \neq \emptyset$ and $B_1 \neq \emptyset$.

Example 2.10 The divide-and-conquer tree for the transformation between ordered trees $T_1 = (V_1, E_1)$ and $T_2 = (V_2, E_2)$ of Fig. 2.1 is shown in Fig. 2.11. A node labeled $v_i v_j w_k w_\ell$ corresponds to a recursive call to the divide-and-conquer procedure upon the subtree of T_1 with nodes from v_i down to v_j and the subtree of T_2 with nodes from w_k down to w_ℓ.

The following divide-and-conquer algorithm for the tree edit problem makes an initial call to the recursive divide-and-conquer procedure upon the subtree of T_1 from the first node down to the last node in preorder and the subtree of T_2 from the first node down to the last node in preorder, that is, upon the whole tree T_1 and the whole tree T_2 and returns the cost of a least-cost transformation of T_1 to T_2, together with the corresponding mapping M.

> **function** *divide_and_conquer_tree_edit*(T_1, T_2)
>   *preorder_tree_traversal*(T_1)
>   *preorder_tree_traversal*(T_2)
>   let M be an empty dictionary (of nodes to nodes)
>   let v_1 and v_2 be the first node and the last node of T_1 in preorder
>   let w_1 and w_2 be the first node and the last node of T_2 in preorder
> $(d, M) \leftarrow$ *tree_edit*$(T_1, v_1, v_2, T_2, w_1, w_2, M)$
>   **return** M

The actual computation of a least-cost transformation M of T_1 to T_2 is done by the following recursive function which, in the general case, takes the predecessor k_1 in preorder of the last child of node v_1, takes the predecessor k_2 in preorder of the last child of node w_1, divides the subtree of T_1 with nodes from v_1 down to v_2, into a subtree A_1 with nodes from v_1 down to k_1 and a subtree B_1 with nodes from the successor of node k_1 in preorder down to v_2, and also divides the subtree of T_2 with nodes from w_1 down to w_2, into a subtree A_2 with nodes from w_1 down to k_2 and a subtree B_2 with nodes from the successor of node k_2 in preorder down to w_2.

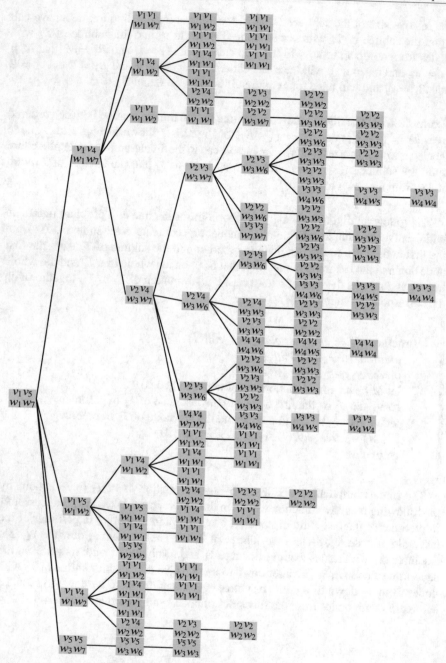

Fig. 2.11 Divide-and-conquer tree for the tree edit problem

```
function tree_edit(T₁, v₁, v₂, T₂, w₁, w₂, M)
    if v₁ = v₂ then
        if w₁ = w₂ then
            M[v₁] = w₁
            dist ← 1
        else
            k₂ ← predecessor_of_last_child(T₂, w₁, w₂)
            (dist, M) ← tree_edit(T₁, v₁, v₂, T₂, w₁, k₂, M)
    else
        if w₁ = w₂ then
            k₁ ← predecessor_of_last_child(T₁, v₁, v₂)
            (dist, M) ← tree_edit(T₁, v₁, k₁, T₂, w₁, w₂, M)
        else
            k₁ ← predecessor_of_last_child(T₁, v₁, v₂)
            let M₁ be a copy of M
            (del, M₁) ← tree_edit(T₁, v₁, k₁, T₂, w₁, w₂, M₁)
            k₂ ← predecessor_of_last_child(T₂, w₁, w₂)
            let M₂ be a copy of M
            (ins, M₂) ← tree_edit(T₁, v₁, v₂, T₂, w₁, k₂, M₂)
            let M₃ be a copy of M
            (pre, M₃) ← tree_edit(T₁, v₁, k₁, T₂, w₁, k₂, M₃)
            let k₁' be the next node after k₁ in T₁ in preorder
            let k₂' be the next node after k₂ in T₂ in preorder
            (pos, M₃) ← tree_edit(T₁, k₁', v₂, T₂, k₂', w₂, M₃)
            dist ← max(del, ins, pre + pos)
            if dist = del then
                M ← M₁
            else if dist = ins then
                M ← M₂
            else
                M ← M₃
    return (dist, M)
```

In the subtree of an ordered tree $T = (V, E)$ with nodes from $v_1 \in V$ down to $v_2 \in V$, where node v_1 is either equal to node v_2 or it is a predecessor of v_2 in preorder, the predecessor $k_1 \in V$ in preorder of the last child of node v_1 is, as a matter of fact, the predecessor in preorder of the last child of node v_1 that lies between nodes v_1 and v_2, for otherwise node k_1 would not belong to the subtree of T_1 with nodes from v_1 down to v_2.

> **function** *predecessor_of_last_child*(T, v_1, v_2)
>   **for all** children k of node v_1 in T in reverse order **do**
>   **if** $k.order \leqslant v_2.order$ **then**
>   break
>   **return** the previous node before k in T in preorder

2.5 Dynamic Programming

The divide-and-conquer technique can be used for finding a solution to combinatorial problems that exhibit the *independence principle*, where a problem instance can be divided into a series of smaller problem instances that are independent of each other. When applied to finding a least-cost solution, the independence principle is often called *optimality principle*, meaning that a least-cost solution to a problem instance can be decomposed into a series of least-cost solutions to smaller problem instances which are independent of each other.

It is often the case, however, that these independent smaller problem instances are overlapping, and a divide-and-conquer algorithm will spend unnecessary effort in recomputing a least-cost solution to overlapping problem instances. For instance, in the divide-and-conquer tree for the tree edit problem instance given in Example 2.11, there are 88 independent smaller problem instances solved for finding a least-cost transformation of tree T_1 to tree T_2, of which 32 are solved exactly once, and the remaining 56 independent smaller problem instances are overlapping. In particular, a least-cost transformation of the subtree of T_1 with nodes from v_2 down to v_3 to the subtree of T_2 with nodes from w_3 down to w_6 is solved three times, during the search of a least-cost transformation of the subtree of T_1 with nodes from v_2 down to v_4 to the subtree of T_2 with nodes from w_3 down to w_7.

The dynamic programming technique improves on the divide-and-conquer technique for optimization problems that exhibit the optimality principle by storing solutions to subproblems, avoiding thus the recomputation of solutions to overlapping subproblems—smaller problems which are common to larger independent problems—during the search for an optimal solution to the problem. Since solutions to common smaller problems are shared among instances of larger problems, the dynamic programming tree corresponding to the decomposition of an optimization problem in smaller problems is, in fact, a directed acyclic graph and is smaller than the corresponding divide-and-conquer tree, as illustrated in Fig. 2.12.

The very name of *dynamic programming* does not refer to a programming methodology but to a tabular method, by which optimal solutions to smaller problems are computed only once and stored in a dictionary for later lookup, whenever the smaller problem is encountered again during the search for an optimal solution to the original problem.

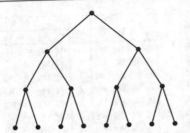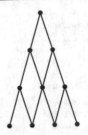

Fig. 2.12 Divide-and-conquer tree (left) and dynamic programming directed acyclic graph (right) for the decomposition of an optimization problem in smaller problems

Essentially, there are two ways in which such a tabular method can be implemented. In top-down dynamic programming, also known as *memoization*, a recursive, divide-and-conquer algorithm for an optimization problem is enhanced by first looking up whether a smaller problem was already solved (and the solution to the smaller problem is already available in a dictionary), and solving it by divide-and-conquer otherwise. In bottom-up dynamic programming, by contrast, computation proceeds in a bottom-up fashion, and optimal solutions to smaller problems are thus readily available, when solving a larger problem.

Consider, for instance, the tree edit problem. Let $T_1 = (V_1, E_1)$ and $T_2 = (V_2, E_2)$ be ordered trees and recall that a solution to the problem of finding a valid transformation of T_1 into T_2 of least cost is given by an assignment $M \subseteq V_1 \times V_2 \cup \{\lambda\}$ of nodes of T_2 to the nodes of T_1 of least cost, where the dummy node λ represents the deletion of a node from T_1, all nodes in $\{w \in V_2 \mid \nexists v \in V_1, (v, w) \in M\}$ are inserted into T_2, and the cost of a solution $M \subseteq V_1 \times V_2 \cup \{\lambda\}$ is given by the number of nodes deleted from T_1, that is, $cost[M] = |\{(v, w) \in M \mid w = \lambda\}|$.

Again, it is more convenient to state the least-cost tree transformation problem as a greatest-benefit problem, where the *benefit* of a least-cost solution $M \subseteq V_1 \times V_2 \cup \{\lambda\}$ is given by the number of nodes of T_1 substituted by nodes of T_2, that is, $benefit[M] = |\{(v, w) \in M \mid w \neq \lambda\}|$.

In a top-down dynamic programming algorithm based on the divide-and-conquer algorithm given in Sect. 2.4 for finding a least-cost transformation between ordered trees T_1 and T_2, a least-cost transformation of the subtree of T_1 with nodes from v_i down to v_j to the subtree of T_2 with nodes from w_j down to w_ℓ can be memoized with the help of a dictionary S of quadruples of nodes (two nodes of T_1 and two nodes of T_2) to pairs consisting of an integer edit distance or cost and a dictionary M of nodes (of tree T_1) to nodes (of tree T_2).

Example 2.11 The execution of the top-down dynamic programming procedure for finding a least-cost transformation between ordered trees $T_1 = (V_1, E_1)$ and $T_2 = (V_2, E_2)$ of Fig. 2.1 is illustrated in Fig. 2.13. The benefit *dist* and the node assignment $M \subseteq V_1 \times V_2 \cup \{\lambda\}$ corresponding to a least-cost transformation of the subtree of T_1 with nodes from v_i down to v_j to the subtree of T_2 with nodes from w_k down to w_ℓ is shown for each smaller problem solved, in the order in which solutions are completed.

T_1		T_2		dist	M				
v_i	v_j	w_k	w_ℓ		v_1	v_2	v_3	v_4	v_5
v_1	v_1	w_1	w_1	1	w_1	λ	λ	λ	λ
v_1	v_1	w_1	w_2	1	w_1	λ	λ	λ	λ
v_1	v_1	w_1	w_7	1	w_1	λ	λ	λ	λ
v_1	v_4	w_1	w_1	1	w_1	λ	λ	λ	λ
v_2	v_2	w_2	w_2	1	w_1	w_2	λ	λ	λ
v_2	v_3	w_2	w_2	1	w_1	w_2	λ	λ	λ
v_2	v_4	w_2	w_2	1	w_1	w_2	λ	λ	λ
v_1	v_4	w_1	w_2	2	w_1	w_2	λ	λ	λ
v_2	v_2	w_3	w_3	1	w_1	w_3	λ	λ	λ
v_2	v_2	w_3	w_6	1	w_1	w_3	λ	λ	λ
v_2	v_2	w_3	w_7	1	w_1	w_3	λ	λ	λ
v_2	v_3	w_3	w_3	1	w_1	w_3	λ	λ	λ
v_3	v_3	w_4	w_4	1	w_1	w_3	w_4	λ	λ
v_3	v_3	w_4	w_5	1	w_1	w_3	w_4	λ	λ
v_3	v_3	w_4	w_6	1	w_1	w_3	w_4	λ	λ
v_2	v_3	w_3	w_6	2	w_1	w_3	w_4	λ	λ
v_3	v_3	w_7	w_7	1	w_1	w_3	w_7	λ	λ

T_1		T_2		dist	M				
v_i	v_j	w_k	w_ℓ		v_1	v_2	v_3	v_4	v_5
v_2	v_3	w_3	w_7	2	w_1	w_3	w_4	λ	λ
v_2	v_4	w_3	w_3	1	w_1	w_3	λ	λ	λ
v_4	v_4	w_4	w_4	1	w_1	w_3	λ	w_4	λ
v_4	v_4	w_4	w_5	1	w_1	w_3	λ	w_4	λ
v_4	v_4	w_4	w_6	1	w_1	w_3	λ	w_4	λ
v_2	v_4	w_3	w_6	2	w_1	w_3	w_4	λ	λ
v_4	v_4	w_7	w_7	1	w_1	w_3	w_4	w_7	λ
v_2	v_4	w_3	w_7	3	w_1	w_3	w_4	w_7	λ
v_1	v_4	w_1	w_7	4	w_1	w_3	w_4	w_7	λ
v_1	v_5	w_1	w_1	1	w_1	λ	λ	λ	λ
v_5	v_5	w_2	w_2	1	w_1	λ	λ	λ	w_2
v_1	v_5	w_1	w_2	2	w_1	w_2	λ	λ	λ
v_5	v_5	w_3	w_3	1	w_1	w_2	λ	λ	w_3
v_5	v_5	w_3	w_6	1	w_1	w_2	λ	λ	w_3
v_5	v_5	w_3	w_7	1	w_1	w_2	λ	λ	w_3
v_1	v_5	w_1	w_7	4	w_1	w_3	w_4	w_7	λ

Fig. 2.13 Memoization of optimal solutions to smaller tree edit distance problems in top-down dynamic programming

The following top-down dynamic programming algorithm for the tree edit problem makes an initial call to the recursive top-down dynamic programming procedure upon the subtree of T_1 from the first node down to the last node in preorder and the subtree of T_2 from the first node down to the last node in preorder, that is, upon the whole tree T_1 and the whole tree T_2 and returns the cost of a least-cost transformation of T_1 to T_2, together with the corresponding mapping M.

function *top_down_dynamic_programming_tree_edit*(T_1, T_2)
  preorder_tree_traversal(T_1)
  preorder_tree_traversal(T_2)
  let M be an empty dictionary (of nodes to nodes)
  let S be an empty dictionary (of quadruples of nodes to pairs
    consisting of an integer and a dictionary of nodes to nodes)
  let v_1 and v_2 be the first node and the last node of T_1 in preorder
  let w_1 and w_2 be the first node and the last node of T_2 in preorder
  $(dist, M, S) \leftarrow tree\_edit(T_1, v_1, v_2, T_2, w_1, w_2, M, S)$
  return M

The actual computation of a least-cost transformation M of T_1 to T_2 is done by the following recursive function which, in the general case, takes the predecessor k_1 in preorder of the last child of node v_1, takes the predecessor k_2 in preorder of the last child of node w_1, divides the subtree of T_1 with nodes from v_1 down to v_2, into a subtree A_1 with nodes from v_1 down to k_1 and a subtree B_1 with nodes from the successor of node k_1 in preorder down to v_2, and also divides the subtree of T_2 with nodes from w_1 down to w_2, into a subtree A_2 with nodes from w_1 down to k_2 and a subtree B_2 with nodes from the successor of node k_2 in preorder down to w_2.

The function is identical to the divide-and-conquer function given in Sect. 2.4, except that in a recursive call upon the subtree of T_1 with nodes from v_i down to v_j and the subtree of T_2 with nodes from w_k down to w_ℓ, the quadruple $\langle v_i, v_j, w_k, w_\ell \rangle$ is first looked up in dictionary S, which is an additional parameter of the recursive function, and the corresponding tree edit problem is decomposed only if no such quadruple is defined in the dictionary. Further, each solution to a smaller problem found is inserted in S.

```
function tree_edit(T₁, v₁, v₂, T₂, w₁, w₂, M, S)
    if ⟨v₁, v₂, w₁, w₂⟩ ∈ S then
        (dist, M) ← S[⟨v₁, v₂, w₁, w₂⟩]
    else
        if v₁ = v₂ then
            if w₁ = w₂ then
                M[v₁] ← w₁
                dist ← 1
            else
                k₂ ← predecessor_of_last_child(T₂, w₁, w₂)
                (dist, M, S) ← tree_edit(T₁, v₁, v₂, T₂, w₁, k₂, M, S)
        else
            if w₁ = w₂ then
                k₁ ← predecessor_of_last_child(T₁, v₁, v₂)
                (dist, M, S) ← tree_edit(T₁, v₁, k₁, T₂, w₁, w₂, M, S)
            else
                k₁ ← predecessor_of_last_child(T₁, v₁, v₂)
                let M₁ be a copy of M
                (del, M₁, S) ← tree_edit(T₁, v₁, k₁, T₂, w₁, w₂, M₁, S)
                k₂ ← predecessor_of_last_child(T₂, w₁, w₂)
                let M₂ be a copy of M
                (ins, M₂, S) ← tree_edit(T₁, v₁, v₂, T₂, w₁, k₂, M₂, S)
                let M₃ be a copy of M
                (pre, M₃, S) ← tree_edit(T₁, v₁, k₁, T₂, w₁, k₂, M₃, S)
                let k₁′ be the next node after k₁ in T₁ in preorder
                let k₂′ be the next node after k₂ in T₂ in preorder
                (pos, M₃, S) ← tree_edit(T₁, k₁′, v₂, T₂, k₂′, w₂, M₃, S)
                dist ← max(del, ins, pre + pos)
                if dist = del then
                    M ← M₁
                else if dist = ins then
                    M ← M₂
                else
                    M ← M₃
        S[⟨v₁, v₂, w₁, w₂⟩] ← (dist, M)
    return (dist, M, S)
```

Consider now a bottom-up dynamic programming algorithm for the tree edit problem. A bottom-up order of computation ensures that optimal solutions to smaller problems are computed before an optimal solution to a larger problem is computed. For the tree edit problem, a natural way to achieve such a bottom-up order of computation consists in finding a least-cost transformation of the subgraph of T_1 with nodes from v_i down to v_j to the subgraph of T_2 with nodes from w_k down to w_ℓ, where $1 \leqslant i \leqslant j \leqslant n_1$ and $1 \leqslant k \leqslant \ell \leqslant n_2$, for decreasing values of i and k starting off with $i = j = n_1$ and $k = \ell = n_2$. However, most of these least-cost transformations are not really needed, and their computation can be avoided as follows.

Recall that the division of T_1 and T_2 into subtrees is made on the basis of a node $v \in V_1$ and a node $w \in V_2$ of same depth. Therefore, only solutions to the smaller problems of finding a least-cost transformation of the subgraph of T_1 with nodes from v_i down to v_j to the subgraph of T_2 with nodes from w_k down to w_ℓ, with $depth[v_i] = depth[w_k]$, need to be computed. Notice that, unlike the top-down dynamic programming algorithm for the tree edit problem, smaller problems are now subgraphs, not necessarily subtrees, of T_1 and T_2.

Then, a least-cost transformation of the subtree of T_1 rooted at node $v \in V_1$ to the subtree of T_2 rooted at node $w \in V_2$, with $depth[v] = depth[w]$, can be found by taking the transformation of lesser cost among

$$
\begin{array}{cccc}
\langle v_i, w_k \rangle & \langle v_i, w_k w_{k+1} \rangle & \cdots & \langle v_i, w_k \ldots w_\ell \rangle \\
\langle v_i v_{i+1}, w_k \rangle & \langle v_i v_{i+1}, w_k w_{k+1} \rangle & \cdots & \langle v_i v_{i+1}, w_k \ldots w_\ell \rangle \\
\vdots & \vdots & \ddots & \vdots \\
\langle v_i \ldots v_j, w_k \rangle & \langle v_i \ldots v_j, w_k w_{k+1} \rangle & \cdots & \langle v_i \ldots v_j, w_k \ldots w_\ell \rangle
\end{array}
$$

where $v_i, \ldots, v_j$ are the children of node v, $w_k, \ldots, w_\ell$ are the children of node w, and $\langle v_i \ldots v_j, w_k \ldots w_\ell \rangle$ denotes a least-cost transformation of the subgraph of T_1 consisting of the subtrees rooted at nodes $v_i, \ldots, v_j$ into the subgraph of T_2 consisting of the subtrees rooted at nodes $w_k, \ldots, w_\ell$. The bottom-up order of computation of a least-cost transformation of the subtree of T_1 rooted at node v into the subtree of T_2 rooted at node w corresponds to computing the entries of the previous matrix of smaller problems in row order. Further, a least-cost transformation of the subtree of T_1 rooted at node v_1 to the subtree of T_2 rooted at node w_1 gives a least-cost transformation of T_1 into T_2.

Again, it is more convenient to state the least-cost tree transformation problem as a greatest-benefit problem, where the *benefit* of a least-cost transformation is given by the number of node substitutions in the transformation.

Example 2.12 The execution of the bottom-up dynamic programming procedure for finding a least-cost transformation between ordered trees $T_1 = (V_1, E_1)$ and $T_2 = (V_2, E_2)$ of Fig. 2.1 is illustrated in Fig. 2.14. The benefit *dist* and the node assignment $M \subseteq V_1 \times V_2 \cup \{\lambda\}$ corresponding to a least-cost transformation of the subgraph of T_1 with nodes from v_i down to v_j to the subgraph of T_2 with nodes from w_k down to w_ℓ is shown for each smaller problem solved, in the order in which solutions are completed.

v	w	T_1		T_2		dist	M				
		v_i	v_j	w_k	w_ℓ		v_1	v_2	v_3	v_4	v_5
v_2	w_2	v_2	v_4	w_2	w_2	1	λ	w_2	λ	λ	λ
v_3	w_4	v_3	v_3	w_4	w_6	1	λ	λ	w_4	λ	λ
v_3	w_7	v_3	v_3	w_7	w_7	1	λ	λ	w_7	λ	λ
v_4	w_4	v_4	v_4	w_4	w_6	1	λ	λ	λ	w_4	λ
v_4	w_7	v_4	v_4	w_7	w_7	1	λ	λ	λ	w_7	λ
v_2	w_3	v_2	v_4	w_3	w_7	3	λ	w_3	w_4	w_7	λ
v_5	w_2	v_5	v_5	w_2	w_2	1	λ	λ	λ	λ	w_2
v_5	w_3	v_5	v_5	w_3	w_7	1	λ	λ	λ	λ	w_3
v_1	w_1	v_1	v_5	w_1	w_7	4	w_1	w_3	w_4	w_7	λ

		w_2	w_2 w_3
	0	0	0
v_2	0	1	3
$v_2 v_5$	0	1	3

		w_4	w_4 w_7
	0	0	0
v_3	0	1	1
$v_3 v_4$	0	1	2

Fig. 2.14 Finding optimal solutions to smaller tree edit distance problems in bottom-up dynamic programming

Further, the matrices of smaller problems are also shown where, for instance, a least-cost transformation of the subgraph of T_1 consisting of the subtrees rooted at nodes v_2, v_5, into the subgraph of T_2 consisting of the subtrees rooted at nodes w_2, w_3 has benefit 3, corresponding to the substitution of nodes v_2, v_3, v_4 by nodes w_3, w_4, w_7, respectively, and a least-cost transformation of the subtree of T_1 rooted at node v_1 into the subtree of T_2 rooted at node w_1 has thus benefit $3 + 1 = 4$, corresponding to the substitution of nodes v_1, v_2, v_3, v_4 by nodes w_1, w_3, w_4, w_7, respectively. Notice that matrices of smaller problems have an additional left column and an additional top row, in order to allow for edit operations on the initial subtree of T_1 and the initial subtree of T_2.

The following bottom-up dynamic programming algorithm for the tree edit problem makes an initial call to the recursive bottom-up dynamic programming procedure upon the subtree of T_1 rooted at node $root[T_1]$ and the subtree of T_2 rooted at node $root[T_2]$, in order to extract a least-cost mapping M of T_1 to T_2 from the list of tree edit operations L computed by the recursive function.

> **function** *bottom_up_dynamic_programming_tree_edit*(T_1, T_2)
>   let L be an empty list (of pairs of nodes)
> $(dist, L) \leftarrow$ *tree_edit*$(T_1, root[T_1], T_2, root[T_2], L)$
>   let M be an empty dictionary (of nodes to nodes)
>   **for all** (v, w) in L **do**
>   **if** $v \neq nil$ **then**
> $M[v] \leftarrow w$
>   **return** M

The actual computation of a least-cost transformation of T_1 to T_2 is done by the following recursive function, which finds the least-cost transformation of the subtree of T_1 rooted at node v, with children nodes $v_i, \ldots, v_j$, into the subtree of T_2 rooted at node w, with children nodes $w_k, \ldots, w_\ell$, by taking the transformation of lesser cost among the least-cost transformations of the subgraph of T_1 consisting of the subtrees rooted in turn at nodes $\{v_i\}, \{v_i, v_{i+1}\}, \ldots, \{v_i, v_{i+1}, \ldots, v_j\}$ into the subgraph of T_2 consisting of the subtrees rooted in turn at nodes $\{w_k\}, \{w_k, w_{k+1}\}, \ldots, \{w_k, w_{k+1}, \ldots, w_\ell\}$.

The least-cost transformation is represented by a list L of tree edit operations, and the matrix of smaller problems is represented by two matrices, an integer matrix D of benefits and a matrix E of lists of edit operations. Both matrices have an additional left column and an additional top row, in order to allow for edit operations on the initial subtrees of T_1 and T_2.

function $tree\_edit(T_1, v, T_2, w, L)$
    let n_1 be the number of children of node r_1 in T_1
    let n_2 be the number of children of node r_2 in T_2
    let $D[0..m, 0..n]$ be a new matrix (of integers)
    let $E[0..m, 0..n]$ be a new matrix (of lists of pairs of nodes)
    $D[0, 0] \leftarrow 0$
    let $E[0, 0]$ be an empty list (of pairs of nodes)
    for $i = 1$ **to** n_1 **do**
        $D[i, 0] \leftarrow D[i - 1, 0]$
        $E[i, 0] \leftarrow E[i - 1, 0]$
        let v' be the i-th child of v in T_1
        append $\langle v', nil \rangle$ to $E[i, 0]$
    for $j = 1$ **to** n_1 **do**
        $D[0, j] \leftarrow D[0, j - 1]$
        $E[0, j] \leftarrow E[0, j - 1]$
        let w' be the j-th child of w in T_2
        append $\langle nil, w' \rangle$ to $E[0, j]$
    for $i = 1$ **to** n_1 **do**
        let v' be the i-th child of v in T_1
        for $j = 1$ **to** n_2 **do**
            let w' be the j-th child of w in T_2
            $del \leftarrow D[i - 1, j]$
            $ins \leftarrow D[i, j - 1]$
            $(sub, L) \leftarrow tree\_edit(T_1, v', T_2, w')$
            if $del \geqslant D[i - 1, j - 1] + sub$ **then**
                $D[i, j] \leftarrow del$
                $E[i, j] \leftarrow E[i - 1, j]$
                append $\langle v', nil \rangle$ to $E[i, j]$
            else
                if $ins \geqslant D[i - 1, j - 1] + sub$ **then**
                    $D[i, j] \leftarrow ins$
                    $E[i, j] \leftarrow E[i, j - 1]$
                    append $\langle nil, w' \rangle$ to $E[i, j]$
                else
                  $D[i, j] \leftarrow D[i - 1, j - 1] + sub$
                  $E[i, j] \leftarrow E[i - 1, j - 1]$
                  concatenate L to $E[i, j]$
    append $\langle v, w \rangle$ to $E[n_1, n_2]$
    return $(D[n_1, n_2] + 1, E[n_1, n_2])$

Lemma 2.6 *The bottom-up dynamic programming algorithm for finding a least-cost transformation of an ordered tree T_1 to an ordered tree T_2 with, respectively, n_1 and n_2 nodes runs in $O(n_1n_2)$ time using $O(n_1n_2)$ additional space.*

Proof Let $T_1 = (V_1, E_1)$ and $T_2 = (V_2, E_2)$ be ordered trees with, respectively, n_1 and n_2 nodes. The algorithm makes one recursive call to the bottom-up dynamic programming procedure upon each pair of nodes $v \in V_1$ and $w \in V_2$ with $depth[v] = depth[w]$. Let also $m = \min\{depth[T_1], depth[T_2]\}$, and let $a_i = |\{v \in V_1 \mid depth[v] = i\}|$ and $b_i = |\{w \in V_2 \mid depth[w] = i\}|$ for $0 \leqslant i \leqslant m$. The number of recursive calls is thus $\sum_{i=0}^{m} a_i b_i \leqslant \sum_{i=0}^{m} a_i \sum_{i=0}^{m} b_i \leqslant n_1n_2$, because $a_i \geqslant 0$ for $0 \leqslant i \leqslant m$. Therefore, the algorithm makes $O(n_1n_2)$ recursive calls. Further, $O(n_1n_2)$ additional space is used. $\square$

Summary

Some of the fundamental algorithmic techniques used for solving combinatorial problems on trees and graphs are reviewed in this chapter. The techniques of backtracking, branch-and-bound, divide-and-conquer, and dynamic programming are discussed in detail. Further, these techniques are illustrated by the problem of computing the edit distance and finding a least-cost transformation between two ordered trees.

Bibliographic Notes

The backtracking technique was introduced in [11,21,57]. The average-case analysis of backtracking algorithms is discussed in [29,42]. A thorough treatment of backtracking and branch-and-bound for exhaustive search can be found in [43, Chap. 4]. The branch-and-bound technique is further reviewed in [30].

The divide-and-conquer algorithm for finding a least-cost transformation between two ordered trees is based on [37]. A thorough exposition of techniques for solving general recurrences can be found in [15, Chap. 4] and also in [22,42,46]. Techniques for the solution of divide-and-conquer recurrences are further discussed in [10,25, 44,55,56,59].

A dynamic programming formulation of several combinatorial problems is given in [23]. A thorough treatment of dynamic programming can be found in [8,9,17]. See also [15, Chap. 15]. The dynamic programming algorithm for finding a least-cost transformation between two ordered trees is based on [47,63]. See also [24,64].

The edit distance between ordered trees constrained by insertion and deletion of leaves only was introduced in [37,47,63]. Further algorithms for computing the edit distance between ordered trees include [3,6,16,20,28,31,32,39,40,48,49,51, 58,60,66,69], where algorithms for computing the edit distance between unordered trees include [4,35,51,62,67,70], and [50,71] between free unordered trees, that is, connected acyclic graphs. See also [1,2,12,13,18,19,27,34,38,41,45,52,54,65,68].

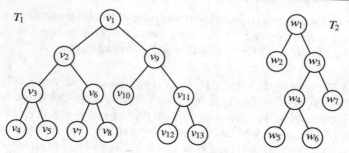

Fig. 2.15 Ordered trees for Problem 2.1. Nodes are numbered according to the order in which they are visited during a preorder traversal

Edit graphs were used by several efficient algorithms for comparing strings and sequences [33, 36, 61] and were first used for tree comparison in [14]. The tree edit distance problem is closely related with isomorphism problems on trees and with the tree inclusion problem [5, 7, 26], in which the elementary edit operation of insertion is forbidden. See also [53] and the bibliographic notes for Chap. 4.

Problems

2.1 Give all valid transformations and a least-cost transformation between the trees of Fig. 2.15, under the assumption that $\gamma(v, w) = 1$ for all edit operations of the form $v \mapsto w$ with $v \in V_1 \cup \{\lambda\}$, $w \in V_2 \cup \{\lambda\}$, and either $v = \lambda$ or $w = \lambda$.

2.2 Give a formulation of the general tree edit distance problem, without the constraint that deletions and insertions be made on leaves only. In a general transformation between two ordered trees, the parent of a deleted node becomes the parent of the children (if any) of the node, and the relative order among sibling nodes is preserved (left siblings of the deleted node become left siblings of the children of the deleted node which, in turn, become left siblings of the right siblings of the deleted node). The same applies to the insertion of non-leaves.

2.3 Extend the formulation of the tree edit distance problem to deal with labeled trees. Recall that in a tree $T = (V, E)$, the information attached to a node $v \in V$ is denoted by $label[v]$.

2.4 Give a formulation of the tree inclusion problem, without the constraint that deletions be made on leaves only but with the additional constraint that insertions are forbidden.

2.5 In the divide-and-conquer formulation of tree edit, the division of a tree into the subtree with nodes up to a certain node in preorder and the subtree with the remaining nodes has the advantage that subtrees are defined by just giving their first and last node in preorder, but the size of the second subtree can be much smaller than half the size of the tree. Give an alternative divide-and-conquer formulation of tree edit, dividing the trees into subtrees of about the same size.

2.6 Give an upper bound on the number of nodes in the backtracking tree, the branch-and-bound tree, and the divide-and-conquer tree for a tree edit problem. Give also an upper bound on the number of vertices in the dynamic programming directed acyclic graph for a tree edit problem.

Exercises

Exercise 2.1 Adapt the tree edit algorithm based on the shortest paths in the edit graph of the trees, in order to give the node assignment corresponding to the valid tree transformation found.

Exercise 2.2 Give a backtracking algorithm for the general tree edit distance problem formulated in Problem 2.2.

Exercise 2.3 Extend the backtracking algorithm for tree edit in order to enumerate not only node assignments that correspond to valid tree transformations, but the valid transformations themselves.

Exercise 2.4 Extend the backtracking and branch-and-bound algorithms for tree edit to labeled trees, according to the formulation of Problem 2.3.

Exercise 2.5 Give a branch-and-bound algorithm for the tree inclusion problem formulated in Problem 2.4.

Exercise 2.6 Give a divide-and-conquer algorithm for tree edit implementing the alternative formulation of Problem 2.5, by which trees are divided into subtrees of about the same size.

Exercise 2.7 Perform experiments on random ordered trees to compare the efficiency of the different algorithms for the tree edit problem. Perform additional experiments on complete binary trees, for the certification of the upper bound given in Problem 2.6.

Exercise 2.8 Give a divide-and-conquer algorithm for the tree inclusion problem formulated in Problem 2.4. Use memoization to extend it to a top-down dynamic programming algorithm for the tree inclusion problem.

Exercise 2.9 Give a bottom-up dynamic programming algorithm for the tree inclusion problem formulated in Problem 2.4. Give also the time and space complexity of the algorithm.

Exercise 2.10 Give an alternative formulation of the tree edit problem and a bottom-up dynamic programming algorithm for tree edit based on the postorder traversal of the trees. Assume that the order of the nodes fixed by the representation of each of the trees coincides with the order in which they are visited during a postorder traversal of the tree.

References

1. Akutsu T (2010) Tree edit distance problems: Algorithms and applications to bioinformatics. IEICE Transactions on Information and Systems E93.D(2):208–218
2. Akutsu T, Fukagawa D, Halldórsson MM, Takasu A, Tanaka K (2006) Approximation and parameterized algorithms for common subtrees and edit distance between unordered trees. Theoretical Computer Science 470(28):10–22
3. Akutsu T, Fukagawa D, Takasu A (2010) Approximating tree edit distance through string edit distance. Algorithmica 57(2):325–348
4. Akutsu T, Fukagawa D, Takasu A, Tamura T (2011) Exact algorithms for computing the tree edit distance between unordered trees. Theoretical Computer Science 412(4–5):352–364
5. Akutsu T, Jansson J, Li R, Takasu A, Tamura T (2021) New and improved algorithms for unordered tree inclusion. Theoretical Computer Science In press
6. Alabbas, M., Ramsay, A.: Optimising tree edit distance with subtrees for textual entailment. In: R. Mitkov, G. Angelova, K. Bontcheva (eds.) Proc. Int. Conf. Recent Advances in Natural Language Processing RANLP 2013, pp. 9–17. INCOMA Ltd., Shoumen, Bulgaria (2013)
7. Alonso L, Schott R (2001) On the tree inclusion problem. Acta Informatica 37(9):653–670
8. Bellman R (1957) Dynamic Programming. Princeton University Press, Princeton NJ
9. Bellman R, Dreyfus SE (1962) Applied Dynamic Programming. Princeton University Press, Princeton NJ
10. Bentley JL, Haken D, Saxe JB (1980) A general method for solving divide-and-conquer recurrences. ACM SIGACT News 12(3):36–44
11. Bitner JR, Reingold EM (1975) Backtrack programming techniques. Communications of the ACM 18(11):651–656
12. Boroujeni M, Ghodsi M, Hajiaghayi MT, Seddighin, S.:ACM, (2019) $1 + epsilon$ approximation of tree edit distance in quadratic time. In: Charikar M, Cohen E (eds). pp 709–720
13. Bringmann K, Gawrychowski P, Mozes S, Weimann O (2020) Tree edit distance cannot be computed in strongly subcubic time. ACM Transactions on Algorithms 16(4):48:1–48:22
14. Chawathe, S.S.: Comparing hierarchical data in external memory. In: M.E.O. Malcolm P. Atkinson, P. Valduriez, S.B. Zdonik, M.L. Brodie (eds.) Proc. 25th Int. Conf. Very Large Data Bases, pp. 90–101. Morgan Kaufmann, New York NY (1999)
15. Cormen TH, Leiserson CE, Rivest RL, Stein C (2009) Introduction to Algorithms, 3rd edn. MIT Press, Cambridge MA
16. Demaine ED, Mozes S, Rossman B, Weimann O (2009) An optimal decomposition algorithm for tree edit distance. ACM Transactions on Algorithms 6(1):2:1–2:19
17. Dreyfus SE, Law AM (1977) The Art and Theory of Dynamic Programming. Academic Press, New York NY

18. Dudek, B., Gawrychowski, P.: Edit distance between unrooted trees in cubic time. In: I. Chatzi-giannakis, C. Kaklamanis, D. Marx, D. Sannella (eds.) Proc. 45th Int. Colloq. Automata, Languages, and Programming, *Leibniz International Proceedings in Informatics*, vol. 107, pp. 45:1–45:14 (2018)

19. Dulucq, S., Touzet, H.: Analysis of tree edit distance algorithms. In: R. Baeza-Yates, E. Chávez, M. Crochemore (eds.) Proc. 14th Annual Symp. Combinatorial Pattern Matching, *Lecture Notes in Computer Science*, vol. 2676, pp. 83–95. Springer (2003)

20. Dulucq S, Touzet H (2005) Decomposition algorithms for the tree edit distance problem. Journal of Discrete Algorithms 3(2–4):448–471

21. Golomb SW, Baumert LD (1965) Backtrack programming. Journal of the ACM 12(4):516–524

22. Graham RL, Knuth DE, Patashnik O (1994) Concrete Mathematics: A Foundation for Computer Science, 2nd edn. Addison-Wesley, Reading MA

23. Held M, Karp RM (1962) A dynamic programming approach to sequencing problems. J. SIAM 10(1):196–210

24. Horesh Y, Mehr R, Unger R (2006) Designing an A* algorithm for calculating edit distance between rooted-unordered trees. Journal of Computational Biology 13(6):1165–1176

25. Kao M (1997) Multiple-size divide-and-conquer recurrences. ACM SIGACT News 28(2):67–69

26. Kilpeläinen P, Mannila H (1995) Ordered and unordered tree inclusion. SIAM Journal on Computing 24(2):340–356

27. Klein, P., Tirthapura, S., Sharvit, D., Kimia, B.: A tree-edit-distance algorithm for comparing simple, closed shapes. In: D. Shmoys (ed.) Proc. 11th Annual ACM-SIAM Symp. Discrete Algorithms, pp. 696–704. SIAM (2000)

28. Klein, P.N.: Computing the edit-distance between unrooted ordered trees. In: G. Bilardi, G.F. Italiano, A. Pietracaprina, G. Pucci (eds.) Proc. 6th Annual European Symp. Algorithms, *Lecture Notes in Computer Science*, vol. 1461, pp. 91–102. Springer, Berlin Heidelberg (1998)

29. Knuth DE, Moore RW (1975) Estimating the efficiency of backtrack programs. Mathematics of Computation 29(1):121–136

30. Lawler EL, Wood DE (1966) Branch-and-bound methods: A survey. Operations Research 14(1):699–719

31. Lu SY (1979) A tree-to-tree distance and its applications to cluster analysis. IEEE Transactions on Pattern Analysis and Machine Intelligence 1(2):219–224

32. Micheli A, Rossin D (2006) Edit distance between unlabeled ordered trees. RAIRO Theoretical Informatics and Applications 40(4):593–609

33. Miller W, Myers EW (1985) A file comparison program. Software: Practice and Experience 15(11):1025–1040

34. Mlinarić D, Milašinović B, Mornar V (2020) Tree inheritance distance. IEEE. Access 8(1):52489–52504

35. Mori T, Tamura T, Fukagawa D, Takasu A, Tomita E, Akutsu T (2012) A clique-based method using dynamic programming for computing edit distance between unordered trees. Journal of Computational Biology 19(10):1089–1104

36. Myers EW (1986) An $O(nd)$ difference algorithm and its variations. Algorithmica 1(2):251–266

37. Noetzel, A.S., Selkow, S.M.: An analysis of the general tree-editing problem. In: D. Sankoff, J.B. Kruskal (eds.) Time Warps, String Edits, and Macromolecules: The Theory and Practice of Sequence Comparison, chap. 8. Center for the Study of Language and Information, Stanford, California (1999)

38. Page RDM, Valiente G (2005) An edit script for taxonomic classifications. BMC Bioinformatics 6:208

39. Pawlik M, Augsten N (2011) RTED: A robust algorithm for the tree edit distance. Proceedings of the VLDB Endowment 5(4):334–345

40. Pawlik M, Augsten N (2015) Efficient computation of the tree edit distance. ACM Transactions on Database Systems 40(1):3:1–3:40

41. Peng, Z., Ting, H.F.: Guided forest edit distance: Better structure comparisons by using domain-knowledge. In: B. Ma, K. Zhang (eds.) Proc. 18th Annual Symp. Combinatorial Pattern Matching, *Lecture Notes in Computer Science*, vol. 4580, pp. 195–204. Springer (2007)

42. Purdom PW, Brown CA (1985) The Analysis of Algorithms. Holt, Rinehart, and Winston, New York NY

43. Reingold EM, Nievergelt J, Deo NJ (1977) Combinatorial Algorithms: Theory and Practice. Prentice Hall, Englewood Cliffs NJ

44. Roura S (2001) Improved master theorems for divide-and-conquer recurrences. Journal of the ACM 48(2):170–205

45. Schwarz, S., Pawlik, M., Augsten, N.: A new perspective on the tree edit distance. In: C. Beecks, F. Borutta, P. Kröger, T. Seidl (eds.) Proc. 10th Int. Conf. Similarity Search and Applications, *Lecture Notes in Computer Science*, vol. 10609, pp. 156–170. Springer (2017)

46. Sedgewick R, Flajolet P (1996) Analysis of Algorithms. Addison-Wesley, Reading MA

47. Selkow SM (1977) The tree-to-tree editing problem. Information Processing Letters 6(6):184–186

48. Shasha D, Zhang K (1990) Fast algorithms for the unit cost editing distance between trees. Journal of Algorithms 11(4):581–621

49. Tai KC (1979) The tree-to-tree correction problem. Journal of the ACM 26(3):422–433

50. Tanaka E (1994) A metric between unrooted and unordered trees and its bottom-up computing method. IEEE Transactions on Pattern Analysis and Machine Intelligence 16(12):1233–1238

51. Tanaka E, Tanaka K (1988) The tree-to-tree editing problem. International Journal of Pattern Recognition and Artificial Intelligence 2(2):221–240

52. Tsur D (2008) Faster algorithms for guided tree edit distance. Information Processing Letters 108(4):251–254

53. Valiente G (2005) Constrained tree inclusion. Journal of Discrete Algorithms 3(2–4):431–447

54. Valiente, G.: Efficient algorithms on trees and graphs with unique node labels. In: A. Kandel, H. Bunke, M. Last (eds.) Applied Graph Theory in Computer Vision and Pattern Recognition, *Studies in Computational Intelligence*, vol. 52, pp. 137–149. Springer (2007)

55. Verma RM (1994) A general method and a master theorem for divide-and-conquer recurrences with applications. Journal of Algorithms 16(1):67–79

56. Verma RM (1997) General techniques for analyzing recursive algorithms with applications. SIAM Journal on Computing 26(2):568–581

57. Walker RJ (1960) An enumerative technique for a class of combinatorial problems. In: Bellman R, Hall M (eds) Combinatorial Analysis, vol 10, chap. 7. American Mathematical Society, Providence, RI, pp 91–94

58. Wang L, Zhang K (2008) Space efficient algorithms for ordered tree comparison. Algorithmica 51(3):283–297

59. Wang X, Fu Q (1996) A frame for general divide-and-conquer recurrences. Information Processing Letters 59(1):45–51

60. Wilhelm R (1981) A modified tree-to-tree correction problem. Information Processing Letters 12(3):127–132

61. Wu S, Manber U, Myers G, Miller W (1990) $O(NP)$ sequence comparison algorithm. Information Processing Letters 35(6):317–323

62. Yamamoto Y, Hirata K, Kuboyama T (2014) Tractable and intractable variations of unordered tree edit distance. International Journal of Foundations of Computer Science 25(3):307–329

63. Yang W (1991) Identifying syntactic differences between two programs. Software: Practice and Experience 21(7):739–755

64. Yoshino T, Higuchi S, Hirata K (2013). In: Hirokawa S, Hashimoto K (eds) A dynamic programming A* algorithm for computing unordered tree edit distance. IEEE, pp 135–140
65. Yoshino T, Hirata K (2017) Tai mapping hierarchy for rooted labeled trees through common subforest. Theory of Computing Systems 60(4):759–783
66. Zhang K (1995) Algorithms for the constrained editing distance between ordered labeled trees and related problems. Pattern Recognition 28(3):463–474
67. Zhang K (1996) A constrained edit distance between unordered labeled trees. Algorithmica 15(3):205–222
68. Zhang K, Jiang T (1994) Some MAX SNP-hard results concerning unordered labeled trees. Information Processing Letters 49(5):249–254
69. Zhang K, Shasha D (1989) Simple fast algorithms for the editing distance between trees and related problems. SIAM Journal on Computing 18(6):1245–1262
70. Zhang K, Statman R, Shasha D (1992) On the editing distance between unordered labeled trees. Information Processing Letters 42(3):133–139
71. Zhang K, Wang JTL, Shasha D (1996) On the editing distance between undirected acyclic graphs. International Journal of Foundations of Computer Science 7(1):43–57

Part II
Algorithms on Trees

Tree Traversal

3.1 Preorder Traversal of a Tree

A traversal of a tree $T = (V, E)$ on n nodes is just a bijection $order : V \to \{1, \ldots, n\}$. In an operational view, a traversal of a tree consists of visiting first the node v with $order[v] = 1$, then the node w with $order[w] = 2$, and so on, until visiting last the node z with $order[z] = n$.

In a preorder traversal of a tree, the root of the tree is visited first, followed by a preorder traversal of the subtree rooted in turn at each of the children of the root. The order in which the children nodes are considered is significant for an ordered tree. The subtree rooted at the first child is traversed first, followed by the subtree rooted at the next sibling, etc. Such an order among the children of a node is also fixed by the representation adopted for both unordered and ordered trees.

Recall from Sect. 1.1 that in a tree $T = (V, E)$, $first[v]$ denotes the first child of node v, $next[v]$ denotes the next sibling of node v, $last[v]$ denotes the last child of node v, and $size[v]$ denotes the number of nodes in the subtree of T rooted at node v, for all nodes $v \in V$.

Definition 3.1 Let $T = (V, E)$ be a tree on n nodes rooted at node r. A bijection $order : V \to \{1, \ldots, n\}$ is a **preorder traversal** of T if $order[r] = 1$ and

- $order[first[v]] = order[v] + 1$ (if v is not a leaf),
- $order[next[v]] = order[v] + size[v]$ (if v is not a last child)

for all nodes $v \in V$.

Example 3.1 The preorder traversal of a rooted tree is illustrated in Fig. 3.1. The root has order 1, the first child of a non-leaf node with order k has order $k + 1$, and the next sibling of a node with order k has order k plus the number of nodes in the

© The Author(s), under exclusive license to Springer Nature Switzerland AG 2021
G. Valiente, *Algorithms on Trees and Graphs*, Texts in Computer Science,
https://doi.org/10.1007/978-3-030-81885-2_3

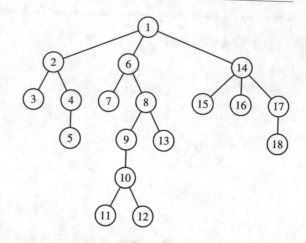

Fig. 3.1 Preorder traversal of a rooted tree. Nodes are numbered according to the order in which they are visited during the traversal

subtree rooted at the node. For instance, the first child of the node with order 6 has order $7 = 6 + 1$, and the next sibling of the same node has order $14 = 6 + 8$, because the subtree rooted at the node with order 6 has 8 nodes.

Now, the preorder traversal of a tree can be constructed from the preorder traversals of the subtrees rooted at the children of the root of the tree. This decomposition property allows the application of the divide-and-conquer technique, yielding a simple recursive algorithm.

The following algorithm performs a preorder traversal of a rooted tree. The order in which nodes are visited is stored in the attribute $v.order$, for all nodes $v \in V$. Thus, nothing has to be done in order to combine the traversals of the subtrees into a traversal of the whole tree.

procedure *preorder_tree_traversal*(T)
   $T.num \leftarrow 0$
   preorder_tree_traversal$(T, root[T])$

procedure *preorder_tree_traversal*(T, v)
   $T.num \leftarrow T.num + 1$
   $v.order \leftarrow T.num$
   for all children w of node v in T **do**
      preorder_tree_traversal(T, w)

Lemma 3.1 *The recursive algorithm for preorder traversal of a rooted tree runs in* $O(n)$ *time using* $O(n)$ *additional space, where n is the number of nodes in the tree.*

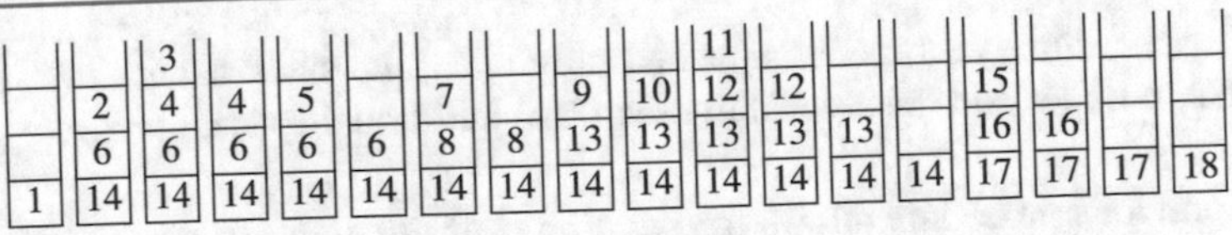

Fig. 3.2 Evolution of the stack of nodes during execution of the preorder traversal procedure, upon the tree of Fig. 3.1

Proof Let $T = (V, E)$ be a tree on n nodes. The algorithm makes $O(n)$ recursive calls, one for each node of the tree. Further, $O(n)$ additional space is used. □

The stack implicit in the recursive algorithm for preorder tree traversal can be made explicit, yielding a simple iterative algorithm in which a stack of nodes holds those nodes, the subtrees rooted at which are waiting to be traversed. Initially, the stack contains only the root of the tree. Every time a node is popped and visited, the children of the popped node are pushed, one after the other, starting with the last child, following with the previous sibling of the last child, and so on, until having pushed the first child of the popped node. The procedure terminates when the stack has been emptied and no children nodes remain to be pushed.

procedure *preorder_tree_traversal*(T)
    let S be an empty stack (of nodes)
    push *root*[T] onto S
    $num \leftarrow 0$
    while S is not empty **do**
        pop from S the top node v
        $num \leftarrow num + 1$
        $v.order \leftarrow num$
        for all children w of node v in T in reverse order **do**
            push w onto S

Example 3.2 Consider the execution of the preorder traversal procedure with the help of a stack of nodes, illustrated in Fig. 3.2. Starting off with a stack containing only the root of the tree, the evolution of the stack of nodes right before a node is popped shows that nodes are, as a matter of fact, popped in preorder traversal order: $1, 2, \ldots, 18$.

Lemma 3.2 *The iterative algorithm for preorder traversal of a rooted tree runs in* $O(n)$ *time using* $O(n)$ *additional space, where* n *is the number of nodes in the tree.*

Proof Let $T = (V, E)$ be a tree on n nodes. Since each node of the tree is pushed and popped exactly once, the loop is executed n times and the algorithm runs in $O(n)$

time. Further, since each node is pushed only once, the stack cannot ever contain more than n nodes and the algorithm uses $O(n)$ additional space. □

The previous procedure allows for iteration over the nodes of a tree in preorder, by building a list L of the nodes of T in the order in which they are visited during the traversal.

```
function preorder_tree_list_traversal(T)
    let L be an empty list (of nodes)
    let S be an empty stack (of nodes)
    push root[T] onto S
    while S is not empty do
        pop from S the top node v
        append v to L
        for all children w of node v in T in reverse order do
            push w onto S
    return L
```

Then,

```
for all nodes v of T in preorder do
    ...
```

is equivalent to

```
L ← preorder_tree_list_traversal(T)
for all v in L do
    ...
```

3.2 Postorder Traversal of a Tree

In a postorder traversal of a tree, a postorder traversal of the subtree rooted in turn at each of the children of the root is performed first, and the root of the tree is visited last. The order in which the children nodes are considered is significant. The subtree rooted at the first child is traversed first, followed by the subtree rooted at the next sibling, etc.

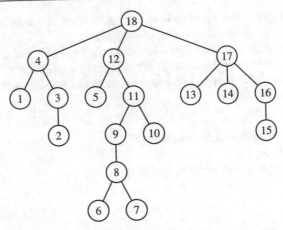

Fig. 3.3 Postorder traversal of the rooted tree of Fig. 3.1. Nodes are numbered according to the order in which they are visited during the traversal

		16	15														
		14	14	14				10			7						
	17	13	13	13	13		11	9	9	8	6	6					
	12	12	12	12	12	12	5	5	5	5	5	5	5		3	2	
18	4	4	4	4	4	4	4	4	4	4	4	4	4	4	1	1	1

Fig. 3.4 Evolution of the stack of nodes during execution of the postorder traversal procedure, upon the tree of Fig. 3.3

Definition 3.2 Let $T = (V, E)$ be a tree on n nodes rooted at node r. A bijection $order : V \rightarrow \{1, \ldots, n\}$ is a **postorder traversal** of T if $order[r] = n$ and

- $order[last[v]] = order[v] - 1$ (if v is not a leaf),
- $order[next[v]] = order[v] + size[next[v]]$ (if v is not a last child)

for all nodes $v \in V$.

Example 3.3 The postorder traversal of a rooted tree is illustrated in Fig. 3.3. The root has order n, the last child of a non-leaf node with order k has order $k - 1$, and the next sibling of a node with order k has order k plus the number of nodes in the subtree rooted at the next sibling node. For instance, the last child of the node with order 12 has order $11 = 12 - 1$, and the next sibling of the same node has order $17 = 12 + 5$, because the subtree rooted at the node with order 17 has 5 nodes.

As with the preorder traversal, the postorder traversal of a tree can be constructed from the postorder traversals of the subtrees rooted at the children of the root of the tree. This decomposition property allows the application of the divide-and-conquer technique, yielding again a simple recursive algorithm.

The following algorithm performs a postorder traversal of a rooted tree. The order in which nodes are visited is stored in the attribute $v.order$, for all nodes $v \in V$. Thus, nothing has to be done in order to combine the traversals of the subtrees into a traversal of the whole tree.

> **procedure** $postorder\_tree\_traversal(T)$
> $T.num \leftarrow 0$
> $postorder\_tree\_traversal(T, root[T])$

> **procedure** $postorder\_tree\_traversal(T, v)$
>   **for all** children w of node v in T **do**
> $postorder\_tree\_traversal(T, w)$
> $T.num \leftarrow T.num + 1$
> $v.order \leftarrow T.num$

Lemma 3.3 *The recursive algorithm for postorder traversal of a rooted tree runs in* $O(n)$ *time using* $O(n)$ *additional space, where n is the number of nodes in the tree.*

Proof Let $T = (V, E)$ be a tree on n nodes. The algorithm makes $O(n)$ recursive calls, one for each node of the tree. Further, $O(n)$ additional space is used. □

The stack implicit in the recursive algorithm for postorder tree traversal can also be made explicit, yielding a simple iterative algorithm in which a stack of nodes holds those nodes the subtrees rooted at which are waiting to be traversed. Initially, the stack contains only the root of the tree. Every time a node is popped and visited, the children of the popped node are pushed, one after the other, starting with the first child, following with the next sibling of the first child, and so on, until having pushed the last child of the popped node.

The procedure terminates when the stack has been emptied and no children nodes remain to be pushed. However, nodes are popped in reverse postorder traversal order and then, the postorder number assigned to node v when visiting it is $n - order[v] + 1$, instead of just $order[v]$.

```
procedure postorder_tree_traversal(T)
    let S be an empty stack (of nodes)
    push root[T] onto S
    num ← 0
    while S is not empty do
        pop from S the top node v
        num ← num + 1
        let n be the number of nodes of T
        v.order ← n − num + 1
        for all children w of node v in T do
            push w onto S
```

Example 3.4 Consider the execution of the postorder traversal procedure with the help of a stack of nodes, illustrated in Fig. 3.4. Starting off with a stack containing only the root of the tree, the evolution of the stack of nodes right before a node is popped shows that nodes are, as a matter of fact, popped in reverse postorder traversal order: 18, 17, ..., 1.

Lemma 3.4 *The iterative algorithm for postorder traversal of a rooted tree runs in* $O(n)$ *time using* $O(n)$ *additional space, where* n *is the number of nodes in the tree.*

Proof Let $T = (V, E)$ be a tree on n nodes. Since each node of the tree is pushed and popped exactly once, the loop is executed n times and the algorithm runs in $O(n)$ time. Further, since each node is pushed only once, the stack cannot ever contain more than n nodes and the algorithm uses $O(n)$ additional space. □

The previous procedure allows for iteration over the nodes of a tree in postorder, by building a list L of the nodes of T in the order in which they are visited during the traversal.

```
function postorder_tree_list_traversal(T)
    let L be an empty list (of nodes)
    let S be an empty stack (of nodes)
    push root[T] onto S
    while S is not empty do
        pop from S the top node v
        append v to L
        for all children w of node v in T do
            push w onto S
    return L reversed
```

Then,

> **for all** nodes v of T in postorder **do**
> ...

is equivalent to

> $L \leftarrow postorder\_tree\_list\_traversal(T)$
> **for all** v in L **do**
> ...

3.3   Top-Down Traversal of a Tree

In a top-down traversal of a tree, nodes are visited in order of non-decreasing depth, and nodes at the same depth are visited in left-to-right order.

Recall from Sect. 1.1 that in a tree $T = (V, E)$, $depth[v]$ denotes the depth of node v, that is, the length of the unique path from the root of T to node v, for all nodes $v \in V$. Let also $rank[v]$ denote the order in which node v is visited during a preorder traversal of T, for all nodes $v \in V$.

Definition 3.3 Let $T = (V, E)$ be a tree on n nodes rooted at node r. A bijection $order : V \rightarrow \{1, \dots, n\}$ is a **top-down traversal** of T if the following conditions

- if $order[v] < order[w]$, then $depth[v] \leqslant depth[w]$
- if $depth[v] = depth[w]$ and $rank[v] < rank[w]$, then $order[v] < order[w]$

are satisfied, for all nodes $v, w \in V$.

It follows that in a top-down traversal of a tree, the root is the first node to be visited.

Lemma 3.5 *Let $T = (V, E)$ be a tree on n nodes rooted at node r, and let order : $V \rightarrow \{1, \dots, n\}$ be a top-down traversal of T. Then, $order[r] = 1$.*

Proof Let $T = (V, E)$ be a tree on n nodes rooted at node r, let $order : V \rightarrow \{1, \dots, n\}$ be a top-down traversal of T, and suppose that $order[r] > 1$. Then, there is a node $v \in V$, $v \neq r$ with $order[v] = 1$, because $order : V \rightarrow \{1, \dots, n\}$ is a bijection and then, $depth[v] \leqslant depth[r]$ by Definition 3.3. But since $depth[r] = 0$, it must also be $depth[v] = 0$, that is, $v = r$, yielding a contradiction. Therefore, $order[r] = 1$. $\qquad\square$

Fig. 3.5 Top-down traversal
of the rooted tree of Fig. 3.1.
Nodes are numbered
according to the order in
which they are visited during
the traversal

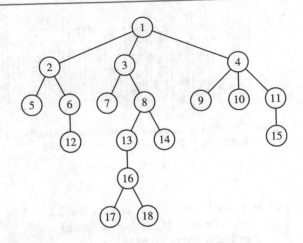

Example 3.5 Figure 3.5 illustrates the top-down traversal of a rooted tree. The root has order 1, nodes are visited in order of non-decreasing depth, and nodes at the same depth are visited by the increasing rank of preorder traversal. For instance, the node with order 8 at depth 2 and preorder rank 8 is visited before the node with order 13 at depth 3, and also before the node with order 11 at depth 2, which has preorder rank 17.

Unlike the preorder and the postorder traversal, the top-down traversal of a tree *cannot* be constructed from the traversals of the subtrees rooted at the children of the root of the tree. A top-down traversal of a tree can be easily realized with the help of a queue of nodes, though, which holds those nodes the subtrees rooted at which are waiting to be traversed.

Initially, the queue contains only the root of the tree. Every time a node is dequeued and visited, the children of the dequeued node are enqueued, one after the other, starting with the first child, following with the next sibling of the first child, and so on, until having enqueued the last child of the dequeued node. The procedure terminates when the queue has been emptied and no children nodes remain to be enqueued.

Example 3.6 Consider the execution of the top-down traversal procedure with the help of a queue of nodes, illustrated in Fig. 3.6. Starting off with a queue containing only the root of the tree, the evolution of the queue of nodes right before a node is dequeued shows that nodes are, as a matter of fact, dequeued in top-down traversal order: 1, 2, . . . , 18.

The following algorithm performs a top-down traversal of a rooted tree $T = (V, E)$, using a queue of nodes Q to implement the previous procedure. The order in which nodes are visited during the top-down traversal is stored in the attribute $v.order$, for all nodes $v \in V$.

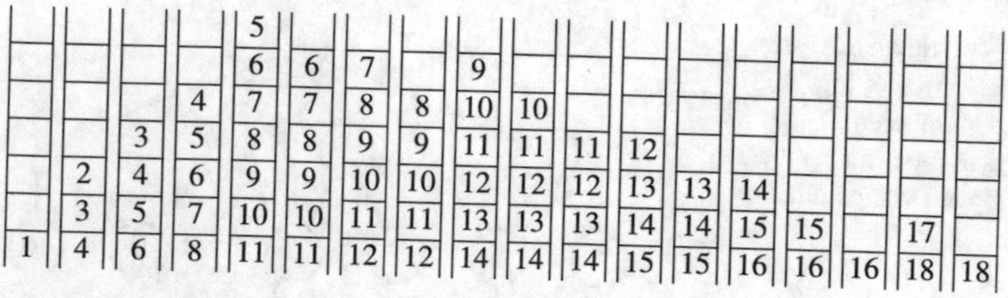

				5													
				6	6	7		9									
			4	7	7	8	8	10	10								
		3	5	8	8	9	9	11	11	11	12						
	2	4	6	9	9	10	10	12	12	12	13	13	14				
	3	5	7	10	10	11	11	13	13	13	14	14	15	15		17	
1	4	6	8	11	11	12	12	14	14	14	15	15	16	16	16	18	18

Fig. 3.6 Evolution of the queue of nodes during execution of the top-down traversal procedure, upon the tree of Fig. 3.5

---

procedure *top_down_tree_traversal*(T)
   let Q be an empty queue (of nodes)
   enqueue *root*[T] into Q
   num $\leftarrow 0$
   while Q is not empty **do**
      dequeue from Q the front node v
      num $\leftarrow$ *num* $+ 1$
      v.*order* $\leftarrow$ *num*
      for all children w of node v in T **do**
         enqueue w into Q

Remark 3.1 Notice that the top-down tree traversal algorithm and the iterative pre-order tree traversal algorithm differ only on the data structure used for holding those nodes the subtrees rooted at which are waiting to be traversed (a stack for the pre-order traversal, and a queue for the top-down traversal) and in the order in which the children of a node are inserted into the data structure (children nodes are enqueued in left-to-right order, but pushed into the stack in right-to-left order).

Lemma 3.6 *The algorithm for top-down traversal of a rooted tree runs in $O(n)$ time using $O(n)$ additional space, where n is the number of nodes in the tree.*

Proof Let $T = (V, E)$ be a tree on n nodes. Since each node of T is enqueued and dequeued exactly once, the loop is executed n times. Processing each dequeued node takes time linear in the number of children of the node, and then, processing all dequeued nodes takes $O(n)$ time. Therefore, the algorithm runs in $O(n)$ time. Further, since each node is enqueued only once, the queue cannot ever contain more than n nodes and the algorithm uses $O(n)$ additional space. ∎

The previous procedure allows for iteration over the nodes of a tree in top-down order, by building a list L of the nodes of T in the order in which they are visited during the traversal.

> **function** $top\_down\_order\_tree\_list\_traversal(T)$
>   let L be an empty list (of nodes)
>   let Q be an empty queue (of nodes)
>   enqueue $root[T]$ into Q
>   **while** Q is not empty **do**
>   dequeue from Q the front node v
>   append v to L
>   **for all** children w of node v in T **do**
>   enqueue w into Q
>   **return** L

Then,

> **for all** nodes v of T in top-down order **do**
>   ...

is equivalent to

> $L \leftarrow top\_down\_order\_tree\_list\_traversal(T)$
> **for all** v in L **do**
>   ...

or just

> **for all** v in $top\_down\_order\_tree\_list\_traversal(T)$ **do**
>   ...

3.4   Bottom-Up Traversal of a Tree

In a bottom-up traversal of a tree, nodes are visited in order of non-decreasing height. Nodes at the same height are visited in order of non-decreasing depth, and nodes of the same height and depth are visited in left-to-right order.

Recall from Sect. 1.1 that in a tree $T = (V, E)$, $height[v]$ denotes the height of node v, that is, the length of a longest path from node v to any node in the subtree of

T rooted at node v, for all nodes $v \in V$. Let also $rank[v]$ denote the order in which node v is visited during a preorder traversal of T, for all nodes $v \in V$.

Definition 3.4 Let $T = (V, E)$ be a tree on n nodes rooted at node r. A bijection $order : V \to \{1, \ldots, n\}$ is a **bottom-up traversal** of T if the following conditions

- if $order[v] < order[w]$, then $height[v] \leqslant height[w]$
- if $height[v] = height[w]$ and $depth[v] < depth[w]$, then $order[v] < order[w]$
- if $height[v] = height[w]$ and $depth[v] = depth[w]$ and $rank[v] < rank[w]$, then $order[v] < order[w]$

are satisfied, for all nodes $v, w \in V$.

The first condition in Definition 3.4 guarantees that in a bottom-up traversal of a tree, nodes are visited in order of non-decreasing height, while the second condition ensures that nodes of the same height are visited in order of non-decreasing depth, and the third condition ensures that nodes of the same height and depth are visited in left-to-right order.

It follows that in a bottom-up traversal of a tree, the root is the last node to be visited.

Lemma 3.7 *Let $T = (V, E)$ be a tree on n nodes rooted at node r, and let order : $V \to \{1, \ldots, n\}$ be a bottom-up traversal of T. Then, $order[r] = n$.*

Proof Let $T = (V, E)$ be a tree on n nodes rooted at node r, let $order : V \to \{1, \ldots, n\}$ be a bottom-up traversal of T, and suppose that $order[r] < n$. Then, there is a node $v \in V$, $v \neq r$ with $order[v] = n$, because $order : V \to \{1, \ldots, n\}$ is a bijection and then, $height[r] \leqslant height[v]$ by Definition 3.4. But since $height[r] = \max_{w \in V} height[w]$, it must also be $height[v] = height[r]$, that is, $v = r$, yielding a contradiction. Therefore, $order[r] = n$. □

Example 3.7 The bottom-up traversal of a tree is illustrated in Fig. 3.7. The root has order n, nodes are visited in order of non-decreasing height, nodes at the same height are visited in order of non-decreasing depth, and nodes of the same height and depth are visited by the increasing rank of preorder traversal. For instance, the node with order 13, which has height 2, depth 1, and preorder rank 2, is visited before the node with order 16, which has height 3, and also before the node with order 15, which has height 2 and depth 3, and before the node with order 14, which has height 2, depth 1, and preorder rank 14.

Unlike the preorder and the postorder traversal, but as with the top-down traversal, the bottom-up traversal of a tree *cannot* be constructed from the traversals of the subtrees rooted at the children of the root of the tree. A bottom-up traversal of a tree

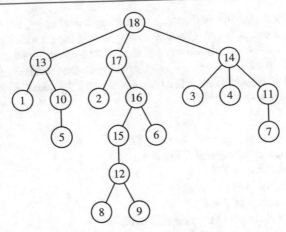

Fig. 3.7 Bottom-up traversal of the tree of Fig. 3.1. Nodes are numbered according to the order in which they are visited during the traversal

1																	
2	2																
3	3	3															
4	4	4	4														
5	5	5	5	5	6												
6	6	6	6	6	7	7	8										
7	7	7	7	7	8	8	9	9	10	11	12	13					
8	8	8	8	8	9	9	10	10	11	12	13	14	14				
9	9	9	9	9	10	10	11	11	12	13	14	15	15	15	16	17	18

Fig. 3.8 Evolution of the queue of nodes during the execution of the bottom-up traversal procedure, upon the tree of Fig. 3.7

can be easily realized with the help of a queue of nodes, though, which holds those nodes which are waiting to be traversed and the subtrees rooted at the children of which have already been traversed.

Initially, the queue contains all leaves of the tree in top-down traversal order. Every time a node is dequeued and visited, the number of non-visited children of the parent of the node is decreased by one, and, as soon as it reaches zero, the parent of the node is enqueued. The procedure terminates when the queue has been emptied and no nodes remain to be enqueued.

Example 3.8 Consider the execution of the bottom-up traversal procedure with the help of a queue of nodes, illustrated in Fig. 3.8. Starting off with a queue containing all leaves of the tree in top-down traversal order, the evolution of the queue of nodes right before a node is dequeued shows that nodes are, as a matter of fact, dequeued in bottom-up traversal order: 1, 2, ..., 18.

The following algorithm performs a bottom-up traversal of a rooted tree $T = (V, E)$, using queues of nodes Q and R to implement the previous procedure. The order in which nodes are visited during the bottom-up traversal is stored in the attribute $v.order$, for all nodes $v \in V$. During a top-down traversal of the tree using a second queue R of nodes, all leaves are enqueued into queue Q in the required top-to-bottom, left-to-right order.

```
procedure bottom_up_tree_traversal(T)
    let Q be an empty queue (of nodes)
    for all nodes v of T do
        v.children ← 0
    for all nodes v of T do
        for all children w of node v in T do
            v.children ← v.children + 1
    let R be an empty queue (of nodes)
    enqueue root[T] into R
    while R is not empty do
        dequeue from R the front node v
        for all children w of node v in T do
            if w is a leaf node of T then
                enqueue w into Q
            else
                enqueue w into R
    num ← 0
    while Q is not empty do
        dequeue from Q the front node v
        num ← num + 1
        v.order ← num
        if v is not the root of T then
            parent[v].children ← parent[v].children − 1
            if parent[v].children = 0 then
                enqueue parent[v] into Q
```

Lemma 3.8 *The algorithm for bottom-up traversal of a rooted tree runs in $O(n)$ time using $O(n)$ additional space, where n is the number of nodes in the tree.*

Proof Let $T = (V, E)$ be a tree on n nodes. All leaves of T are enqueued in $O(n)$ time using $O(n)$ additional space during a top-down traversal. Further, since each node of T is enqueued and dequeued exactly once, the loop is executed n times, and given that processing each dequeued node takes $O(1)$ time, the algorithm runs in $O(n)$ time. Further, since each node is enqueued only once, the queue cannot ever contain more than n nodes and the algorithm uses $O(n)$ additional space. $\square$

The previous procedure allows for iteration over the nodes of a tree in bottom-up order, by building a list L of the nodes of T in the order in which they are visited during the traversal.

```
function bottom_up_tree_list_traversal(T)
    let L be an empty list (of nodes)
    let Q be an empty queue (of nodes)
    for all nodes v of T do
        v.children ← 0
    for all nodes v of T do
        for all children w of node v in T do
            v.children ← v.children + 1
    let R be an empty queue (of nodes)
    enqueue root[T] into R
    while R is not empty do
        dequeue from R the front node v
        for all children w of node v in T do
            if w is a leaf node of T then
                enqueue w into Q
            else
                enqueue w into R
    while Q is not empty do
        dequeue from Q the front node v
        append v to L
        if v is not the root of T then
            parent[v].children ← parent[v].children − 1
            if parent[v].children = 0 then
                enqueue parent[v] into Q
    return L
```

Then,

```
for all nodes v of T in bottom-up order do
    ...
```

is equivalent to

```
L ← bottom_up_tree_list_traversal(T)
for all v in L do
    ...
```

or just

> **for all** v in *bottom_up_tree_list_traversal*(T) **do**
> ...

3.5 Applications

The different methods for exploring a tree can be used, for instance, to compute the depth and the height of the tree, as well as for producing simple tree layouts. Further applications of tree traversal will be addressed in Chap. 4.

The depth of the nodes of a rooted tree can be computed in $O(n)$ time by a preorder traversal of the tree. The depth of the tree is equal to the depth of the deepest node in the tree.

In the following algorithm, the depth of the nodes is stored in the attribute $v.depth$, for all nodes $v \in V$, and the depth of the tree is stored in the attribute $T.depth$.

> **procedure** *preorder_tree_depth*(T)
>   **let** *deepest* be the first node of T
>   **for all** nodes v of T in preorder **do**
>   **if** v is the root of T **then**
> $v.depth \leftarrow 0$
>   **else**
> $v.depth \leftarrow parent[v].depth + 1$
>   **if** $v.depth > deepest.depth$ **then**
>   *deepest* $\leftarrow v$
> $T.depth \leftarrow deepest.depth$

The height of the nodes of a tree can be computed in $O(n)$ time by a bottom-up traversal of the tree, in which the order among leaves is not significant. The height of a leaf is equal to zero, and the height of a non-leaf node is equal to one plus the largest height among the children of the node. For a rooted tree, the height of the tree is equal to the height of the root.

In the following algorithm, the height of the nodes is stored in the attribute $v.height$, for all nodes $v \in V$, and the height of the tree is stored in the attribute $T.height$.

```
procedure bottom_up_tree_height(T)
    for all nodes v of T in bottom-up order do
        v.height ← 0
        if v is not a leaf node of T then
            for all children w of node v in T do
                v.height ← max(v.height, w.height)
            v.height ← v.height + 1
    T.height ← root[T].height
```

As a further application of tree traversal, consider the problem of producing a layout of a rooted tree. A straight-line *drawing* or *layout* of a graph is a mapping of vertices of the graph to distinct points in the plane, and it induces a mapping of edges of the graph to straight line segments, usually drawn as arrows, joining the points corresponding to their source and target vertices.

A good drawing of a graph should provide some intuition to understand the model of a problem represented by the graph. When it comes to rooted trees, one of the most important sources of intuition is given by the hierarchical structure of the tree. Therefore, rooted trees are usually drawn using a *layered* layout, in which nodes are arranged into vertical layers according to node depth and, since rooted trees in computer science are most often drawn downwards with the root at the top, each node is drawn at a vertical coordinate proportional to the opposite of the depth of the node. The tree drawings shown in Figs. 3.1, 3.3, 3.5, and 3.7 are all straight-line, layered drawings.

The intuition conveyed by a tree drawing can be backed up by a series of aesthetic criteria. As a matter of fact, the following aesthetic criteria are often adopted for drawing rooted trees:

1. Since rooted trees impose a distance on the nodes, no node should be closer to the root than any of its ancestors.
2. Nodes at the same depth should lie along a straight line, and the straight lines corresponding to the depths should be parallel.
3. The relative order of nodes on any level should be the same as in the top-down traversal of the tree.
4. For a binary tree, a left child should be positioned to the left of its parent node and a right child to the right.
5. A parent node should be centered over its children nodes.
6. A subtree of a given tree should be drawn the same way regardless of where it occurs in the tree.

The first three aesthetic rules guarantee that the drawing is planar, that is, that no two distinct, non-adjacent edges cross or intersect. The sixth aesthetic rule guarantees that isomorphic subtrees are *congruent*, helping thus to visually identify repeated patterns in the drawing of a tree.

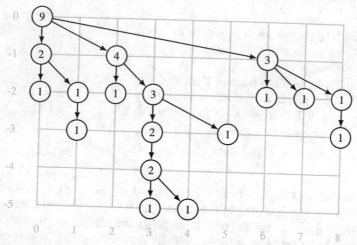

Fig. 3.9 Computing initial node coordinates in a simple layered layout for the rooted tree of Fig. 3.1. Nodes are labeled by their breadth

Further criteria usually adopted for drawing trees and graphs arise from physical constraints. For instance, as long as trees are drawn on media having a drawing surface of bounded width, a layered drawing of a rooted tree should be as narrow as possible.

Now, a simple layered layout of a rooted tree can be obtained in $O(n)$ time using $O(n)$ additional space by a series of traversals of the tree, as follows. Let the *breadth* of a rooted tree be the total breadth of the children nodes of the root, where the breadth of a node in a rooted tree be the breadth of the subtree rooted at the node, and the breadth of a leaf node is equal to 1.

1. The depth of all nodes in a tree can be computed during either a preorder or a top-down traversal, and the breadth of all nodes can be computed during either a postorder or a bottom-up traversal of the tree.
2. Given the depth and breadth of all nodes in the tree, the horizontal coordinate of a node can be initially set proportional to the horizontal coordinate plus the breadth of the previous sibling (or, if the node is a first child, to the horizontal coordinate of the parent), and vertical coordinates can be set proportional to opposite the depth of the nodes, during either a preorder or a top-down traversal of the tree.
3. Parent nodes can be centered over their children nodes during either a postorder or a bottom-up traversal of the tree.

The initial horizontal coordinates computed in the second step of the previous procedure are alright for leaves, as can be seen in Fig. 3.9. Horizontal coordinates are improved for non-leaves in the third step. The resulting layered tree layout is illustrated in Fig. 3.10.

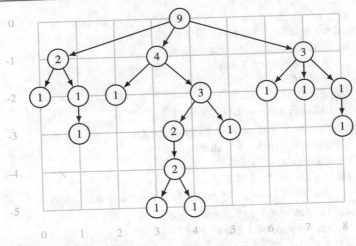

Fig. 3.10 Improving initial node coordinates by centering parent nodes over their children nodes, in the simple layered layout of Fig. 3.9. Nodes are labeled by their breadth

Despite the simplicity of the procedure, all six aesthetic criteria for drawing rooted trees are satisfied. See the bibliographic notes below for more complex algorithms producing tidier tree drawings.

The following algorithm computes a layered layout of a rooted tree $T = (V, E)$, performing a preorder traversal, a bottom-up traversal, a top-down traversal, and a second bottom-up traversal of T to implement the previous procedure. The layout is stored in the form of horizontal coordinates $v.x$ and vertical coordinates $v.y$, for all nodes $v \in V$. First, the depth of all nodes in the tree is computed during an iterative preorder traversal of the tree, and it is stored in the attribute $v.depth$, for all nodes $v \in V$. Second, the breadth of all nodes in the tree is easily computed during a bottom-up traversal of the tree, in which the top-down order among leaves is not significant. The breadth of a leaf node is equal to one, and the breadth of a non-leaf node is equal to the sum of the breadth of the children of the node. The breadth of the nodes is stored in the attribute $v.breadth$, for all nodes $v \in V$.

Next, initial coordinates for all nodes in the tree are computed during a top-down traversal of the tree. The horizontal coordinate of the root node is set to zero. The horizontal coordinate $v.x$ of a non-root node v is set to either $v.x = parent[v].x$, if v is a first child, or $v.x = u.x + u.breadth$, where node u is the previous sibling of node v. The vertical coordinate $v.y$ of a node v is set to $v.y = -v.depth$. Finally, parent nodes are centered over their first and last children nodes during a bottom-up traversal of the tree, in which the top-down order among leaves is not significant.

procedure *layered_tree_layout(T)*
  for all nodes v of T in preorder **do**
    if v is the root of T **then**
      $v.depth \leftarrow 0$
    else
      $v.depth \leftarrow parent[v].depth + 1$
  for all nodes v of T in bottom-up order **do**
    if v is a leaf node of T **then**
      $v.breadth \leftarrow 1$
    if v is not the root of T **then**
      $parent[v].breadth \leftarrow parent[v].breadth + v.breadth$
  for all nodes v of T in top-down order **do**
    if v is the root of T **then**
      $v.x \leftarrow 0$
    else
      if v is a first child in T **then**
        $v.x \leftarrow parent[v].x$
      else
        let w be the previous sibling of node v in T
        $v.x \leftarrow w.x + w.breadth$
      $v.y \leftarrow -v.depth$
  for all nodes v of T in bottom-up order **do**
    if v is not a leaf node of T **then**
      let w be the first child of node v in T
      let z be the last child of node v in T
      $v.x \leftarrow (w.x + z.x)/2$

Summary

The most common methods for exploring a general, rooted tree were addressed in this chapter. Simple algorithms are given in detail (or proposed as exercises) for the different methods of tree traversal: depth-first prefix leftmost (preorder), depth-first prefix rightmost, depth-first postfix leftmost (postorder), depth-first postfix rightmost, breadth-first leftmost (top-down), breadth-first rightmost, and, finally, bottom-up traversal. References to more space-efficient traversal algorithms on *threaded trees* are given in the bibliographic notes below, together with references to further methods for the traversal of ordered binary trees. Application of the different tree traversal methods to tree drawing is also discussed in detail.

Bibliographic Notes

All the tree traversal algorithms discussed in this chapter require, in the worst case, $O(n)$ additional space. More space-efficient algorithms are known [3,4,9,11,12] which visit all nodes of a binary tree on n nodes in $O(n)$ time using $O(1)$ additional space. Space efficiency is achieved by representing trees as *threaded trees* and temporarily modifying the threads during the traversal.

There is still a further common method of exploring a tree. In a *symmetric order* traversal of a binary tree, also known as *inorder* traversal, the left subtree is traversed before visiting the root of the tree, followed by a traversal of the right subtree. A detailed account of the symmetric order traversal of a binary tree and its significance in solving searching and sorting problems can be found, for instance, in [6,7,10,16].

A comprehensive treatment of tree and graph drawing can be found in [1]. A rather simple algorithm for drawing binary trees was given in [5] which consists of setting the vertical coordinate of a node proportional to opposite the depth of the node and the horizontal coordinate proportional to the symmetric traversal order of the node and which, despite its simplicity, still satisfies the first four aesthetic criteria. Several algorithms have been since proposed for producing tidier drawings of rooted trees, all of them also running in $O(n)$ time using $O(n)$ additional space. The first five aesthetic criteria for drawing rooted trees are due to [17], and algorithms for drawing binary trees satisfying these criteria were presented in [14,17]. The sixth aesthetic criteria were introduced in [8], together with a divide-and-conquer algorithm for drawing binary trees, which was improved in [2] in order to produce different drawings of non-isomorphic subtrees and extended in [15] for drawing general, ordered rooted trees. The problem of producing a drawing of a binary tree satisfying all six aesthetic criteria and having minimum width can be solved in polynomial time by means of linear programming, but becomes NP-hard if a grid drawing, that is, a drawing in which all nodes have integral coordinates, is required [13].

Problems

3.1 Give the order in which nodes are visited during a preorder traversal and during a top-down traversal of the tree in Fig. 3.11.

3.2 Give the order in which nodes are visited during a postorder traversal and during a bottom-up traversal of the tree in Fig. 3.11.

3.3 Show the evolution of the stack of nodes during execution of the iterative preorder traversal algorithm upon the tree in Fig. 3.11.

3.4 Show the evolution of the queue of nodes during execution of the top-down traversal algorithm upon the tree in Fig. 3.11.

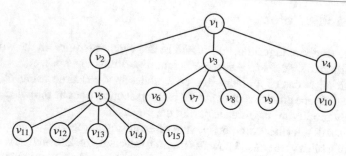

Fig. 3.11 Rooted tree for Problems 3.1–3.4

3.5 Give a full binary tree on $2k + 1$ nodes whose preorder traversal and top-down traversal coincide.

3.6 Give a full binary tree on $2k + 1$ nodes whose postorder traversal and bottom-up traversal coincide.

Exercises

Exercise 3.1 The preorder traversal of a tree is also called *depth-first prefix leftmost* traversal, because parents are visited before children and siblings are visited in left-to-right order. In a *depth-first prefix rightmost* traversal, parents are still visited before children but siblings are visited in right-to-left order instead. The depth-first prefix rightmost traversal of a tree is illustrated in Fig. 3.12. Adapt the iterative preorder tree traversal algorithm to perform a depth-first prefix rightmost traversal of a tree.

Exercise 3.2 The postorder traversal of a tree is also called *depth-first postfix leftmost* traversal, because parents are visited after children and siblings are visited in left-to-right order. In a *depth-first postfix rightmost* traversal, parents are still visited after children but siblings are visited in right-to-left order instead. The depth-first postfix rightmost traversal of a tree is illustrated in Fig. 3.12. Adapt the iterative postorder tree traversal algorithm to perform a depth-first postfix rightmost traversal of a tree.

Exercise 3.3 The depth of a rooted tree was computed in $O(n)$ time in Sect. 3.5 as a by-product of computing the depth of each node, during an iterative preorder traversal of the tree. Extend the recursive preorder traversal algorithm to compute the depth of a rooted tree.

Exercise 3.4 The depth of a rooted tree was computed in $O(n)$ time in Sect. 3.5 as a by-product of computing the depth of each node, during an iterative preorder traversal of the tree. The depth of a rooted tree can also be computed in $O(n)$ time by a top-down traversal of the tree. Extend the top-down traversal algorithm to compute the depth of a rooted tree.

Exercise 3.5 The height of a tree was computed in $O(n)$ time in Sect. 3.4 as a by-product of computing the height of each node, during an iterative postorder traversal of the tree and also in Sect. 3.5, during a bottom-up traversal of the tree. Extend the recursive postorder traversal algorithm to compute the height of a rooted tree.

Exercise 3.6 The depth and the breadth of all nodes in a rooted tree were computed in $O(n)$ time in Sect. 3.5 during an iterative preorder and a bottom-up traversal of the tree, respectively. Show that both the depth and the breadth of all nodes in a rooted tree can be computed in $O(n)$ time during a single, recursive preorder traversal of the tree.

Exercise 3.7 The inductive definition of binary trees yields a simple recursive algorithm for the symmetric order traversal of a binary tree. Give instead an iterative algorithm for the symmetric order traversal of a *full* binary tree.

Exercise 3.8 In the iterative algorithm for the top-down traversal of a tree, a queue is used to hold children nodes while the sibling and cousin nodes of the parent node are visited, because children nodes are only accessible from their parent node. Give instead a recursive algorithm for the top-down traversal of a tree, without using an auxiliary queue. Give also the time and space complexity of the recursive top-down tree traversal algorithm.

Exercise 3.9 Implement the simple algorithm for drawing *full* binary trees, in which the vertical coordinate of a node is set proportional to opposite the depth of the node and the horizontal coordinate of a node is set proportional to the symmetric traversal order of the node.

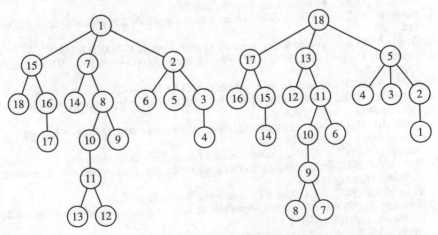

Fig. 3.12 Depth-first prefix (left) and postfix (right) rightmost traversal of a rooted tree. Nodes are numbered according to the order in which they are visited during the traversal

Fig. 3.13 Tidier drawing of
the simple layered layout of
Fig. 3.10. Nodes are labeled
by their breadth

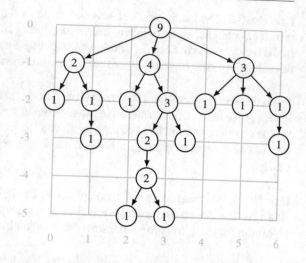

Exercise 3.10 Extend the layered tree layout algorithm to produce tidier drawings,
in which siblings are set as close to each other as possible while keeping a horizontal
distance between nodes of at least one. For instance, the simple layered layout of
Fig. 3.10 could be improved as in Fig. 3.13. Give also the time and space complexity
of the improved algorithm.

References

1. Battista GD, Eades P, Tamassia R, Tollis IG (1999) Graph drawing: algorithms for the visualization of graphs. Prentice Hall, Englewood Cliffs NJ
2. Brüggemann-Klein A, Wood D (1989) Drawing trees nicely with TeX. Electron Publ 2(2):101–
   115
3. Burkhard WA (1975) Nonrecursive traversals of trees. Comput J 18(3):227–230
4. Hirschberg DS, Seiden SS (1993) A bounded-space tree traversal algorithm. Inf Process Lett
   47(4):215–219
5. Knuth DE (1971) Optimum binary search trees. Acta Inf 1(1):14–25
6. Knuth DE (1997) Fundamental algorithms, the art of computer programming, vol 1, 3rd edn.
   Addison-Wesley, Reading MA
7. Knuth DE (1998) Sorting and searching, the art of computer programming, vol 3, 3rd edn.
   Addison-Wesley, Reading MA
8. Reingold EM, Tilford JS (1981) Tidier drawing of trees. IEEE Trans Softw Eng 7(2):223–228
9. Robson JM (1973) An improved algorithm for traversing binary trees without auxiliary stack.
   Inf Process Lett 2(1):12–14
10. Sedgewick R (1992) Algorithms in C++. Addison-Wesley, Reading MA
11. Siklòssy L (1972) Fast and read-only algorithms for traversing trees without an auxiliary stack.
    Inf Process Lett 1(4):149–152
12. Soule S (1977) A note on the nonrecursive traversal of binary trees. Comput J 20(4):350–352
13. Supowit KJ, Reingold EM (1983) The complexity of drawing trees nicely. Acta Inf 18(4):377–
    392

14. Vaucher JG (1980) Pretty-printing of trees. Softw: Practice Exp 10(5):553–561
15. Walker JQ II (1990) A node-positioning algorithm for general trees. Softw: Practice Exp 20(7):685–705
16. Weiss MA (1999) Data structures and algorithm analysis in C++, 2nd edn. Addison-Wesley, Reading MA
17. Wetherell C, Shannon A (1979) Tidy drawing of trees. IEEE Trans Softw Eng 5(5):514–520

Tree Isomorphism

4.1 Tree Isomorphism

Tree isomorphism is the problem of determining whether a tree is isomorphic to another tree and, beside being a fundamental problem with a variety of applications, is also the basis of simple solutions to the more general problems of subtree isomorphism and maximum common subtree isomorphism.

Since trees can be either ordered or unordered, there are different notions of tree isomorphism.

4.1.1 Ordered Tree Isomorphism

Two ordered trees are isomorphic if there is a bijective correspondence between their node sets which preserves and reflects the structure of the ordered trees—that is, such that the node corresponding to the root of one tree is the root of the other tree, a node v_1 is the parent of a node v_2 if and only if the node corresponding to v_1 is the parent of the node corresponding to v_2 in the other tree, and a node v_3 is the next sibling of a node v_2 if and only if the node corresponding to v_3 is also the next sibling of the node corresponding to v_2 in the other tree.

In most applications, however, further information is attached to nodes and edges in the form of node and edge labels. Then, two ordered labeled trees are isomorphic if the underlying ordered trees are isomorphic, and furthermore, corresponding nodes and edges share the same label. Given that a node in a tree can have at most one incoming edge, though the information attached to an edge can be attached to the target node of the edge instead, and edge labels are not really needed. The algorithms discussed in this chapter deal, thus with trees which are either unlabeled or whose nodes are labeled.

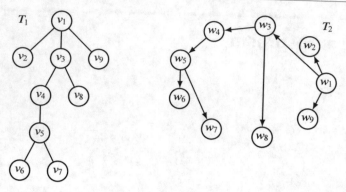

Fig. 4.1 Isomorphic ordered trees. Nodes are numbered according to the order in which they are visited during a preorder traversal

Definition 4.1 Two ordered trees $T_1 = (V_1, E_1)$ and $T_2 = (V_2, E_2)$ are **iso-morphic**, denoted by $T_1 \cong T_2$, if there is a bijection $M \subseteq V_1 \times V_2$ such that $(root[T_1], root[T_2]) \in M$ and the following conditions

- $(first[v], first[w]) \in M$ for all non-leaves $v \in V_1$ and $w \in V_2$ with $(v, w) \in M$,
- $(next[v], next[w]) \in M$ for all non-last children nodes $v \in V_1$ and $w \in V_2$ with $(v, w) \in M$

are satisfied. In such a case, M is an ordered tree isomorphism of T_1 to T_2.

Isomorphism expresses what, in less formal language, is meant when two trees are said to be the same tree. Two isomorphic trees may be depicted in such a way that they look very different—they are differently labeled, perhaps also differently drawn, and it is for this reason that they look different.

Example 4.1 The ordered trees $T_1 = (V_1, E_1)$ and $T_2 = (V_2, E_2)$ shown in Fig. 4.1 are isomorphic. The bijection $M \subseteq V_1 \times V_2$ given by $M = \{(v_1, w_1), (v_2, w_2), (v_3, w_3), (v_4, w_4), (v_5, w_5), (v_6, w_6), (v_7, w_7), (v_8, w_8), (v_9, w_9)\}$ is an ordered tree isomorphism of T_1 to T_2.

An Algorithm for Ordered Tree Isomorphism

A straightforward procedure for testing isomorphism of two ordered trees consists of performing a traversal of both trees using a same traversal method, and then testing whether the node mapping induced by the traversal is an isomorphism between the ordered trees.

Let $T_1 = (V_1, E_1)$ and $T_2 = (V_2, E_2)$ be ordered trees on n nodes, let $order_1 : V_1 \to \{1, \ldots, n\}$ be a traversal of T_1, and let $order_2 : V_2 \to \{1, \ldots, n\}$ be a traversal of T_2. The node mapping $M \subseteq V_1 \times V_2$ induced by $order_1$ and $order_2$ is given by $M[v] = w$ if and only if $order_1[v] = order_2[w]$, for all nodes $v \in V_1$ and $w \in V_2$. Since both $order_1$ and $order_2$ are bijections, M is also a bijection.

The following algorithm implements the previous procedure for testing isomorphism of two ordered trees, based on a top-down traversal of each of the trees. The bijection $M \subseteq V_1 \times V_2$ computed by the procedure upon two isomorphic ordered trees $T_1 = (V_1, E_1)$ and $T_2 = (V_2, E_2)$ is stored in the attribute $v.M$, for all nodes $v \in V_1$.

Recall first that for a tree $T = (V, E)$, the label of a node $v \in V$ is stored in the attribute $v.label$. Unlabeled trees are dealt with in the following algorithms as labeled trees with all node labels set to nil, that is, undefined.

function *simple_ordered_tree_isomorphism*(T_1, T_2)
    if T_1 and T_2 do not have the same number of nodes **then**
        return false
    top_down_tree_traversal(T_1)
    top_down_tree_traversal(T_2)
    let n be the number of nodes of T_1
    let *disorder*$[1..n]$ be a new array
    for all nodes w of T_2 **do**
        disorder$[w.order] \leftarrow w$
    for all nodes v of T_1 **do**
        $v.M \leftarrow$ *disorder*$[v.order]$
    for all nodes v of T_1 **do**
        $w \leftarrow v.M$
        if $v.label \neq w.label$ **then**
            return false
        if v is not a leaf node of T_1 and (w is not a leaf node of T_2 or
                first$[v].M \neq$ *first*$[w]$) **then**
            return false
        if v is not a last child in T_1 and (w is not a last child in T_2 or
                next$[v].M \neq$ *next*$[w]$) **then**
            return false
    return true

Although the previous procedure is correct, non-isomorphism of two ordered trees could be detected earlier by performing a simultaneous traversal of the two trees, instead of performing a traversal of each of the trees and then testing the induced node mapping for ordered tree isomorphism.

The following algorithm performs a simultaneous preorder traversal of the two ordered trees T_1 and T_2, collecting also a mapping M of nodes of T_1 to nodes of T_2 stored in the attribute $v.M$, for all nodes v of T_1.

```
function ordered_tree_isomorphism(T_1, T_2)
    if T_1 and T_2 do not have the same number of nodes then
        return false
    return map_ordered_tree(T_1, root[T_1], T_2, root[T_2])
```

The simultaneous preorder traversal succeeds if T_1 and T_2 are isomorphic, but otherwise fails as soon as either the structure of T_1 and the structure of T_2 differ, or corresponding nodes in T_1 and T_2 do not share the same label.

```
function map_ordered_tree(T_1, v_1, T_2, v_2)
    if v_1.label ≠ v_2.label then
        return false
    v_1.M = v_2
    if nodes v_1 and v_2 do not have the same number of children then
        return false
    for all children w_1 of node v_1 in T_1 and w_2 of node v_2 in T_2 do
        if not map_ordered_tree(T_1, w_1, T_2, w_2) then
            return false
    return true
```

Lemma 4.1 *The algorithm for ordered tree isomorphism runs in $O(n)$ time using $O(n)$ additional space, where n is the number of nodes in the trees.*

Proof Let T_1 and T_2 be ordered trees on n nodes. The algorithm makes $O(n)$ recursive calls, one for each non-leaf node of T_1 and although within a recursive call on some node, the effort spent is not bounded by a constant but is proportional to the number of children of the node, the total effort spent over all non-leaf nodes of T_1 is proportional to n and the algorithm runs in $O(n)$ time. Further, $O(n)$ additional space is used.

□

4.1.2   Unordered Tree Isomorphism

Two unordered trees are isomorphic if there is a bijective correspondence between their node sets which preserves and reflects the structure of the trees—that is, such that the node corresponding to the root of one tree is the root of the other tree, and a node v_1 is the parent of a node v_2 if and only if the node corresponding to v_1 is the parent of the node corresponding to v_2 in the other tree.

Further, two unordered labeled trees are isomorphic if the underlying unordered trees are isomorphic and corresponding nodes share the same label.

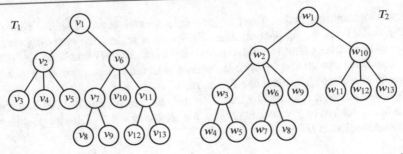

Fig. 4.2 Isomorphic trees. Nodes are numbered according to the order in which they are visited during a preorder traversal

Definition 4.2 Two unordered trees $T_1 = (V_1, E_1)$ and $T_2 = (V_2, E_2)$ are **isomorphic**, denoted by $T_1 \cong T_2$, if there is a bijection $M \subseteq V_1 \times V_2$ such that $(root[T_1], root[T_2]) \in M$ and the following condition

- $(parent[v], parent[w]) \in M$ for all non-root nodes $v \in V_1$ and $w \in V_2$ with $(v, w) \in M$

is satisfied. In such a case, M is an unordered tree isomorphism, or just a tree isomorphism, of T_1 to T_2.

Isomorphism of unordered trees expresses what, in less formal language, is meant when two unordered trees are said to be, up to permuting subtrees rooted at some node, the same tree.

Example 4.2 The unordered trees $T_1 = (V_1, E_1)$ and $T_2 = (V_2, E_2)$ in Fig. 4.2 are isomorphic. The bijection $M \subseteq V_1 \times V_2$ given by $M = \{(v_1, w_1), (v_2, w_{10}),$ $(v_3, w_{11}), (v_4, w_{12}), (v_5, w_{13}), (v_6, w_2), (v_7, w_3), (v_8, w_4), (v_9, w_5), (v_{10}, w_9),$ $(v_{11}, w_6), (v_{12}, w_7), (v_{13}, w_8)\}$ is a tree isomorphism of T_1 to T_2.

Remark 4.1 An ordered tree isomorphism between two ordered trees is also a tree isomorphism between the underlying unordered trees.

The tree isomorphism problem is actually harder for unordered trees than for ordered trees, because two isomorphic unordered trees may be represented in many different ways. Given an (ordered or unordered) tree, any non-identical permutation of the subtrees rooted at any node in the tree yields, in fact, a non-isomorphic ordered tree which is isomorphic as unordered tree to the given tree. That is, many non-isomorphic ordered trees share the same underlying unordered tree.

A simple *certificate* or necessary and sufficient condition for two unordered trees to be isomorphic consists of associating a *unique* isomorphism code to each of the nodes in the two trees. Then, the two trees are isomorphic if and only if their respective root nodes share the same isomorphism code.

The isomorphism code cannot be a single short integer, though. There are $\Omega(2^n/(n\sqrt{n}))$ non-isomorphic unordered trees on n nodes, and encoding such an exponential number of trees would require integers of arbitrary precision, with $\Omega(n)$ bits. Therefore, the isomorphism code for an unordered tree on n nodes is instead a sequence of n integers in the range $1, \ldots, n$.

Recall from Sect. 1.1 that in a tree $T = (V, E)$, $size[v]$ denotes the number of nodes in the subtree of T rooted at node v, for all nodes $v \in V$. Let also $p; q$ denote the concatenation of sequences p and q.

Definition 4.3 Let $T = (V, E)$ be an unordered tree on n nodes. The **isomorphism code** of the root of T is the sequence of n integers in the range $1, \ldots, n$ given by $code[root[T]] = [size[root[T]]], code[w_1], \ldots, code[w_k]$, where nodes $w_1, \ldots, w_k$ are the children of the root of T arranged in non-decreasing lexicographic order of isomorphism code. The isomorphism code of an unordered tree is the isomorphism code of the root of the tree.

It follows from Definition 4.3 that in an unordered tree $T = (V, E)$, all leaves $v \in V$ share the same isomorphism code, $code[v] = [1]$.

Example 4.3 The unique isomorphism codes for the 9 non-isomorphic unordered trees on 5 nodes are shown in Fig. 4.3, together with the isomorphism codes for the subtrees rooted at each of their nodes.

Theorem 4.1 *Let T_1 and T_2 be unordered trees. Then, $T_1 \cong T_2$ if and only if $code[T_1] = code[T_2]$.*

Proof It is immediate from Definition 4.3 that in order for two unordered trees to be isomorphic, it is necessary that their isomorphism codes be identical. Showing that it is sufficient that the isomorphism codes of two unordered trees be identical in order for the trees to be isomorphic, is equivalent to showing that a unique, up to isomorphism, unordered tree can be reconstructed from its isomorphism code.

The latter can be shown by induction on the length of the isomorphism code. Let $code[T]$ be an isomorphism code. If the length of $code[T]$ is equal to one, then it must be $code[T] = [1]$ and the unordered tree T consists of a single node.

Assume now that a unique, up to isomorphism, unordered tree T can be reconstructed from an isomorphism code $code[T]$ of length i, for all $1 \leqslant i \leqslant n$. After discarding the first integer, the rest of a sequence $code[T]$ of length $n+1$ can be partitioned in a unique way into a sequence of disjoint sequences, because an isomorphism code is a sequence of isomorphism codes arranged in non-decreasing lexicographic order and further, the first integer in an isomorphism code is larger than the others. As a matter of fact, each of these sequences starts with the first integer in the rest of the isomorphism code which is at least as large as the first integer of the previous sequence.

Now, since none of the sequences resulting from partitioning the isomorphism code can have more than n integers, a unique, up to isomorphism, unordered tree

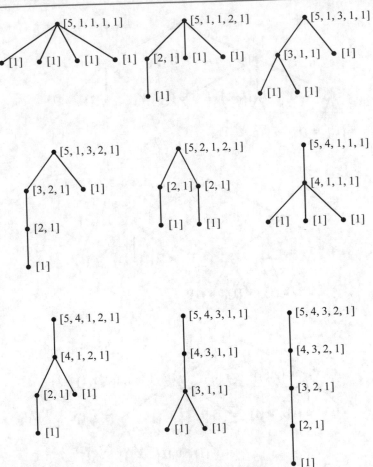

Fig. 4.3 Isomorphism codes of the 9 non-isomorphic unordered trees on 5 nodes. The isomorphism code of each node is shown to the right of the node

can be reconstructed from each of them. Then, the unordered tree T consists of an additional node, the root of the tree, together with these reconstructed unordered trees as subtrees. □

An Algorithm for Unordered Tree Isomorphism

The isomorphism code of an unordered tree can be obtained by performing a series of permutations of the children of the nodes in the tree, transforming the unordered tree into a *canonical* ordered tree in which the left-to-right order of siblings reflects the non-decreasing lexicographic order of their isomorphism codes, as illustrated

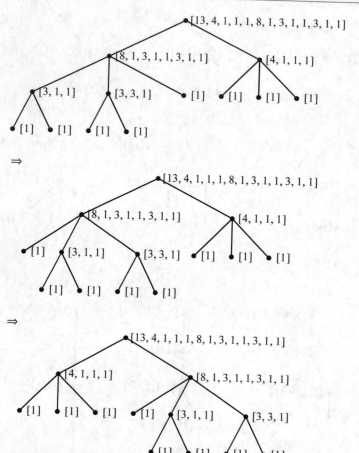

Fig. 4.4 Transformation of an unordered tree into a canonical ordered tree. The isomorphism code of each node is shown to the right of the node

in Fig. 4.4. Then, two unordered trees are isomorphic if and only if their canonical ordered trees are isomorphic.

An alternative, simple procedure for testing isomorphism of two unordered trees consists of computing the isomorphism code for each of the nodes in the trees during either a postorder traversal or a bottom-up traversal, and then comparing the isomorphism codes of their root nodes.

Remark 4.2 Correctness of the simple procedure for unordered tree isomorphism follows from Theorem 4.1.

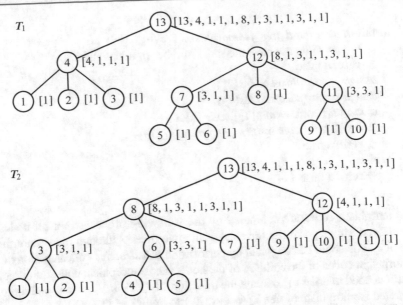

Fig. 4.5 Execution of the simple unordered tree isomorphism procedure upon the trees of Fig. 4.2. Nodes are numbered according to the order in which they are visited during a postorder traversal, and the isomorphism code of each node is also shown to the right of the node

Example 4.4 Consider the execution of the simple isomorphism procedure for unordered trees, illustrated in Fig. 4.5. During a postorder traversal of a tree, the isomorphism codes for the children of a node are computed before computing the isomorphism code of the node. For instance, the isomorphism code of the eighth node in postorder of tree T_2, [8, 1, 3, 1, 1, 3, 1, 1], is obtained by arranging in nondecreasing lexicographic order the isomorphism codes of the third, sixth, and seventh node of the tree, concatenating them, and preceding them by the size of the subtree of T_2 rooted at the eighth node.

Notice that testing unordered tree isomorphism based on isomorphism codes applies to unlabeled trees only. However, the previous isomorphism procedure for unordered unlabeled trees with n nodes can be extended to unordered trees whose nodes are labeled by integers in the range $1, \ldots, n$. See Exercise 4.3.

The following algorithm determines whether two unordered trees T_1 and T_2 are isomorphic, computing isomorphism codes during a postorder traversal to implement the previous procedure. The bijection $M \subseteq V_1 \times V_2$ computed by the procedure upon two isomorphic unordered trees $T_1 = (V_1, E_1)$ and $T_2 = (V_2, E_2)$ is stored in the attribute $v.M$, for all nodes $v \in V_1$.

function *unordered_tree_isomorphism*(T_1, T_2)
   if T_1 and T_2 do not have the same number of nodes **then**
      return false
   assign isomorphism codes(T_1)
   assign isomorphism codes(T_2)
   if $root[T_1].code = root[T_2].code$ **then**
      build_tree_isomorphism_mapping(T_1, T_2)
      return true
   else
      return false

Isomorphism codes are computed by the following procedure for all nodes of T_1 and T_2, during a postorder traversal of the two trees. Although the isomorphism code of a node could be computed by performing a *destructive* concatenation of the isomorphism codes of the children of the node, the isomorphism codes of all nodes are still needed in order to compute the bijection $M \subseteq V_1 \times V_2$ corresponding to an unordered isomorphism of tree $T_1 = (V_1, E_1)$ to tree $T_2 = (V_2, E_2)$.

procedure *assign isomorphism codes*(T)
   for all nodes v of T in postorder **do**
      let $v.code$ be an empty list (of integers)
      if v is a leaf node of T **then**
         append 1 to $v.code$
      else
         let L be an empty list (of lists of integers)
         $size \leftarrow 1$
         for all children w of node v in T **do**
            let x be the first element of $w.code$
            $size \leftarrow size + x$
            append $w.code$ to L
         sort L in lexicographic order
         append $size$ to $v.code$
         for all x in L **do**
            for all y in x **do**
               append y to $v.code$

If the unordered trees $T_1 = (V_1, E_1)$ and $T_2 = (V_2, E_2)$ are isomorphic, the corresponding bijection $M \subseteq V_1 \times V_2$ can be computed from the isomorphism codes of the nodes of T_1 and T_2. Node $v \in V_1$ is mapped to node $w \in T_2$, that is, $(v, w) \in M$, if and only if $code_1[v] = code_2[w]$ and $(parent[v], parent[w]) \in M$.

```
procedure build_tree_isomorphism_mapping(T₁, T₂)
    for all nodes w of T₂ do
        w.mapped_to ← false
    root[T₁].M ← root[T₂]
    root[T₂].mapped_to ← true
    for all nodes v of T₁ in preorder do
        if v is not the root of T₁ then
            for all children w of node parent[v].M in T₂ do
                if v.code = w.code and not w.mapped_to then
                    if w is the root of T₂ then
                        v.M ← w
                        w.mapped_to ← true
                        break
                    else
                        if parent[v].M = parent[w] then
                            v.M ← w
                            w.mapped_to ← true
                            break
```

Theorem 4.2 *The algorithm for unordered tree isomorphism runs in $O(n^2)$ time using $O(n)$ additional space, where n is the number of nodes in the trees.*

Proof Let $T_1 = (V_1, E_1)$ and $T_2 = (V_2, E_2)$ be unordered trees on n nodes. The effort spent in sorting, in lexicographic order, the isomorphism codes of the children of a node is proportional not only to the number of children of the node, but also to the length of the isomorphism codes to be sorted, that is, to the size of the subtree rooted at each of the children nodes.

Now, radix sorting a list of arrays of integers takes time linear in the total length of the arrays, even if the arrays to be sorted are of different lengths. Then, the total effort spent in radix sorting the isomorphism codes of the children of all nodes in the trees is proportional to the sum over all nodes in the trees, of the size of the subtree rooted at the node. This sum is bounded by the recurrence $T(1) = 1, T(n) = T(n-1) + n$, which solves to $T(n) = O(n^2)$.

Therefore, the algorithm runs in $O(n^2)$ time using $O(n)$ additional space. (Taking the construction of bijection $M \subseteq V_1 \times V_2$ into account, though, the algorithm runs in $O(n^2)$ time using $O(n^2)$ additional space.) □

References to more efficient algorithms for unordered tree isomorphism are given in the bibliographic notes, at the end of the chapter.

4.2 Subtree Isomorphism

An important generalization of tree isomorphism is known as subtree isomorphism. Subtree isomorphism is the problem of determining whether a tree is isomorphic to a subtree of another tree, and is a fundamental problem with a variety of applications in engineering and life sciences. Trees can be either ordered or unordered, and further, there are several different notions of subtree.

In the most general sense, a subtree of a given unordered tree is a connected subgraph of the tree, while in a more restricted sense, a *bottom-up subtree* of a given unordered tree is the whole subtree rooted at some node of the tree. Further, a connected subgraph is called a *top-down subtree* if the parent of all nodes in the subtree also belong to the subtree.

Definition 4.4 Let $T = (V, E)$ be an unordered tree, and let $W \subseteq V$. Let also *children*$[v]$ denote the set of children of node v, for all nodes $v \in V$. An unordered tree (W, S) is a **subtree** of T if $S \subseteq E$. A subtree (W, S) is a **top-down subtree** if *parent*$[v] \in W$, for all nodes $v \in W$ different from the root, and it is a **bottom-up subtree** if *children*$[v] \subseteq W$, for all non-leaves $v \in W$.

For ordered trees, the different notions of subtree become those of *leftmost* subtrees. An ordered subtree is a subtree in which the previous sibling (if any) of each of the nodes in the subtree also belong to the subtree, and the same holds for top-down and bottom-up ordered subtrees.

Recall from Sect. 1.1 that in an ordered tree $T = (V, E)$, *previous*$[v]$ denotes the previous sibling of node v, for all non-first child nodes $v \in V$.

Definition 4.5 Let $T = (V, E)$ be an ordered tree, and let $W \subseteq V$. Let also *children*$[v]$ denote the set of children of node v, for all nodes $v \in V$. An ordered tree (W, S) is an **ordered subtree** of T if $S \subseteq E$, and furthermore, *previous*$[v] \in W$ for all non-first child nodes $v \in W$. An ordered subtree (W, S) is a **top-down ordered subtree** if *parent*$[v] \in W$, for all nodes $v \in W$ different from the root, and it is a **bottom-up ordered subtree** if *children*$[v] \subseteq W$, for all non-leaves $v \in W$.

The difference between a connected subgraph, a top-down subtree, and a bottom-up subtree of an ordered and an unordered tree is illustrated in Fig. 4.6. Notice that a subtree (connected subgraph) of an ordered tree is also a subtree of the underlying unordered tree, and a top-down subtree of an ordered tree is also a top-down subtree of the underlying unordered tree, but the converse does not always hold. The notion of bottom-up subtree, however, is the same for both ordered and unordered trees, because a bottom-up subtree of a given tree contains all of the nodes in the subtree rooted at some node of the tree.

Remark 4.3 A top-down subtree of a tree is a connected subgraph of the tree which is rooted at the root of the tree.

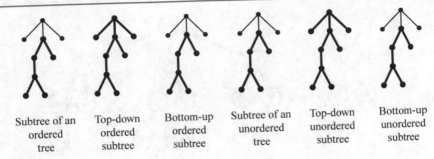

Fig. 4.6 A connected subgraph, a top-down subtree, and a bottom-up subtree of an ordered and an unordered tree

Because of the previous remark, connected subgraph isomorphism problems on trees can be reduced to top-down subtree isomorphism problems upon the subtrees (of enough size and height) rooted at each node of the given tree. Therefore, more attention will be paid in the rest of this chapter to top-down and bottom-up subtree isomorphism problems.

4.2.1   Top-Down Subtree Isomorphism

An ordered tree T_1 is isomorphic to a top-down subtree of another ordered tree T_2 if there is an injective correspondence of the node set of T_1 into the node set of T_2 which preserves the ordered structure of T_1—that is, such that the node of T_2 corresponding to the root of T_1 is the root of T_2, the node of T_2 corresponding to the parent in T_1 of a node v_1 is the parent in T_2 of the node corresponding to v_1, and the node of T_2 corresponding to the next sibling in T_1 of a node v_2 is also the next sibling in T_2 of the node corresponding to v_2.

Definition 4.6 An ordered tree $T_1 = (V_1, E_1)$ is **isomorphic to a top-down subtree** of another ordered tree $T_2 = (V_2, E_2)$ if there is an injection $M \subseteq V_1 \times V_2$ such that the following conditions

- $(root[T_1], root[T_2]) \in M$,
- $(v, w) \in M$ for all non-leaves $v \in V_1$ and $w \in V_2$ such that $(first[v], first[w]) \in M$, and
- $(v, w) \in M$ for all non-last children $v \in V_1$ and $w \in V_2$ such that $(next[v], next[w]) \in M$

are satisfied. In such a case, M is a top-down ordered subtree isomorphism of T_1 into T_2.

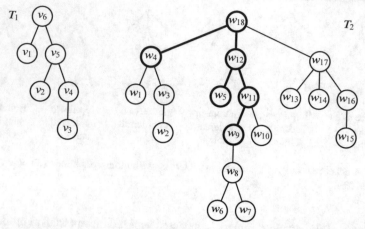

Fig. 4.7 An ordered tree which is isomorphic to a top-down subtree of another ordered tree. Nodes are numbered according to the order in which they are visited during a postorder traversal. The subtree of T_2 which is isomorphic to T_1 is shown highlighted

Example 4.5 The ordered tree $T_1 = (V_1, E_1)$ in Fig. 4.7 is isomorphic to a top-down subtree of the ordered tree $T_2 = (V_2, E_2)$. The injection $M \subseteq V_1 \times V_2$ given by $M = \{(v_1, w_4), (v_2, w_5), (v_3, w_9), (v_4, w_{11}), (v_5, w_{12}), (v_6, w_{18})\}$ is a top-down ordered subtree isomorphism of T_1 into T_2.

An Algorithm for Top-Down Ordered Subtree Isomorphism

A top-down subtree isomorphism of an ordered tree T_1 into another ordered tree T_2 can be obtained by performing a simultaneous traversal of the two trees in the same way as done in Sect. 4.1 for testing isomorphism of two ordered trees.

The following algorithm performs a simultaneous preorder traversal of the ordered tree T_1 and a top-down subtree of the ordered tree T_2, collecting also a mapping M of nodes of T_1 to nodes of T_2 stored in the attribute $v.M$, for all nodes v of T_1.

> **function** *top_down_ordered_subtree*(T_1, T_2)
>   **if** T_1 has more nodes than T_2 **then**
>   **return** false
>   **return** *map_ordered_subtree*$(T_1, root[T_1], T_2, root[T_2])$

The simultaneous preorder traversal succeeds if T_1 is isomorphic to a top-down ordered subtree of T_2, but otherwise fails as soon as the structure of T_1 and the top-down ordered structure of T_2 differ.

```
function map_ordered_subtree(T_1, v_1, T_2, v_2)
    if v_1.label ≠ v_2.label then
        return false
    v_1.M ← v_2
    if v_1 has more children than v_2 then
        return false
    for all children w_1 of node v_1 in T_1 and w_2 of node v_2 in T_2 do
        if not map_ordered_subtree(T_1, w_1, T_2, w_2) then
            return false
    return true
```

Lemma 4.2 *Let T_1 and T_2 be ordered trees with, respectively, n_1 and n_2 nodes, where $n_1 \leqslant n_2$. The algorithm for top-down ordered subtree isomorphism runs in $O(n_1)$ time using $O(n_1)$ additional space.*

Proof Let T_1 and T_2 be ordered trees with, respectively, n_1 and n_2 nodes, where $n_1 \leqslant n_2$. The algorithm makes $O(n_1)$ recursive calls, one for each non-leaf node of T_1 and although within a recursive call on some node, the effort spent is not bounded by a constant but is proportional to the number of children of the node, the total effort spent over all non-leaf nodes of T_1 is proportional to n_1 and the algorithm runs in $O(n_1)$ time. Further, $O(n_1)$ additional space is used. □

4.2.2 Top-Down Unordered Subtree Isomorphism

An unordered tree T_1 is isomorphic to a top-down subtree of another unordered tree T_2 if there is an injective correspondence of the node set of T_1 into the node set of T_2 which preserves the structure of T_1—that is, such that the node of T_2 corresponding to the root of T_1 is the root of T_2, and the node of T_2 corresponding to the parent in T_1 of a node v_1 is the parent in T_2 of the node corresponding to v_1.

Definition 4.7 An unordered tree $T_1 = (V_1, E_1)$ is **isomorphic to a top-down subtree** of another unordered tree $T_2 = (V_2, E_2)$ if there is an injection $M \subseteq V_1 \times V_2$ such that the following conditions

- $(root[T_1], root[T_2]) \in M$,
- $(parent[v], parent[w]) \in M$ for all non-root nodes $v \in V_1$ and $w \in V_2$ with $(v, w) \in M$

are satisfied. In such a case, M is a top-down unordered subtree isomorphism of T_1 into T_2.

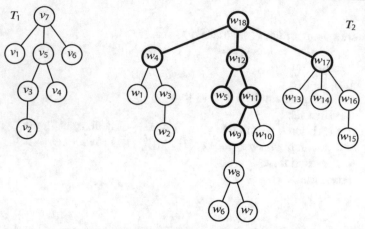

Fig. 4.8 An unordered tree which is isomorphic to a top-down subtree of another unordered tree. Nodes are numbered according to the order in which they are visited during a postorder traversal. The subtree of T_2 which is isomorphic to T_1 is shown highlighted

Example 4.6 The unordered tree $T_1 = (V_1, E_1)$ in Fig. 4.8 is isomorphic to a top-down subtree of the unordered tree $T_2 = (V_2, E_2)$. The injection $M \subseteq V_1 \times V_2$ given by $M = \{(v_1, w_4), (v_2, w_9), (v_3, w_{11}), (v_4, w_5), (v_5, w_{12}), (v_6, w_{17}), (v_7, w_{18})\}$ is a top-down unordered subtree isomorphism of T_1 into T_2.

An Algorithm for Top-Down Unordered Subtree Isomorphism

A top-down subtree isomorphism of an unordered tree T_1 into another unordered tree T_2 can be constructed from subtree isomorphisms of each of the subtrees of T_1 rooted at the children of node v into each of the subtrees of T_2 rooted at the children of node w, provided that these isomorphic subtrees do not overlap. This decomposition property allows application of the divide-and-conquer technique, yielding a simple recursive algorithm.

Consider first the problem of determining whether or not there is a top-down unordered subtree isomorphism of T_1 into T_2, postponing for a while the discussion about construction of an actual subtree isomorphism mapping M of T_1 into T_2. If node v is a leaf in T_1, then it can be mapped to node w in T_2, that is, (v, w) belongs to some top-down unordered subtree isomorphism mapping of T_1 into T_2, provided that $v.label = w.label$.

Otherwise, let p be the number of children of node v in T_1, and let q be the number of children of node w in T_2. Let also $v_1, \dots, v_p$ and $w_1, \dots, w_q$ be the children of nodes v and w, respectively. Build a bipartite graph $G = (\{v_1, \dots, v_p\}, \{w_1, \dots, w_q\}, E)$ on $p + q$ vertices, with an edge $(v_i, w_j) \in E$ if and only if node v_i can be mapped to node w_j. Then, node v can be mapped to node w if and only if G has a maximum cardinality bipartite matching with p edges.

Example 4.7 Consider the execution of the top-down unordered subtree isomorphism procedure, upon the unordered trees of Fig. 4.8. In order to decide if node v_7 can be mapped to node w_{18}, that is, if T_1 is isomorphic to a top-down unordered subtree of T_2, the following maximum cardinality bipartite matching problem is solved:

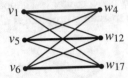

Since there is a solution of cardinality equal to 3, the number of children of node v_7, node v_7 can, in fact, be mapped to node w_{18} and there is a top-down unordered subtree isomorphism of T_1 into T_2. However, stating this bipartite matching problem involves (recursively) solving further maximum cardinality bipartite matching problems.

First, in order to decide if node v_5 can be mapped to node w_4, the following maximum cardinality bipartite matching problem is also solved:

which, in turn, requires solving the following maximum cardinality bipartite matching problem, in order to decide if node v_3 can be mapped to node w_3:

$$v_2 \bullet\!\!\!-\!\!\!\bullet w_2$$

Since the latter (trivial) bipartite matching problem has a solution of cardinality equal to 1, the number of children of node v_3, node v_3 can be mapped to node w_3, and furthermore, node v_5 can be mapped to node w_4, because the former bipartite matching problem has a solution of cardinality equal to 2, the number of children of node v_5.

Next, in order to decide if node v_5 can be mapped to node w_{12}, the following maximum cardinality bipartite matching problem is also solved:

which, in turn, requires solving the following maximum cardinality bipartite matching problem, in order to decide if node v_3 can be mapped to node w_{11}:

Since the latter (trivial) bipartite matching problem has a solution of cardinality equal to 1, the number of children of node v_3, node v_3 can be mapped to node w_{11}, and furthermore, node v_5 can be mapped to node w_{12}, because the former bipartite matching problem has a solution of cardinality equal to 2, the number of children of node v_5.

Finally, in order to decide if node v_5 can be mapped to node w_{17}, the following maximum cardinality bipartite matching problem is also solved:

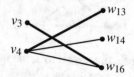

which, in turn, requires solving the following maximum cardinality bipartite matching problem, in order to decide if node v_3 can be mapped to node w_{16}:

Again, since the latter (trivial) bipartite matching problem has a solution of cardinality equal to 1, the number of children of node v_3, node v_3 can be mapped to node w_{16}, and furthermore, node v_5 can be mapped to node w_{17}, because the former bipartite matching problem has a solution of cardinality equal to 2, the number of children of node v_5.

Now, the previous decision procedure for top-down unordered subtree isomorphism can be extended with the construction of an actual unordered subtree isomorphism mapping M of T_1 into T_2, based on the following result. Notice first that the solution to a maximum cardinality bipartite matching problem on a subtree (W_1, S_1) of T_1 and a subtree (W_2, S_2) of T_2 is a set of edges of the corresponding bipartite graph, that is, a set of ordered pairs of nodes $B \in W_1 \times W_2$. Since $W_1 \subseteq V_1$ and $W_2 \subseteq V_2$, it also holds that $B \in V_1 \times V_2$.

Lemma 4.3 *Let $T_1 = (V_1, E_1)$ be an unordered tree isomorphic to a top-down subtree of an unordered tree $T_2 = (V_2, E_2)$, and let $B \subseteq V_1 \times V_2$ be the solutions to all the maximum cardinality bipartite matching problems solved during the top-down unordered subtree isomorphism procedure upon T_1 and T_2. Then, there is a unique top-down unordered subtree isomorphism $M \in V_1 \times V_2$ such that $M \subseteq B$.*

Proof Let $T_1 = (V_1, E_1)$ be an unordered tree isomorphic to a top-down subtree of an unordered tree $T_2 = (V_2, E_2)$, and let $B \subseteq V_1 \times V_2$ be the corresponding solutions of maximum cardinality bipartite matching problems. Showing existence and uniqueness of a top-down unordered subtree isomorphism $M \subseteq B$ is equivalent to showing that, for each node $v \in V_1$ with $(parent(v), z) \in B$ for some node $z \in V_2$, there is a unique $(v, w) \in B$ such that $parent(w) = z$, because a unique top-down subtree of T_2 isomorphic to T_1 can then be reconstructed by order of non-decreasing depth. Therefore, it suffices to show the weaker condition that, for all $(v, w_1), (v, w_2) \in B$ with $w_1 \neq w_2$, it holds that $parent(w_1) \neq parent(w_2)$.

Let $(v, w_1), (v, w_2) \in B$ with $w_1 \neq w_2$. Then, (v, w_1) and (v, w_2) must belong to the solution to different bipartite matching problems, for otherwise the corresponding edges in the bipartite graph would not be independent. But if (v, w_1) and (v, w_2) belong to the solution to different bipartite matching problems, then nodes w_1 and w_2 cannot be siblings, because in each bipartite matching problem, all children of some node of T_1 are matched against all children of some node of T_2. Therefore, $parent(w_1) \neq parent(w_2)$. $\qquad\square$

Given the solutions $B \subseteq V_1 \times V_2$ to all the maximum cardinality bipartite matching problems solved during the top-down unordered subtree isomorphism procedure upon unordered trees $T_1 = (V_1, E_1)$ and $T_2 = (V_2, E_2)$, the corresponding subtree isomorphism mapping $M \subseteq V_1 \times V_2$ can be reconstructed as follows. Set $root[T_1].M$ to $root[T_2]$ and, for all nodes $v \in V_1$ during a preorder traversal, set $v.M$ to the unique node w with $(v, w) \in B$ and $(parent(v), parent(w)) \in B$.

Example 4.8 Consider the execution shown in Example 4.7 of the top-down unordered subtree isomorphism procedure, upon the unordered trees of Fig. 4.8. The solutions $B \subseteq V_1 \times V_2$ to all the maximum cardinality bipartite matching problems solved are as follows:

$v_1 : \boxed{w_4}$
$v_2 : w_2 \quad \boxed{w_9} \quad w_{15}$
$v_3 : w_3 \quad \boxed{w_{11}} \quad w_{16}$
$v_4 : w_1 \quad \boxed{w_5} \quad w_{13}$

$v_5 : \boxed{w_{12}}$
$v_6 : \boxed{w_{17}}$
$v_7 : \boxed{w_{18}}$

The unique subtree isomorphism mapping $M \subseteq B \subseteq V_1 \times V_2$ is indicated by the highlighted nodes of T_2.

The following result yields a simple improvement of the algorithm. The recursive solution to top-down unordered subtree isomorphism problems on subtrees that actually cannot have a solution, because the subtree of the first tree either does not have enough children, is not deep enough, or is not large enough, can be avoided even before constructing the corresponding maximum cardinality bipartite matching problems.

Lemma 4.4 *Let $M \subseteq V_1 \times V_2$ be a top-down subtree isomorphism of an unordered tree $T_1 = (V_1, E_1)$ into another unordered tree $T_2 = (V_2, E_2)$. Then*

- $children(v) \leqslant children(w)$
- $height[v] \leqslant height[w]$
- $size[v] \leqslant size[w]$

for all $(v, w) \in M$.

Proof Let $M \subseteq V_1 \times V_2$ be a top-down subtree isomorphism of an unordered tree $T_1 = (V_1, E_1)$ into an unordered tree $T_2 = (V_2, E_2)$. Suppose that $children(v) > children(w)$ for some $(v, w) \in M$. Let also $X_1 = (W_1, S_1)$ and $X_2 = (W_2, S_2)$ be the subtrees of T_1 and T_2 induced, respectively, by $W_1 = \{v\} \cup \{x \in V_1 \mid (v, x) \in E_1\}$ and $W_2 = \{w\} \cup \{y \in V_2 \mid (w, y) \in E_2\}$, and let $X = \{(x, y) \in M \mid x \in W_1, y \in W_2\}$. Since $|W_1| > |W_2|$, $X \subseteq W_1 \times W_2$ is not an injection, and then M is not an injection either, contradicting the hypothesis that M is a top-down subtree isomorphism of T_1 into T_2. Therefore, it must be $children(v) \leqslant children(w)$.

Suppose now that $height[v] > height[w]$ for some $(v, w) \in M$. Let $X_1 = (W_1, S_1)$ and $X_2 = (W_2, S_2)$ be the subtrees of T_1 and T_2 induced, respectively, by $W_1 = \{v\} \cup \{parent[x] \mid x \in W_1, x \neq root[T_1]\}$ and $W_2 = \{w\} \cup \{parent[y] \mid y \in W_2, y \neq root[T_2]\}$, and let $X = \{(x, y) \in M \mid x \in W_1, y \in W_2\}$. Since $height[v] + 1 = |W_1| > |W_2| = height[w] + 1$, $X \subseteq W_1 \times W_2$ is not an injection, and then M is not an injection either, contradicting the hypothesis that M is a top-down subtree isomorphism of T_1 into T_2. Therefore, it must be $height[v] \leqslant height[w]$.

Finally, suppose that $size[v] > size[w]$ for some $(v, w) \in M$. Let $X_1 = (W_1, S_1)$ and $X_2 = (W_2, S_2)$ be the subtrees of T_1 and T_2 rooted, respectively, at nodes v and w, and let $X = \{(x, y) \in M \mid x \in W_1, y \in W_2\}$. Since $|W_1| > |W_2|$, $X \subseteq W_1 \times W_2$ is not an injection, and then M is not an injection either, contradicting the hypothesis that M is a top-down subtree isomorphism of T_1 into T_2. Therefore, it must be $size[v] \leqslant size[w]$. $\square$

The following algorithm implements the previous procedure for top-down unordered subtree isomorphism. The injection $M \subseteq V_1 \times V_2$ computed by the function upon two unordered trees $T_1 = (V_1, E_1)$ and $T_2 = (V_2, E_2)$ is stored in the attribute $v.M$, for all nodes $v \in V_1$.

Notice first that an unordered labeled tree is isomorphic to a top-down subtree of another unordered labeled tree if the unlabeled tree underlying the former labeled tree is isomorphic to the unlabeled tree underlying the subtree of the latter labeled tree, and furthermore, corresponding nodes share the same label.

The correspondence between children nodes in the trees and the vertices of the bipartite graph is stored in the attribute $v.GT$, for all nodes v of G, and in the opposite direction in the attributes $v_1.T1G$ and $v_2.T2G$, for all nodes v_1 of T_1 and v_2 of T_2.

```
function top_down_unordered_subtree(T₁, v₁, T₂, v₂)
    if v₁.label ≠ v₂.label then
        return false
    if v₁ is a leaf node of T₁ then
        return true
    if T₁ has more nodes than T₂ or v₁.height > v₂.height or
            v₁.size > v₂.size then
        return false
    let G be a new graph
    for all children w₁ of node v₁ in T₁ do
        add a new vertex v to G
        v.GT ← w₁
        w₁.T1G ← v
    for all children w₂ of node v₂ in T₂ do
        add a new vertex w to G
        w.GT ← w₂
        w₂.T2G ← w
    for all children w₁ of node v₁ in T₁ do
        for all children w₂ of node v₂ in T₂ do
            if top_down_unordered_subtree(T₁, w₁, T₂, w₂) then
                add a new edge to G labeled (w₁.T1G, w₂.T2G)
    let L be a maximum cardinality bipartite matching of G
    if the length of L is equal to the number of children of v₁ in T₁ then
        for all edges e = (v, w) in L do
            v.GT.B ← v.GT.B ∪ {w.GT}
        return true
    else
        return false
```

The following algorithm implements the previous procedure for top-down subtree isomorphism of an unordered tree T_1 into another unordered tree T_2, calling the previous recursive function upon the root of T_1 and the root of T_2. The solutions $B \subseteq V_1 \times V_2$ to all the maximum cardinality bipartite matching problems are stored in the attribute $v.B$, for all vertices $v \in V_1$.

```
function top_down_unordered_subtree(T₁, T₂)
    compute_height_and_size_of_all_nodes(T₁)
    compute_height_and_size_of_all_nodes(T₂)
    for all nodes v of T₁ do
        let v.B be an empty set (of nodes)
    if top_down_unordered_subtree(T₁, root[T₁], T₂, root[T₂]) then
        reconstruct_unordered_subtree_isomorphism(T₁, T₂)
    return iso
```

The following procedure reconstructs the top-down unordered subtree isomorphism mapping $M \subseteq V_1 \times V_2$ included in the solutions $B \subseteq V_1 \times V_2$ to all the maximum cardinality bipartite matching problems. The root of T_1 is mapped to the root of T_2 and, during a preorder traversal, each non-root node $v \in V_1$ is mapped to the unique node $w \in V_2$ with $(v, w) \in B$ and $(parent(v)), parent(w)) \in B$.

```
procedure reconstruct_unordered_subtree_isomorphism(T₁, T₂)
    root[T₁].M ← root[T₂]
    for all nodes v of T₁ in preorder do
        if v ≠ root[T₁] then
            for all nodes w in v.B do
                if parent[v].M = parent[w] then
                    v.M ← w
                    break
```

Remark 4.4 Correctness of the algorithm for top-down unordered subtree isomorphism follows from Lemmas 4.3 and 4.4.

Recall that the maximum cardinality bipartite matching algorithm of Sect. 5.3 runs in $O(nm)$ time using $O(m)$ additional space, where n is the number of vertices and m is the number of edges in the bipartite graph, with $n \leq m$. This is $O(n^3)$ time using $O(n^2)$ additional space.

Lemma 4.5 *Let T_1 and T_2 be unordered trees with, respectively, n_1 and n_2 nodes, and let $n = n_1 + n_2$. The algorithm for top-down unordered subtree isomorphism runs in $O(n^3)$ time using $O(n^2)$ additional space.*

Proof Let $T_1 = (V_1, E_1)$ and $T_2 = (V_2, E_2)$ be unordered trees on respectively n_1 and n_2 nodes, let $n = n_1 + n_2$, and let $k(v)$ denote the number of children of node v. The effort spent on a leaf node of T_1 is $O(1)$, and the total effort spent on the leaves of T_1 is thus bounded by $O(n_1)$. The effort spent on a non-leaf node v of T_1 and a non-leaf node w of T_2, on the other hand, is dominated by solving a maximum cardinality bipartite matching problem on a bipartite graph with $k(v)+k(w)$ vertices, and is thus bounded by $O((k(v) + k(w))^3)$. The total effort spent on non-leaves of T_1 and T_2 is thus bounded by $O(\sum_{v \in V_1} \sum_{w \in V_2} (k(v) + k(w))^3)$. Reconstructing the actual subtree isomorphism mapping M of T_1 into T_2 takes $O(n_1 n_2)$ time. Therefore, the algorithm runs in $O(n^3)$ time.

A similar argument shows that the algorithm uses $O(n^2)$ additional space. $\square$

Remark 4.5 Notice that a maximum cardinality bipartite matching can be computed in $O(n\sqrt{n}\sqrt{m}/\log n)$ time using $O(m)$ additional space, see the bibliographic notes for Chap. 5. On a bipartite graph with $n = p+q$ vertices and $m \leq pq$ edges, this is $O((p+q)\sqrt{p+q}\sqrt{pq}/\log(p+q))$ time using $O(pq)$ additional space. Therefore, the algorithm for top-down unordered subtree isomorphism can be implemented to run in $O(n^2 \sqrt{n}/\log n)$ time using $O(n^2)$ additional space.

The algorithm solves the problem of finding a top-down subtree isomorphism of an unordered tree into another unordered tree. The problem of enumerating all top-down unordered subtree isomorphisms can be solved with the subgraph isomorphism algorithms given in Sect. 7.3.

A few details still need to be filled in. The height and size of all nodes in the trees is computed by the following procedure during an iterative postorder traversal of the trees:

procedure *compute_height_and_size_of_all_nodes(T)*
    for all nodes v of T in postorder **do**
        v.height $\leftarrow 0$
        v.size $\leftarrow 1$
        if v is not a leaf node of T **then**
            for all children w of node v in T **do**
                v.height $\leftarrow$ max(*v.height*, *w.height*)
                v.size $\leftarrow$ *v.size* + *w.size*
            v.height $\leftarrow$ *v.height* + 1

4.2.3 Bottom-Up Subtree Isomorphism

An ordered tree T_1 is isomorphic to a bottom-up subtree of another ordered tree T_2 if there is an injective correspondence of the node set of T_1 into the node set of T_2 which preserves the ordered structure of T_1, and furthermore, reflects the ordered structure of a bottom-up subtree of T_2—that is, such that the node of T_2 corresponding to the first child in T_1 of a node v_1 is the first child in T_2 of the node corresponding to v_1, the node of T_2 corresponding to the next sibling in T_1 of a node v_2 is also the next sibling in T_2 of the node corresponding to v_2, and all nodes of T_2 corresponding to leaves of T_1 are also leaves of T_2.

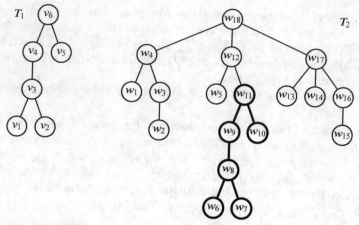

Fig. 4.9 An ordered tree is isomorphic to a bottom-up subtree of another ordered tree. Nodes are numbered according to the order in which they are visited during a postorder traversal. The subtree of T_2, which is isomorphic to T_1, is highlighted

Definition 4.8 An ordered tree $T_1 = (V_1, E_1)$ is **isomorphic to a bottom-up subtree** of another ordered tree $T_2 = (V_2, E_2)$ if there is an injection $M \subseteq V_1 \times V_2$ such that the following conditions

- $(first[v], first[w]) \in M$ for all non-leaves $v \in V_1$ and $w \in V_2$ with $(v, w) \in M$,
- $(next[v], next[w]) \in M$ for all non-last children nodes $v \in V_1$ and $w \in V_2$ with $(v, w) \in M$, and
- for all leaves $v \in V_1$ and nodes $w \in V_2$ with $(v, w) \in M$, w is a leaf in T_2

are satisfied. In such a case, M is a bottom-up ordered subtree isomorphism of T_1 into T_2.

Example 4.9 The ordered tree $T_1 = (V_1, E_1)$ in Fig. 4.9 is isomorphic to a bottom-up subtree of the ordered tree $T_2 = (V_2, E_2)$. The injection $M \subseteq V_1 \times V_2$ given by $M = \{(v_1, w_6), (v_2, w_7), (v_3, w_8), (v_4, w_9), (v_5, w_{10}), (v_6, w_{11})\}$ is a bottom-up ordered subtree isomorphism of T_1 into T_2.

An Algorithm for Bottom-Up Ordered Subtree Isomorphism

A bottom-up subtree isomorphism of an ordered tree T_1 into another ordered tree T_2 can be obtained by repeated invocation of the algorithm for ordered tree isomorphism, upon the root of T_1 and each of the nodes of T_2 in turn.

The following result yields a simple improvement of the procedure. Notice that it holds for bottom-up subtrees only, both ordered and unordered, but it does not hold for top-down subtrees.

Lemma 4.6 Let $T_1 = (V_1, E_1)$ and $T_2 = (V_2, E_2)$ be trees with, respectively, n_1 and n_2 nodes, where $n_1 \leqslant n_2$ and $V_1 \cap V_2 = \emptyset$. Then, for all bottom-up subtrees $X_1 = (W_1, S_1)$ and $X_2 = (W_2, S_2)$ of T_2 which are isomorphic to T_1

- $height[X_1] = height[X_2] = height[T_1]$,
- $size[X_1] = size[X_2] = n_1$,
- $W_1 \cap W_2 = \emptyset$ if $X_1 \neq X_2$.

Proof Let $T_1 = (V_1, E_1)$ and $T_2 = (V_2, E_2)$ be trees with, respectively, n_1 and n_2 nodes, where $n_1 \leqslant n_2$ and $V_1 \cap V_2 = \emptyset$. Let also $X_1 = (W_1, S_1)$ and $X_2 = (W_2, S_2)$ be bottom-up subtrees of T_2 isomorphic to T_1, and let $M_1 \subseteq V_1 \times W_1$ and $M_2 \subseteq V_1 \times W_2$ be bottom-up subtree isomorphisms of T_1 into X_1 and into X_2, respectively. Being M_1 and M_2 bijections, it must be $|V_1| = |X_1|$ and $|V_1| = |X_2|$, that is, $n_1 = size[X_1] = size[X_2]$.

Suppose now that $height[T_1] \neq height[X_1]$. Let $v \in V_1$ be a leaf of largest depth in T_1, and let $w \in W_1$ be a leaf of largest depth in X_1. Let also $T_1' = (V_1', E_1')$ and $X_1' = (W_1', S_1')$ be the subtrees of T_1 and X_1 induced, respectively, by $V_1' = \{v\} \cup \{parent[x] \mid x \in V_1', x \neq root[T_1]\}$ and $W_1' = \{w\} \cup \{parent[y] \mid y \in W_1', y \neq root[X_1]\}$, and let $X = \{(x, y) \in M_1 \mid x \in V_1', y \in W_1'\}$. Since $depth[v] + 1 = height[T_1'] + 1 = |V_1'| \neq |W_1'| = height[X_1'] + 1 = depth[w] + 1$, $X \subseteq V_1' \times W_1'$ is not an injection, and then M_1 is not an injection either, contradicting the hypothesis that M_1 is a bottom-up subtree isomorphism of T_1 into X_1. Therefore, it must be $height[T_1] = height[X_1]$. A similar argument shows that $height[T_1] = height[X_2]$.

Let now $X_1 = (W_1, S_1)$ and $X_2 = (W_2, S_2)$ be distinct bottom-up subtrees of T_2 which are isomorphic to T_1. Suppose $W_1 \cap W_2 = \emptyset$, and let $v \in W_1 \cap W_2$ be a node of largest height in $W_1 \cap W_2$. If $v = root[X_1]$ then $v = root[X_2]$ as well, because $height[X_1] = height[X_2]$ and then $X_1 = X_2$, because X_1 and X_2 are bottom-up subtrees, contradicting the hypothesis that X_1 and X_2 are distinct. Otherwise, there are nodes $u \in W_1$ and $w \in W_2$ such that $(u, v) \in S_1$ and $(w, v) \in S_2$ and then, $parent[v]$ is not well-defined in T_2. Therefore, $W_1 \cap W_2 = \emptyset$ if $X_1 \neq X_2$. $\quad\square$

Now, the repeated invocation of the algorithm for ordered tree isomorphism upon the root of T_1 and each of the nodes v of T_2 with $height[T_1] = height[v]$ and $size[v] = n_1$, is sufficient in order to find all bottom-up subtrees of T_2 which are isomorphic to T_1.

The following algorithm implements the previous procedure for bottom-up subtree isomorphism of an ordered tree T_1 into another ordered tree T_2, collecting also a list of mappings M of nodes of T_1 to nodes of T_2.

```
function bottom_up_ordered_subtree(T₁, T₂)
    compute_height_and_size_of_all_nodes(T₁)
    compute_height_and_size_of_all_nodes(T₂)
    let L be an empty list (of lists of nodes)
    for all nodes v₂ of T₂ do
        if root[T₁].height = v₂.height and root[T₁].size = v₂.size then
            if map_ordered_subtree(T₁, root[T₁], T₂, v₂) then
                let M be the list of pairs (v₁, v₁.M) for all nodes v₁ of T₁
                append M to L
    return L
```

Remark 4.6 Correctness of the algorithm for bottom-up ordered subtree isomorphism follows from Lemma 4.6.

Lemma 4.7 *Let T_1 and T_2 be ordered trees with, respectively, n_1 and n_2 nodes, where $n_1 \leqslant n_2$. The algorithm for bottom-up ordered subtree isomorphism runs in $O(n_2)$ time using $O(n_1)$ additional space.*

Proof Let T_1 and T_2 be ordered trees with, respectively, n_1 and n_2 nodes, where $n_1 \leqslant n_2$, and let k be the number of distinct bottom-up ordered subtrees of T_2 of height $height[T_1]$ and size n_1. The algorithm makes k calls to the ordered tree isomorphism procedure, and thus takes $O(kn_1)$ time using $O(n_1)$ additional space. Since by Lemma 4.6 these k bottom-up ordered subtrees of T_2 are pairwise node-disjoint, it follows that $kn_1 \leqslant n_2$. Therefore, the algorithm runs in $O(n_2)$ time using $O(n_1)$ additional space. ☐

4.2.4 Bottom-Up Unordered Subtree Isomorphism

An unordered tree T_1 is isomorphic to a bottom-up subtree of another unordered tree T_2 if there is an injective correspondence of the node set of T_1 into the node set of T_2 which preserves the structure of T_1, and furthermore, reflects the structure of a bottom-up subtree of T_2—that is, such that the node of T_2 corresponding to the parent in T_1 of a node v is the parent in T_2 of the node corresponding to v, and all nodes of T_2 corresponding to leaves of T_1 are also leaves of T_2.

Definition 4.9 An unordered tree $T_1 = (V_1, E_1)$ is **isomorphic to a bottom-up subtree** of another unordered tree $T_2 = (V_2, E_2)$ if there is an injection $M \subseteq V_1 \times V_2$ such that the following conditions

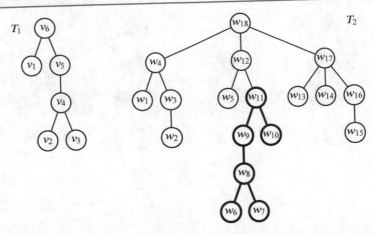

Fig. 4.10 An unordered tree which is isomorphic to a bottom-up subtree of another unordered tree. Nodes are numbered according to the order in which they are visited during a postorder traversal. The subtree of T_2 which is isomorphic to T_1 is shown highlighted

- for all non-root nodes $v \in V_1$ such that $(parent[v], parent[y]) \in M$ for some non-root node $y \in V_2$, $(v, w) \in M$ for some node non-root $w \in V_2$ and, furthermore, $parent[w] = parent[y]$,
- for all leaves $v \in V_1$ and nodes $w \in V_2$ with $(v, w) \in M$, w is a leaf in T_2

are satisfied. In such a case, M is a bottom-up unordered subtree isomorphism of T_1 into T_2.

Example 4.10 The unordered tree $T_1 = (V_1, E_1)$ in Fig. 4.10 is isomorphic to a bottom-up subtree of the unordered tree $T_2 = (V_2, E_2)$. The injection $M \subseteq V_1 \times V_2$ given by $M = \{(v_1, w_{10}), (v_2, w_6), (v_3, w_7), (v_4, w_8), (v_5, w_9), (v_6, w_{11})\}$ is a bottom-up unordered subtree isomorphism of T_1 into T_2.

The problem of finding all bottom-up subtrees of an unordered tree $T_2 = (V_2, E_2)$ on n_2 nodes which are isomorphic to an unordered tree $T_1 = (V_1, E_1)$ on n_1 nodes, where $n_1 \leqslant n_2$, can be reduced to the problem of partitioning $V_1 \cup V_2$ into equivalence classes of bottom-up subtree isomorphism. Two nodes (in the same or in different trees) are equivalent if and only if the bottom-up unordered subtrees rooted at them are isomorphic. Then, T_1 is isomorphic to the bottom-up subtree of T_2 rooted at node $w \in V_2$ if and only if nodes $root[T_1]$ and w belong to the same equivalence class of bottom-up subtree isomorphism.

Example 4.11 The partition of the unordered trees $T_1 = (V_1, E_1)$ and $T_2 = (V_2, E_2)$ of Fig. 4.10 in equivalence classes of bottom-up subtree isomorphism is illustrated in Fig. 4.11. The nine equivalence classes are numbered from 1 to 9.

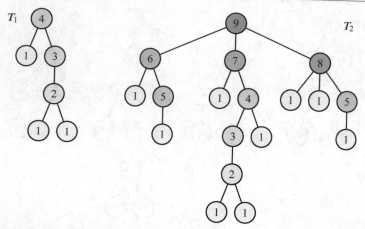

Fig. 4.11 Bottom-up unordered subtree isomorphism equivalence classes for the unordered trees in Fig. 4.10. Nodes are numbered according to the equivalence class to which they belong, and the equivalence classes are shown highlighted in different shades of gray

Tree T_1 is isomorphic to the bottom-up subtree of T_2 rooted at node w_{11}, because nodes v_6 (the root of T_1) and w_{11} both belong to the same equivalence class, the one numbered 4.

Remark 4.7 Notice that partitioning the nodes of two unordered trees in equivalence classes of bottom-up subtree isomorphism is different from assigning isomorphism codes to the nodes in the trees, although isomorphism codes do yield a partition of an unordered tree in bottom-up subtree isomorphism equivalence classes. While the isomorphism code for an unordered tree in Definition 4.3 allows the reconstruction of a unique (up to isomorphism) unordered tree, the bottom-up subtree isomorphism equivalence classes of the nodes of a tree do not convey enough information to make reconstruction of the tree possible. As a matter of fact, the partition of the nodes of a tree (say, T_2) in bottom-up subtree isomorphism equivalence classes is not a function of T_2 alone, but of tree T_1 as well.

An Algorithm for Bottom-Up Unordered Subtree Isomorphism

The bottom-up subtree isomorphisms of an unordered tree $T_1 = (V_1, E_1)$ into another unordered tree $T_2 = (V_2, E_2)$ can be obtained by first partitioning the set of nodes $V_1 \cup V_2$ in bottom-up subtree isomorphism equivalence classes, and then testing equivalence of the root of T_1 and each of the nodes of T_2 in turn.

A simple procedure for partitioning an unordered tree in equivalence classes of bottom-up subtree isomorphism consists of maintaining a dictionary of known equivalence classes during a postorder traversal (or a bottom-up traversal) of the trees. Let $T_1 = (V_1, E_1)$ and $T_2 = (V_2, E_2)$ be unordered trees with, respectively, n_1 and n_2

key	element
[1, 1]	2
[2]	3
[1, 3]	4
[1]	5
[1, 5]	6
[1, 4]	7
[1, 1, 5]	8
[6, 7, 8]	9

Fig. 4.12 Contents of the dictionary of known equivalence classes upon partitioning the unordered trees $T_1 = (V_1, E_1)$ and $T_2 = (V_2, E_2)$ of Fig. 4.10 in equivalence classes of bottom-up subtree isomorphism

nodes, where $n_1 \leqslant n_2$. The k equivalence classes of bottom-up subtree isomorphism of T_1 into T_2 can be numbered from 1 to k, where $1 \leqslant k \leqslant n_1 + n_2$, as follows.

Let the number of known equivalence classes be initially equal to 1, corresponding to the equivalence class of all leaves in the trees. For all nodes v of T_1 and T_2 in postorder, set the equivalence class of v to 1 if node v is a leaf. Otherwise, look up in the dictionary the ordered list of equivalence classes to which the children of node v belong. If the ordered list (key) is found in the dictionary, set the equivalence class of node v to the value (element) found. Otherwise, increment by one the number of known equivalence classes, insert the ordered list together with the number of known equivalence classes in the dictionary, and set the equivalence class of node v to the number of known equivalence classes.

Example 4.12 After partitioning the unordered trees $T_1 = (V_1, E_1)$ and $T_2 = (V_2, E_2)$ of Fig. 4.10 in equivalence classes of bottom-up subtree isomorphism, as illustrated in Fig. 4.11, the dictionary of known equivalence classes contains the entries shown in Fig. 4.12.

Notice that testing bottom-up unordered subtree isomorphism based on equivalence classes of subtree isomorphism applies to unlabeled trees only. However, the previous bottom-up subtree isomorphism procedure for unordered unlabeled trees with n nodes can be extended to unordered trees whose nodes are labeled by integers in the range $1, \ldots, n$. See Exercise 4.7.

The following algorithm implements the previous procedure for bottom-up subtree isomorphism of an unordered tree T_1 into another unordered tree T_2, collecting a list of mappings M of nodes of T_1 to nodes of T_2 stored in the attribute $v.M$, for all nodes v of T_1.

function *bottom_up_unordered_subtree*(T_1, T_2)
   $num \leftarrow 1$
   let D be an empty dictionary (of lists of integers to integers)
   partition_tree_in_isomorphism_equivalence_classes(T_1, num, D)
   partition_tree_in_isomorphism_equivalence_classes(T_2, num, D)
   let L be an empty list (of pairs of nodes)
   for all nodes v_2 of T_2 in postorder **do**
      if $root[T_1].code = v_2.code$ **then**
         $root[T_1].M \leftarrow v_2$
         map_unordered_subtree(T_1, $root[T_1]$, T_2, v_2)
         let M be the list of pairs $(v, v.M)$ for all nodes v of T_1
         append M to L
   return L

Partitioning an unordered tree in equivalence classes of bottom-up subtree isomorphism is done with the help of a variable global to the whole procedure, the number
num of known equivalence classes, and a dictionary D, also global to the whole
procedure, with lists of integers (equivalence classes of children nodes) as keys and
integers (equivalence classes of nodes) as elements. After all nodes of tree T_1 are
partitioned in equivalence classes of bottom-up subtree isomorphism, the nodes of
tree T_2 are also partitioned, sharing the same variable *num* and dictionary D.

Given that the equivalence classes of all nodes in the trees are integers falling
in a fixed range, the ordered list of equivalence classes to which the children of a
node belong can be built in time linear in the number of children plus the range
of equivalence classes of the children nodes, by bucket sorting the corresponding
unordered list of equivalence classes.

procedure *partition_tree_in_isomorphism_equivalence_classes*(T, num, D)
   for all nodes v of T in postorder **do**
      if v is a leaf node of T **then**
         $v.code \leftarrow 1$
      else
         let L be an empty list (of integers)
         for all children w of node v in T **do**
            append $w.code$ to L
         sort L in lexicographic order
         if $L \in D$ **then**
            $v.code \leftarrow D[L]$
         else
            $num \leftarrow num + 1$
            $D[L] \leftarrow num$
            $v.code \leftarrow num$

Now, an actual bottom-up subtree isomorphism mapping $M \subseteq V_1 \times V_2$ of tree $T_1 = (V_1, E_1)$ into the subtree of $T_2 = (V_2, E_2)$ rooted at node v can be constructed by mapping the root of T_1 to node v, and then mapping the remaining nodes in T_1 to the remaining nodes in the subtree of T_2 rooted at node v, such that mapped nodes belong to the same equivalence class of bottom-up subtree isomorphism.

Mapping the nodes of T_1 to equivalent nodes in the subtree of T_2 during a pre-order traversal of T_1, guarantees that the bottom-up subtree isomorphism mapping preserves the structure of tree T_1. In the following recursive procedure, each of the children of node $r_1 \in V_1$ is mapped to some unmapped child of node $r_2 \in V_2$ belonging to the same equivalence class.

procedure *map_unordered_subtree*(T_1, v_1, T_2, v_2)
    let L_2 be an empty list (of nodes)
    for all children w_2 of v_2 in T_2 **do**
        append w_2 to L_2
    for all children w_1 of v_1 in T_1 **do**
        for all nodes w_2 in L_2 **do**
            if $w_1.code = w_2.code$ **then**
                $w_1.M \leftarrow w_2$
                delete w_2 from L_2
                map_unordered_subtree(T_1, w_1, T_2, w_2)
                break

Remark 4.8 Correctness of the algorithm for bottom-up unordered subtree isomorphism follows from the fact that the equivalence class of bottom-up subtree isomorphism which each node in the trees belongs to, is either equal to 1, if the node is a leaf, or is determined by the equivalence classes which the children of the node belong to. During a postorder traversal of each of the trees in turn, known equivalence classes are looked up in a dictionary and new equivalence classes are inserted into the dictionary.

Theorem 4.3 *Let $T_1 = (V_1, E_1)$ and $T_2 = (V_2, E_2)$ be unordered trees with, respectively, n_1 and n_2 nodes. The algorithm for bottom-up unordered subtree isomorphism runs in $O((n_1 + n_2)^2)$ time using $O(n_1 + n_2)$ additional space.*

Proof Let T_1 and T_2 be unordered trees with, respectively, n_1 and n_2 nodes, and let $k(v)$ denote the number of children of node v. The effort spent on a leaf node of T_1 or T_2 is $O(1)$, and the total effort spent on the leaves of T_1 and T_2 is bounded by $O(n_1 + n_2)$.

The effort spent on a non-leaf v of T_1 or T_2, on the other hand, is dominated by bucket sorting a list of $k(v)$ integers, despite the expected $O(1)$ time taken to look up or insert the sorted list of integers in the dictionary. Since all these integers fall in the range $1, \ldots, n_1 + n_2$, the time taken for bucket sorting the codes of

the children nodes is bounded by $O(k(v) + n_1 + n_2)$. The total effort spent on non-leaves of T_1 and T_2 is thus bounded by $O(\sum_{v \in V_1 \cup V_2} (k(v) + n_1 + n_2)) = O(\sum_{v \in V_1 \cup V_2} k(v)) + O((n_1 + n_2)^2) = O(n_1 + n_2) + O((n_1 + n_2)^2) = O((n_1 + n_2)^2)$. Therefore, the algorithm runs in $O((n_1 + n_2)^2)$ time using $O(n_1 + n_2)$ additional space.
□

Lemma 4.8 *Let $T_1 = (V_1, E_1)$ and $T_2 = (V_2, E_2)$ be unordered trees of degree bounded by a constant k and with, respectively, n_1 and n_2 nodes. The algorithm for bottom-up unordered subtree isomorphism can be implemented to run in expected $O(n_1 + n_2)$ time using $O(n_1 + n_2)$ additional space.*

Proof For trees of degree bounded by a constant k, bucket sorting a list of at most k integers can be replaced with appropriate code performing at most k comparisons, and taking thus $O(1)$ time to sort a list of at most k integers. In this case, the effort spent on a non-leaf v of T_1 or T_2 is dominated instead by the expected $O(1)$ time required to look up or insert the sorted list of k integers in the dictionary, and the algorithm runs in expected $O(n_1 + n_2)$ time, still using $O(n_1 + n_2)$ additional space.
□

References to more efficient algorithms for bottom-up unordered subtree isomorphism are given in the bibliographic notes, at the end of the chapter.

4.3 Maximum Common Subtree Isomorphism

Another important generalization of tree isomorphism which also generalizes subtree isomorphism is known as maximum common subtree isomorphism. The maximum common subtree isomorphism problem consists of finding a largest common subtree between two trees, and is also a fundamental problem with a variety of applications in engineering and life sciences.

Since trees can be either ordered or unordered, and further, there are several different notions of subtree, there are different notions of maximum common subtree isomorphism.

4.3.1 Top-Down Maximum Common Subtree Isomorphism

A top-down common subtree of two ordered trees T_1 and T_2 is an ordered tree T such that there are top-down ordered subtree isomorphisms of T into T_1 and into T_2. A maximal top-down common subtree of two ordered trees T_1 and T_2 is a top-down common subtree of T_1 and T_2 which is not a proper subtree of any other top-down common subtree of T_1 and T_2. A top-down maximum common subtree of two ordered trees T_1 and T_2 is a top-down common subtree of T_1 and T_2 with the largest number of nodes.

Definition 4.10 A **top-down common subtree** of an ordered tree $T_1 = (V_1, E_1)$ to another ordered tree $T_2 = (V_2, E_2)$ is a structure (X_1, X_2, M), where $X_1 = (W_1, S_1)$ is a top-down ordered subtree of T_1, $X_2 = (W_2, S_2)$ is a top-down ordered subtree of T_2, and $M \subseteq W_1 \times W_2$ is an ordered tree isomorphism of X_1 to X_2. A top-down common subtree (X_1, X_2, M) of T_1 to T_2 is **maximal** if there is no top-down common subtree (X'_1, X'_2, M') of T_1 to T_2 such that X_1 is a proper top-down subtree of X'_1 and X_2 is a proper top-down subtree of X'_2, and it is **maximum** if there is no top-down common subtree (X'_1, X'_2, M') of T_1 to T_2 with $size[X_1] < size[X'_1]$.

The following remark applies to top-down and bottom-up common subtrees as well, both ordered and unordered.

Remark 4.9 Let (X_1, X_2, M) be a top-down common subtree of an ordered tree $T_1 = (V_1, E_1)$ to another ordered tree $T_2 = (V_2, E_2)$, where $X_1 = (W_1, S_1)$, $X_2 = (W_2, S_2)$, and $M \subseteq W_1 \times W_2$. Since $W_1 \subseteq V_1$ and $W_2 \subseteq V_2$, the injection $M \subseteq W_1 \times W_2$ is also a partial injection $M \subseteq V_1 \times V_2$.

Example 4.13 For the ordered trees $T_1 = (V_1, E_1)$ and $T_2 = (V_2, E_2)$ in Fig. 4.13, the partial injection $M \subseteq V_1 \times V_2$ given by $M = \{(v_5, w_1), (v_6, w_4), (v_9, w_5), (v_{10}, w_{11}), (v_{11}, w_{12}), (v_{12}, w_{18})\}$ is a top-down maximum common subtree isomorphism of T_1 to T_2.

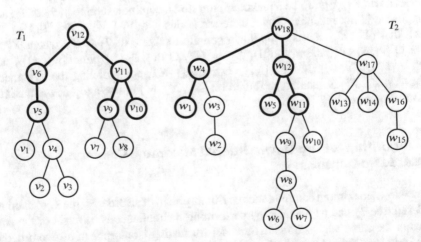

Fig. 4.13 A top-down maximum common subtree of two ordered trees. Nodes are numbered according to the order in which they are visited during a postorder traversal. The common subtree of T_1 and T_2 is shown highlighted in both trees

Now, since a top-down ordered subtree is a top-down subtree in which the previous sibling (if any) of each of the nodes in the subtree also belongs to the subtree, the notions of top-down maximal and maximum common subtree isomorphism coincide for ordered trees.

Lemma 4.9 *Let (X_1, X_2, M) be a top-down common subtree of an ordered tree $T_1 = (V_1, E_1)$ to another ordered tree $T_2 = (V_2, E_2)$. Then, (X_1, X_2, M) is a top-down maximal common subtree of T_1 to T_2 if and only if it is a top-down maximum common subtree of T_1 to T_2.*

Proof Let (X_1, X_2, M) be a maximal top-down common subtree of an ordered tree $T_1 = (V_1, E_1)$ to another ordered tree $T_2 = (V_2, E_2)$, where $X_1 = (W_1, S_1)$, $X_2 = (W_2, S_2)$, and $M \subseteq W_1 \times W_2$, and suppose it is not maximum. Let also (X_1', X_2', M') be a maximum top-down common subtree of T_1 to T_2, where $X_1' = (W_1', S_1')$, $X_2' = (W_2', S_2')$, and $M' \subseteq W_1' \times W_2'$. Suppose $M \not\subseteq M'$, and let $(v, w) \in M \setminus M'$. Let also $W_1'' = \{v\} \cup \{parent[x] \mid x \in W_1'', x \neq root[X_1']\} \cup \{previous[x] \mid x \in W_1'', x \neq root[X_1'], x \neq first[parent[x]]]\}$, let $W_2'' = \{w\} \cup \{parent[y] \mid y \in W_2'', y \neq root[X_2']\} \cup \{previous[y] \mid y \in W_2'', y \neq root[X_2'], y \neq first[parent[y]]]\}$, and let X_1'' and X_2'' be the top-down subtrees of T_1 and T_2 induced, respectively, by W_1'' and W_2''. Let also $M'' = \{(v, w) \in M' \mid v \in W_1'', w \in W_2''\}$.

Then, $(X_1 \cup X_1'', X_2 \cup X_2'', M \cup M'')$ is a top-down common subtree of T_1 to T_2, contradicting the hypothesis that (X_1, X_2, M) is maximal. Therefore, it must be $M \subseteq M'$ and, since (X_1, X_2, M) is maximal, it must be $M = M'$. Therefore, (X_1, X_2, M) is also a top-down maximum common subtree of T_1 to T_2.

Let now (X_1, X_2, M) be a maximum top-down common subtree of T_1 to T_2, and suppose it is not maximal. Then, there are nodes $v \in V_1 \setminus W_1$ and $w \in V_2 \setminus W_2$ such that (X_1', X_2', M') is a top-down common subtree of T_1 to T_2, where $X_1' = (W_1 \cup \{v\}, S_1 \cup \{(parent[v], v)\})$, $X_2' = (W_2 \cup \{w\}, S_2 \cup \{(parent[w], w)\})$, and $M' = M \cup \{(v, w)\}$. But $size[X_1'] = size[X_1] + 1$, contradicting the hypothesis that (X_1, X_2, M) is maximum. Therefore, (X_1, X_2, M) is also a top-down maximal common subtree of T_1 to T_2. $\qquad\square$

An Algorithm for Top-Down Ordered Maximum Common Subtree Isomorphism

A top-down maximum common subtree isomorphism of an ordered tree T_1 to another ordered tree T_2 can be obtained by performing a simultaneous traversal of the two trees in the same way as done in Sect. 4.1 for testing isomorphism of two ordered trees and in in Sect. 4.2 for finding ordered subtree isomorphisms.

The following algorithm performs a simultaneous preorder traversal of a top-down subtree of the ordered tree T_1 and a top-down subtree of the ordered tree T_2, collecting also a partial mapping M of nodes of T_1 to nodes of T_2 stored in the attribute $v.M$, for all nodes v of T_1.

```
procedure top_down_ordered_max_common_subtree(T₁, T₂)
    for all nodes v of T₁ do
        v.M ← nil
        map_ordered_common_subtree(T₁, root[T₁], T₂, root[T₂])
```

The simultaneous preorder traversal proceeds as long as a top-down ordered subtree of T_1 is isomorphic to a top-down ordered subtree of T_2, but otherwise stops as soon as the top-down ordered structures of T_1 and T_2 differ.

```
function map_ordered_common_subtree(T₁, v₁, T₂, v₂)
    if v₁.label ≠ v₂.label then
        return false
    v₁.M ← v₂
    for all children w₁ of node v₁ in T₁ and w₂ of node v₂ in T₂ do
        if not map_ordered_common_subtree(T₁, w₁, T₂, w₂) then
            return false
    return true
```

Lemma 4.10 *Let T_1 and T_2 be ordered trees with, respectively, n_1 and n_2 nodes, where $n_1 \leqslant n_2$. The algorithm for top-down ordered maximum common subtree isomorphism runs in $O(n_1)$ time using $O(n_1)$ additional space.*

Proof Let T_1 and T_2 be ordered trees with, respectively, n_1 and n_2 nodes, where $n_1 \leqslant n_2$. The algorithm makes $O(n_1)$ recursive calls, one for each non-leaf node of T_1 and although within a recursive call on some node, the effort spent is not bounded by a constant but is proportional to the number of children of the node, the total effort spent over all non-leaf nodes of T_1 is proportional to n_1 and the algorithm runs in $O(n_1)$ time. Further, $O(n_1)$ additional space is used.   □

4.3.2 Top-Down Unordered Maximum Common Subtree Isomorphism

A top-down common subtree of two unordered trees T_1 and T_2 is an unordered tree T such that there are top-down unordered subtree isomorphisms of T into T_1 and into T_2. A maximal top-down common subtree of two unordered trees T_1 and T_2 is a top-down common subtree of T_1 and T_2 which is not a proper subtree of any other top-down common subtree of T_1 and T_2. A top-down maximum common subtree of two unordered trees T_1 and T_2 is a top-down common subtree of T_1 and T_2 with the largest number of nodes.

Definition 4.11 A **top-down common subtree** of an unordered tree $T_1 = (V_1, E_1)$ to another unordered tree $T_2 = (V_2, E_2)$ is a structure (X_1, X_2, M), where $X_1 = (W_1, S_1)$ is a top-down unordered subtree of T_1, $X_2 = (W_2, S_2)$ is a top-down unordered subtree of T_2, and $M \subseteq W_1 \times W_2$ is an unordered tree isomorphism of X_1 to X_2. A top-down common subtree (X_1, X_2, M) of T_1 to T_2 is **maximal** if there is no top-down common subtree (X'_1, X'_2, M') of T_1 to T_2 such that X_1 is a proper top-down subtree of X'_1 and X_2 is a proper top-down subtree of X'_2, and it is **maximum** if there is no top-down common subtree (X'_1, X'_2, M') of T_1 to T_2 with $size[X_1] < size[X'_1]$.

Example 4.14 For the unordered trees $T_1 = (V_1, E_1)$ and $T_2 = (V_2, E_2)$ shown in Fig. 4.14, the partial injection $M \subseteq V_1 \times V_2$ given by $M = \{(v_1, w_{10}), (v_2, w_8), (v_4, w_9), (v_5, w_{11}), (v_6, w_{12}), (v_7, w_{15}), (v_9, w_{16}), (v_{10}, w_{13}), (v_{11}, w_{17}), (v_{12}, w_{18})\}$ is a top-down unordered maximum common subtree isomorphism of T_1 and T_2.

Unlike the case of ordered trees, though, the notions of top-down maximal and maximum common subtree isomorphism do not coincide for ordered trees. Nevertheless, top-down unordered maximum common subtree isomorphisms are still maximal.

Lemma 4.11 *Let (X_1, X_2, M) be a maximum top-down common subtree of an unordered tree $T_1 = (V_1, E_1)$ to another unordered tree $T_2 = (V_2, E_2)$. Then, (X_1, X_2, M) is also a top-down maximal common subtree of T_1 to T_2.*

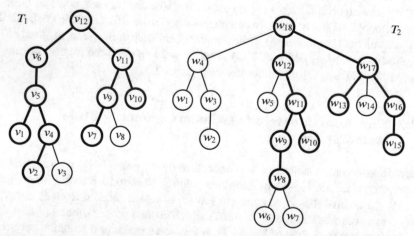

Fig. 4.14 A top-down maximum common subtree of two unordered trees. Nodes are numbered according to the order in which they are visited during a postorder traversal. The common subtree of T_1 and T_2 is shown highlighted in both trees

Proof Let (X_1, X_2, M) be a maximum top-down common subtree of an unordered tree $T_1 = (V_1, E_1)$ to another unordered tree $T_2 = (V_2, E_2)$, where $X_1 = (W_1, S_1)$, $X_2 = (W_2, S_2)$, and $M \subseteq W_1 \times W_2$.

Suppose (X_1, X_2, M) is not maximal. Then, there are nodes $v \in V_1 \setminus W_1$ and $w \in V_2 \setminus W_2$ such that (X_1', X_2', M') is a top-down common subtree of T_1 to T_2, where $X_1' = (W_1 \cup \{v\}, S_1 \cup \{(parent[v], v)\})$, $X_2' = (W_2 \cup \{w\}, S_2 \cup \{(parent[w], w)\})$, and $M' = M \cup \{(v, w)\}$. But $size[X_1'] = size[X_1] + 1$, contradicting the hypothesis that (X_1, X_2, M) is maximum. Therefore, (X_1, X_2, M) is also a top-down maximal common subtree of T_1 to T_2. $\qquad\square$

An Algorithm for Top-Down Unordered Maximum Common Subtree Isomorphism

A top-down maximum common subtree isomorphism of an unordered tree T_1 into another unordered tree T_2 can be constructed from maximum common subtree isomorphisms of each of the subtrees of T_1 rooted at the children of node v into each of the subtrees of T_2 rooted at the children of node w, provided that these common isomorphic subtrees do not overlap. As in the case of top-down unordered subtree isomorphism, this decomposition property allows application of the divide-and-conquer technique, yielding again a simple recursive algorithm.

Consider first the problem of determining the size of a top-down unordered maximum common subtree isomorphism of T_1 and T_2, postponing for a while the discussion about construction of an actual maximum common subtree isomorphism mapping M of T_1 to T_2. If node v is a leaf in T_1 or node w is a leaf in T_2, mapping node v to node w yields a maximum common subtree of size 1, provided that $v.label = w.label$.

Otherwise, let p be the number of children of node v in T_1, and let q be the number of children of node w in T_2. Let also $v_1, \ldots, v_p$ and $w_1, \ldots, w_q$ be the children of nodes v and w, respectively. Build a bipartite graph $G = (\{v_1, \ldots, v_p\}, \{w_1, \ldots, w_q\}, E)$ on $p + q$ vertices, with edge $(v_i, w_j) \in E$ if and only if the size of a maximum common subtree of the subtree of T_1 rooted at node v_i and the subtree of T_2 rooted at node w_j is non-zero and, in such a case, with edge $(v_i, w_j) \in E$ weighted by that non-zero size.

Then, a maximum common subtree of the subtree of T_1 rooted at node v and the subtree of T_2 rooted at node w has size equal to one plus the weight of a maximum weight bipartite matching in G.

Example 4.15 Consider the execution of the top-down unordered maximum common subtree isomorphism procedure, upon the unordered trees of Fig. 4.14. In order to find the size of a top-down unordered maximum common subtree of the subtree of T_1 rooted at node v_{12} and the subtree of T_2 rooted at node w_{18}, the following maximum weight bipartite matching problem is solved:

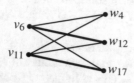

	w_4	w_{12}	w_{17}
v_6	3	5	3
v_{11}	4	5	4

A solution to this maximum weight bipartite matching problem has weight $5+4 = 9$ and then, a maximum common subtree of the subtree of T_1 rooted at node v_{12} and the subtree of T_2 rooted at node w_{18} has $9 + 1 = 10$ nodes. However, stating this bipartite matching problem involves (recursively) solving further maximum weight bipartite matching problems.

First, in order to find the size of a maximum common subtree of the subtree of T_1 rooted at node v_6 and the subtree of T_2 rooted at node w_4, the following maximum weight bipartite matching problem is also solved:

	w_1	w_3
v_5	1	2

which, in turn, requires solving the following maximum weight bipartite matching problem, in order to find the size of a maximum common subtree of the subtree of T_1 rooted at node v_5 and the subtree of T_2 rooted at node w_3:

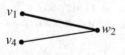

	w_2
v_1	1
v_4	1

Since the latter (trivial) bipartite matching problem has a solution of weight 1, the former bipartite matching problem has a solution of weight 2, and a maximum common subtree of the subtree of T_1 rooted at node v_6 and the subtree of T_2 rooted at node w_4 has, in fact, $2 + 1 = 3$ nodes.

Next, in order to find the size of a maximum common subtree of the subtree of T_1 rooted at node v_6 and the subtree of T_2 rooted at node w_{12}, the following maximum weight bipartite matching problem is also solved

	w_5	w_{11}
v_5	1	4

which, in turn, requires solving the following maximum weight bipartite matching problem, in order to find the size of a maximum common subtree of the subtree of T_1 rooted at node v_5 and the subtree of T_2 rooted at node w_{11}:

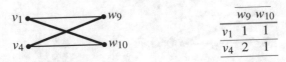

This, in turn, requires solving the following maximum weight bipartite matching problem, in order to find the size of a maximum common subtree of the subtree of T_1 rooted at node v_4 and the subtree of T_2 rooted at node w_9:

Since the latter (trivial) bipartite matching problem has a solution of weight 1, the previous bipartite matching problem has a solution of weight $2 + 1 = 3$, the former bipartite matching problem has a solution of weight 4, and a maximum common subtree of the subtree of T_1 rooted at node v_6 and the subtree of T_2 rooted at node w_{12} has thus $4 + 1 = 5$ nodes.

Now, in order to find the size of a maximum common subtree of the subtree of T_1 rooted at node v_6 and the subtree of T_2 rooted at node w_{17}, the following maximum weight bipartite matching problem is also solved:

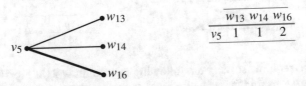

which, in turn, requires solving the following maximum weight bipartite matching problem, in order to find the size of a maximum common subtree of the subtree of T_1 rooted at node v_5 and the subtree of T_2 rooted at node w_{16}:

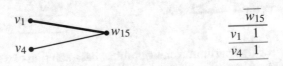

Since the latter (trivial) bipartite matching problem has a solution of weight 1, the former bipartite matching problem has a solution of weight 2, and a maximum common subtree of the subtree of T_1 rooted at node v_6 and the subtree of T_2 rooted at node w_{17} has thus $2 + 1 = 3$ nodes.

In the same way, in order to find the size of a maximum common subtree of the subtree of T_1 rooted at node v_{11} and the subtree of T_2 rooted at node w_4, the following maximum weight bipartite matching problem is solved:

	w_1	w_3
v_9	1	2
v_{10}	1	1

which, in turn, requires solving the following maximum weight bipartite matching problem, in order to find the size of a maximum common subtree of the subtree of T_1 rooted at node v_9 and the subtree of T_2 rooted at node w_3:

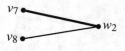

	w_2
v_7	1
v_8	1

Since the latter (trivial) bipartite matching problem has a solution of weight 1, the former bipartite matching problem has a solution of weight $2 + 1 = 3$, and a maximum common subtree of the subtree of T_1 rooted at node v_{11} and the subtree of T_2 rooted at node w_4 has thus $3 + 1 = 4$ nodes.

Now, in order to find the size of a maximum common subtree of the subtree of T_1 rooted at node v_{11} and the subtree of T_2 rooted at node w_{12}, the following maximum weight bipartite matching problem is solved:

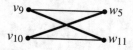

	w_5	w_{11}
v_9	1	3
v_{10}	1	1

which, in turn, requires solving the following maximum weight bipartite matching problem, in order to find the size of a maximum common subtree of the subtree of T_1 rooted at node v_9 and the subtree of T_2 rooted at node w_{11}:

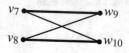

	w_9	w_{10}
v_7	1	1
v_8	1	1

Since the latter (trivial) bipartite matching problem has a solution of weight $1 + 1 = 2$, the former bipartite matching problem has a solution of weight $3 + 1 = 4$, and a maximum common subtree of the subtree of T_1 rooted at node v_{11} and the subtree of T_2 rooted at node w_{12} has thus $4 + 1 = 5$ nodes.

Finally, in order to find the size of a maximum common subtree of the subtree of T_1 rooted at node v_{11} and the subtree of T_2 rooted at node w_{17}, the following maximum weight bipartite matching problem is also solved

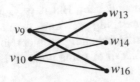

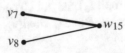

which, in turn, requires solving the following maximum weight bipartite matching problem, in order to find the size of a maximum common subtree of the subtree of T_1 rooted at node v_9 and the subtree of T_2 rooted at node w_{16}:

Since the latter (trivial) bipartite matching problem has a solution of weight 1, the former bipartite matching problem has a solution of weight $2+1 = 3$, and a maximum common subtree of the subtree of T_1 rooted at node v_{11} and the subtree of T_2 rooted at node w_{17} has thus $3 + 1 = 4$ nodes.

Now, the previous procedure can be extended with the construction of an actual top-down unordered maximum common subtree isomorphism mapping M of T_1 into T_2, based on the following result. Notice first that the solution to a maximum weight bipartite matching problem on a subtree (W_1, S_1) of T_1 and a subtree (W_2, S_2) of T_2 is a set of (weighted) edges of the corresponding bipartite graph, that is, a set of ordered pairs of nodes $B \in W_1 \times W_2$. Since $W_1 \subseteq V_1$ and $W_2 \subseteq V_2$, it also holds that $B \in V_1 \times V_2$.

Lemma 4.12 *Let $T_1 = (V_1, E_1)$ and $T_2 = (V_2, E_2)$ be unordered trees, and let $B \subseteq V_1 \times V_2$ be the solutions to all the maximum weight bipartite matching problems solved during the top-down unordered maximum common subtree isomorphism procedure upon T_1 and T_2. Then, there is a unique top-down unordered maximum common subtree isomorphism $M \in V_1 \times V_2$ such that $M \subseteq B$.*

Proof Let $T_1 = (V_1, E_1)$ and $T_2 = (V_2, E_2)$ be unordered trees, and let $B \subseteq V_1 \times V_2$ be the corresponding solutions to maximum weight bipartite matching problems. Showing existence and uniqueness of a top-down unordered maximum common subtree isomorphism $M \subseteq B$ is equivalent to showing that, for each node $v \in V_1$ with $(parent(v), z) \in B$ for some node $z \in V_2$, there is a unique $(v, w) \in B$ such that $parent(w) = z$, because a unique top-down subtree of T_2 isomorphic to a top-down subtree of T_1 can then be reconstructed by order of non-decreasing depth. Therefore, it suffices to show the weaker condition that, for all $(v, w_1), (v, w_2) \in B$ with $w_1 \neq w_2$, it holds that $parent(w_1) \neq parent(w_2)$.

Let $(v, w_1), (v, w_2) \in B$ with $w_1 \neq w_2$. Then, (v, w_1) and (v, w_2) must belong to the solution to different bipartite matching problems, for otherwise the corresponding edges in the bipartite graph would not be independent. But if (v, w_1) and (v, w_2) belong to the solution to different bipartite matching problems, then nodes w_1 and w_2 cannot be siblings, because in each bipartite matching problem, all children of some node of T_1 are matched against all children of some node of T_2. Therefore, $parent(w_1) \neq parent(w_2)$.

$\square$

Given the solutions $B \subseteq V_1 \times V_2$ to all the maximum weight bipartite matching problems solved during the top-down unordered maximum common subtree isomorphism procedure upon unordered trees $T_1 = (V_1, E_1)$ and $T_2 = (V_2, E_2)$, the corresponding maximum common subtree isomorphism mapping $M \subseteq V_1 \times V_2$ can be reconstructed as follows. Set $M[root[T_1]]$ to $root[T_2]$ and, for all nodes $v \in V_1$ in pre-order, set $M[v]$ to the unique node w with $(v, w) \in B$ and $(parent(v), parent(w)) \in B$.

Example 4.16 Consider the execution shown in Example 4.15 of the top-down unordered maximum common subtree isomorphism procedure, upon the unordered trees of Fig. 4.14. The solutions $B \subseteq V_1 \times V_2$ to all the maximum weight bipartite matching problems solved are as follows:

$v_1 : \quad w_2 \quad \boxed{w_{10}} \quad w_{15}$ $v_7 : \quad w_2 \quad w_9 \quad \boxed{w_{15}}$

$v_2 : \quad \boxed{w_8}$ $v_8 : \quad w_{10}$

$v_3 : \quad \boxed{w_8}$ $v_9 : \quad w_3 \quad w_{11} \quad \boxed{w_{16}}$

$v_4 : \quad \boxed{w_9}$ $v_{10} : \quad w_1 \quad w_5 \quad \boxed{w_{13}}$

$v_5 : \quad w_3 \quad \boxed{w_{11}} \quad w_{16}$ $v_{11} : \quad \boxed{w_{17}}$

$v_6 : \quad \boxed{w_{12}}$ $v_{12} : \quad \boxed{w_{18}}$

The unique maximum common subtree isomorphism mapping $M \subseteq B \subseteq V_1 \times V_2$ is indicated by the highlighted nodes of T_2.

The following algorithm implements the previous procedure for top-down maximum common subtree isomorphism of an unordered tree T_1 into another unordered tree T_2. The partial injection $M \subseteq V_1 \times V_2$ computed by the function upon two unordered trees $T_1 = (V_1, E_1)$ and $T_2 = (V_2, E_2)$ is stored in the attribute $v.M$, for all nodes $v \in V_1$.

Notice first that a top-down subtree of an unordered labeled tree is isomorphic to a top-down subtree of another unordered labeled tree if the unlabeled trees underlying the subtrees of the labeled trees are isomorphic, and furthermore, corresponding nodes share the same label.

Further, as in the case of top-down unordered subtree isomorphism, the correspondence between children nodes in the trees and the vertices of the bipartite graph is stored in the attribute $v.GT$, for all nodes v of G, and in the opposite direction in the attributes $v_1.TG$ and $v_2.TG$, for all nodes v_1 of T_1 and v_2 of T_2.

function $max\_common\_subtree\_isomorphism(T_1, r_1, T_2, r_2)$
  if $r_1.label \neq r_2.label$ **then**
    return 0
  if r_1 is a leaf node of T_1 or r_2 is a leaf node of T_2 **then**
    return 1
  let G be a new graph
  for all children v_1 of node r_1 in T_1 **do**
    add a new vertex v to G
    $v.GT \leftarrow v_1$
    $v_1.TG \leftarrow v$
  for all children v_2 of node r_2 in T_2 **do**
    add a new vertex w to G
    $w.GT \leftarrow v_2$
    $v_2.TG \leftarrow w$
  for all children v_1 of node r_1 in T_1 **do**
    for all children v_2 of node r_2 in T_2 **do**
      $res \leftarrow max\_common\_subtree\_isomorphism(T_1, v_1, T_2, v_2)$
      if $res \neq 0$ **then**
        add a new edge e to G labeled $(v_1.TG, v_2.TG)$
        $e.weight \leftarrow res$
  let L be a maximum weight bipartite matching of G
  $res \leftarrow 1$
  for all edges $e = (v, w)$ in L **do**
    $v.GT.B \leftarrow v.GT.B \cup \{w.GT\}$
    $res \leftarrow res + e.weight$
  return res

The following algorithm implements the previous procedure for top-down maximum common subtree isomorphism of an unordered tree T_1 into another unordered tree T_2, calling the previous recursive procedure upon the root of T_1 and the root of T_2 and then reconstructing the top-down unordered maximum common subtree isomorphism mapping $M \subseteq V_1 \times V_2$ included in the solutions $B \subseteq V_1 \times V_2$ to all the maximum weight bipartite matching problems. The root of T_1 is mapped to the root of T_2 and, during a preorder traversal, each non-root node $v \in V_1$ is mapped to the unique node $w \in V_2$ with $(v, w) \in B$ and $(parent[v], parent[w]) \in B$.

procedure *top_down_unordered_max_common_subtree*(T_1, T_2)
    for all nodes v_1 of T_1 **do**
        $v_1.TG = nil$
    for all nodes v_2 of T_2 **do**
        $v_2.TG \leftarrow nil$
    if $root[T_1].label = root[T_2].label$ **then**
        for all nodes v_1 of T_1 **do**
            let $v_1.B$ be an empty set (of nodes)
            $v_1.M \leftarrow nil$
        $root[T_1].B \leftarrow root[T_1].B \cup \{root[T_2]\}$
        $max\_common\_subtree\_isomorphism(T_1.root[T_1], T_2, root[T_2])$
        $root[T_1].M \leftarrow root[T_2]$
        for all nodes v of T_1 in preorder **do**
            if v is not the root of T_1 **then**
                for all nodes w in $v.B$ **do**
                    if $parent[v].M = parent[w]$ **then**
                        $v.M \leftarrow w$
                        break

Remark 4.10 Correctness of the algorithm for top-down unordered maximum common subtree isomorphism follows from Lemma 4.12.

Recall that the maximum weight bipartite matching algorithm of Sect. 5.3 runs in $O(n^3)$ time using $O(n^2)$ additional space, where n is the number of vertices in the bipartite graph.

Lemma 4.13 *Let T_1 and T_2 be unordered trees with, respectively, n_1 and n_2 nodes, and let $n = n_1 + n_2$. The algorithm for top-down unordered maximum common subtree isomorphism runs in $O(n^3)$ time using $O(n^2)$ additional space.*

Proof Let $T_1 = (V_1, E_1)$ and $T_2 = (V_2, E_2)$ be unordered trees on, respectively, n_1 and n_2 nodes, let $n = n_1 + n_2$, and let $k(v)$ denote the number of children of node v. The effort spent on a leaf node of T_1 is $O(1)$, and the total effort spent on the leaves of T_1 is thus bounded by $O(n_1)$. The effort spent on a non-leaf node v of T_1 and a non-leaf node w of T_2, on the other hand, is dominated by solving a maximum weight bipartite matching problem on a complete bipartite graph with $k(v) + k(w)$ vertices, and is thus bounded by $O((k(v) + k(w))^3)$. The total effort spent on non-leaves of T_1 and T_2 is thus bounded by $O(\sum_{v \in V_1} \sum_{w \in V_2} (k(v) + k(w))^3)$. Therefore, the algorithm runs in $O(n^3)$ time.

A similar argument shows that the algorithm uses $O(n^2)$ additional space. □

Remark 4.11 Notice that a maximum weight bipartite matching can be computed in $O(n(m + n \log n))$ time using $O(m)$ additional space, see the bibliographic notes for Chap. 5. On a bipartite graph with $n = p + q$ vertices and $m = pq$ edges, this is $O((p+q)(pq+(p+q) \log (p + q)))$ time using $O(pq)$ additional space. Therefore, the algorithm for top-down unordered maximum common subtree isomorphism can be implemented to run in $O((n_1 + n_2)(n_1 n_2 + (n_1 + n_2) \log (n_1 + n_2)))$ time using $O(n_1 n_2)$ additional space.

The algorithm solves the problem of finding a top-down maximum common subtree isomorphism of an unordered tree into another unordered tree. The problem of enumerating all top-down unordered maximum common subtree isomorphisms can be solved with the maximum common subgraph isomorphism algorithms given in Sect. 7.3.

4.3.3 Bottom-Up Maximum Common Subtree Isomorphism

A bottom-up common subtree of two ordered trees T_1 and T_2 is an ordered tree T such that there are bottom-up ordered subtree isomorphisms of T into T_1 and into T_2. A maximal bottom-up common subtree of two ordered trees T_1 and T_2 is a bottom-up common subtree of T_1 and T_2 which is not a proper subtree of any other bottom-up common subtree of T_1 and T_2. A maximum bottom-up common subtree of two ordered trees T_1 and T_2 is a bottom-up common subtree of T_1 and T_2 with the largest number of nodes.

Definition 4.12 A **bottom-up common subtree** of an ordered tree $T_1 = (V_1, E_1)$ to another ordered tree $T_2 = (V_2, E_2)$ is a structure (X_1, X_2, M), where $X_1 = (W_1, S_1)$ is a bottom-up ordered subtree of T_1, $X_2 = (W_2, S_2)$ is a bottom-up ordered subtree of T_2, and $M \subseteq W_1 \times W_2$ is an ordered tree isomorphism of X_1 to X_2. A bottom-up common subtree (X_1, X_2, M) of T_1 to T_2 is **maximal** if there is no bottom-up common subtree (X_1', X_2', M') of T_1 to T_2 such that X_1 is a proper bottom-up subtree of X_1' and X_2 is a proper bottom-up subtree of X_2', and it is **maximum** if there is no bottom-up common subtree (X_1', X_2', M') of T_1 to T_2 with $size[X_1] < size[X_1']$.

Example 4.17 For the ordered trees $T_1 = (V_1, E_1)$ and $T_2 = (V_2, E_2)$ in Fig. 4.15, the partial injection $M \subseteq V_1 \times V_2$ given by $M = \{(v_9, w_6), (v_{10}, w_7), (v_{11}, w_8),$ $(v_{12}, w_9), (v_{13}, w_{10}), (v_{14}, w_{11})\}$ is a bottom-up maximum common subtree isomorphism of T_1 to T_2.

The problem of finding a bottom-up maximum common subtree of an ordered tree $T_2 = (V_2, E_2)$ to an ordered tree $T_1 = (V_1, E_1)$, can also be reduced to the problem of partitioning $V_1 \cup V_2$ into equivalence classes of bottom-up subtree isomorphism. Two nodes (in the same or in different trees) are equivalent if and only if the bottom-up ordered subtrees rooted at them are isomorphic. Then, the bottom-up subtree of

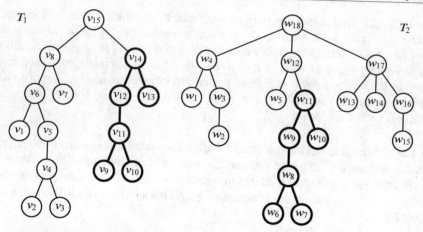

Fig. 4.15 A bottom-up maximum common subtree of two ordered trees. Nodes are numbered according to the order in which they are visited during a postorder traversal. The common subtree of T_1 and T_2 is shown highlighted in both trees

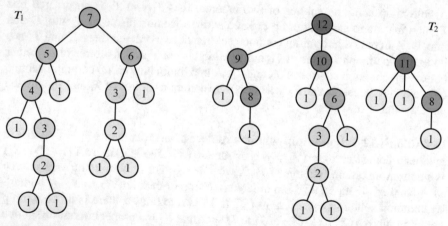

Fig. 4.16 Bottom-up ordered subtree isomorphism equivalence classes for the ordered trees in Fig. 4.15. Nodes are numbered according to the equivalence class to which they belong, and the equivalence classes are shown highlighted in different shades of gray

T_1 rooted at node $v \in V_1$ is isomorphic to the bottom-up subtree of T_2 rooted at node $w \in V_2$ if and only if nodes v and w belong to the same equivalence class of bottom-up subtree isomorphism.

Example 4.18 The partition of the ordered trees $T_1 = (V_1, E_1)$ and $T_2 = (V_2, E_2)$ of Fig. 4.15 in equivalence classes of bottom-up subtree isomorphism is illustrated in Fig. 4.16. The twelve equivalence classes are numbered from 1 to 12. The bottom-up subtree of T_1 rooted at node v_{14} is isomorphic to the bottom-up subtree of T_2 rooted at node w_{11}, because nodes v_{14} and w_{11} both belong to the same equivalence class, the one numbered 6.

An Algorithm for Bottom-Up Ordered Maximum Common Subtree Isomorphism

A bottom-up ordered maximum common subtree isomorphism of an ordered tree $T_1 = (V_1, E_1)$ to another ordered tree $T_2 = (V_2, E_2)$ can be obtained by first partitioning the set of nodes $V_1 \cup V_2$ in bottom-up subtree isomorphism equivalence classes, and then finding equivalent nodes $v \in V_1$ and $w \in V_2$ of largest size.

Notice that finding bottom-up ordered maximum common subtree isomorphisms based on equivalence classes of subtree isomorphism applies to unlabeled trees only. However, the bottom-up maximum common subtree isomorphism procedure for ordered unlabeled trees with n nodes can be extended to ordered trees whose nodes are labeled by integers in the range $1, \ldots, n$.

The following algorithm implements the procedure for bottom-up maximum common subtree isomorphism of an ordered tree T_1 to another ordered tree T_2, collecting a mapping M of nodes of T_1 to nodes of T_2 stored in the attribute $v.M$, for all nodes v of T_1.

procedure *bottom_up_ordered_max_common_subtree*(T_1, T_2)
    num ← 1
    let D be an empty dictionary (of lists of integers to integers)
    partition_tree_in_isomorphism_equivalence_classes(T_1, num, D)
    partition_tree_in_isomorphism_equivalence_classes(T_2, num, D)
    (v, w) ← *find_largest_common_subtree*(T_1, T_2)
    map_ordered_subtree(T_1, v, T_2, w)

procedure *partition_tree_in_isomorphism_equivalence_classes*(T, num, D)
    for all nodes v of T in postorder **do**
        if v is a leaf node of T **then**
            $v.code$ ← 1
        else
            let L be an empty list (of integers)
            for all children w of node v in T **do**
                append $w.code$ to L
            if $L \in D$ **then**
                $v.code$ ← $D[L]$
            else
                num ← *num* + 1
                $D[L]$ ← *num*
                $v.code$ ← *num*

Now, given a partition of the nodes of the ordered trees $T_1 = (V_1, E_1)$ and $T_2 = (V_2, E_2)$ in equivalence classes of bottom-up subtree isomorphism, equivalent nodes $v \in V_1$ and $w \in V_2$ of largest size can be found by a simultaneous traversal of a list of nodes of T_1 and a list of nodes of T_2, sorted by non-increasing order of subtree size and, within size, by (non-decreasing) order of equivalence class of bottom-up subtree isomorphism.

The following function uses instead priority queues of nodes of T_1 and T_2, implementing the simultaneous traversal by selectively deleting nodes of largest subtree size until the nodes with highest priority in both queues belong to the same equivalence class. The required ordering of nodes is accomplished by setting the priority of a node to opposite the subtree size of the node, together with the equivalence class of bottom-up subtree isomorphism to which it belongs.

function *find_largest_common_subtree*(T_1, T_2)
    let Q_1, Q_2 be priority queues (of nodes and pairs of integers)
    for all nodes v of T_1 in postorder **do**
        $v.size \leftarrow 1$
        for all children w of node v in T_1 **do**
            $v.size \leftarrow v.size + w.size$
        enqueue $(v, (-v.size, v.code))$ into Q_1
    for all nodes v of T_2 in postorder **do**
        $v.size \leftarrow 1$
        for all children w of node v in T_2 **do**
            $v.size \leftarrow v.size + w.size$
        enqueue $(v, (-v.size, v.code))$ into Q_2
    while Q_1 is not empty and Q_2 is not empty **do**
        let (v, p) be an element v with the minimum priority p in Q_1
        let (w, q) be an element w with the minimum priority q in Q_2
        if $v.code = w.code$ **then**
            break
        if $p < q$ **then**
            dequeue from Q_1 an element with the minimum priority
        else
            dequeue from Q_2 an element with the minimum priority
    $v.M \leftarrow w$
    return (v, w)

Theorem 4.4 *Let $T_1 = (V_1, E_1)$ and $T_2 = (V_2, E_2)$ be ordered trees with, respectively, n_1 and n_2 nodes, where $n_1 \leqslant n_2$. The algorithm for bottom-up ordered maximum common subtree isomorphism runs in $O(n_2 \log n_2)$ time using $O(n_1 + n_2)$ additional space.*

Proof Let T_1 and T_2 be ordered trees with, respectively, n_1 and n_2 nodes, where $n_1 \leqslant n_2$. Despite the expected $O(n_1 + n_2)$ time needed to partition the nodes of

T_1 and T_2 in equivalence classes of bottom-up subtree isomorphism, the effort spent on the nodes of T_1 and T_2 is dominated by the n_1 and n_2 insertions and $O(n_2)$ deletions from node priority queues with, respectively, n_1 and n_2 elements, which can be implemented by binary heaps. Therefore, the algorithm runs in $O(n_2 \log n_2)$ time using $O(n_1 + n_2)$ additional space. □

4.3.4   Bottom-Up Unordered Maximum Common Subtree Isomorphism

A bottom-up common subtree of two unordered trees T_1 and T_2 is an unordered tree T such that there are bottom-up unordered subtree isomorphisms of T into T_1 and into T_2. A maximal bottom-up common subtree of two unordered trees T_1 and T_2 is a bottom-up common subtree of T_1 and T_2 which is not a proper subtree of any other bottom-up common subtree of T_1 and T_2. A maximum bottom-up common subtree of two unordered trees T_1 and T_2 is a bottom-up common subtree of T_1 and T_2 with the largest number of nodes.

Definition 4.13 A **bottom-up common subtree** of an unordered tree $T_1 = (V_1, E_1)$ to another unordered tree $T_2 = (V_2, E_2)$ is a structure (X_1, X_2, M), where $X_1 = (W_1, S_1)$ is a bottom-up unordered subtree of T_1, $X_2 = (W_2, S_2)$ is a bottom-up unordered subtree of T_2, and $M \subseteq W_1 \times W_2$ is an unordered tree isomorphism of X_1 to X_2. A bottom-up common subtree (X_1, X_2, M) of T_1 to T_2 is **maximal** if there is no bottom-up common subtree (X_1', X_2', M') of T_1 to T_2 such that X_1 is a proper bottom-up subtree of X_1' and X_2 is a proper bottom-up subtree of X_2', and it is **maximum** if there is no bottom-up common subtree (X_1', X_2', M') of T_1 to T_2 with $size[X_1] < size[X_1']$.

Example 4.19 For the unordered trees $T_1 = (V_1, E_1)$ and $T_2 = (V_2, E_2)$ in Fig. 4.17, the partial injection $M \subseteq V_1 \times V_2$ given by $M = \{ (v_1, w_{10}), (v_2, w_6), (v_3, w_7), (v_4, w_8), (v_5, w_9) (v_6, w_{11}), (v_7, w_5), (v_8, w_{12}) \}$ is a bottom-up maximum common subtree isomorphism of T_1 to T_2.

Again, the problem of finding a bottom-up maximum common subtree of an unordered tree $T_2 = (V_2, E_2)$ on n_2 nodes to an unordered tree $T_1 = (V_1, E_1)$ on n_1 nodes, where $n_1 \leqslant n_2$, can be reduced to the problem of partitioning $V_1 \cup V_2$ into equivalence classes of bottom-up subtree isomorphism. Two nodes (in the same or in different trees) are equivalent if and only if the bottom-up unordered subtrees rooted at them are isomorphic. Then, the bottom-up subtree of T_1 rooted at node $v \in V_1$ is isomorphic to the bottom-up subtree of T_2 rooted at node $w \in V_2$ if and only if nodes v and w belong to the same equivalence class of bottom-up subtree isomorphism.

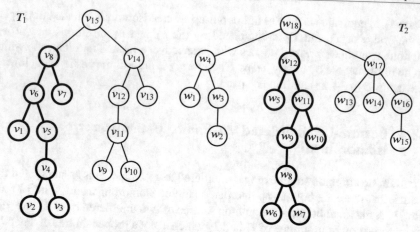

Fig. 4.17 A bottom-up maximum common subtree of two unordered trees. Nodes are numbered according to the order in which they are visited during a postorder traversal. The common subtree of T_1 and T_2 is shown highlighted in both trees

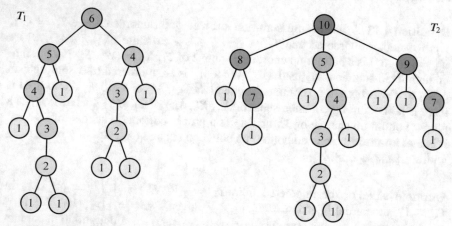

Fig. 4.18 Bottom-up unordered subtree isomorphism equivalence classes for the unordered trees in Fig. 4.17. Nodes are numbered according to the equivalence class to which they belong, and the equivalence classes are shown highlighted in different shades of gray

Example 4.20 The partition of the unordered trees $T_1 = (V_1, E_1)$ and $T_2 = (V_2, E_2)$ of Fig. 4.17 in equivalence classes of bottom-up subtree isomorphism is illustrated in Fig. 4.18. The ten equivalence classes are numbered from 1 to 10. The bottom-up subtree of T_1 rooted at node v_8 is isomorphic to the bottom-up subtree of T_2 rooted at node w_{12}, because nodes v_8 and w_{12} both belong to the same equivalence class, the one numbered 5.

An Algorithm for Bottom-Up Unordered Maximum Common Subtree Isomorphism

A bottom-up unordered maximum common subtree isomorphism of an unordered tree $T_1 = (V_1, E_1)$ to another unordered tree $T_2 = (V_2, E_2)$ can also be obtained by first partitioning the set of nodes $V_1 \cup V_2$ in bottom-up subtree isomorphism equivalence classes, and then finding equivalent nodes $v \in V_1$ and $w \in V_2$ of largest size.

Again, notice that finding bottom-up unordered maximum common subtree isomorphisms based on equivalence classes of subtree isomorphism applies to unlabeled trees only. However, the bottom-up maximum common subtree isomorphism procedure for unordered unlabeled trees with n nodes can be extended to unordered trees whose nodes are labeled by integers in the range $1, \ldots, n$.

The following algorithm implements the procedure for bottom-up maximum common subtree isomorphism of an unordered tree T_1 to another unordered tree T_2, collecting a mapping M of nodes of T_1 to nodes of T_2 stored in the attribute $v.M$, for all nodes v of T_1.

procedure *bottom_up_unordered_max_common_subtree*(T_1, T_2)
    num $\leftarrow$ 1
    let D be an empty dictionary (of lists of integers to integers)
    partition_tree_in_isomorphism_equivalence_classes(T_1, num, D)
    partition_tree_in_isomorphism_equivalence_classes(T_2, num, D)
    for all nodes v of T_1 **do**
        $v.M \leftarrow nil$
    for all nodes w of T_2 **do**
        $w.mapped\_to \leftarrow$ false
    $(v, w) \leftarrow$ *find_largest_common_subtree*(T_1, T_2)
    $v.M = w$
    $w.mapped\_to \leftarrow$ true
    $L_1 \leftarrow$ *preorder_subtree_list_traversal*(T_1, v)
    for all nodes v_1 in L_1 **do**
        if $v_1 \neq v$ **then**
            for all children w_1 of node *parent*$[v_1].M$ **do**
                if $v_1.code = w_1.code$ and not $w_1.mapped\_to$ **then**
                    $v_1.M \leftarrow w_1$
                    $w_1.mapped\_to \leftarrow$ true
                    break

The following function performs an iterative preorder traversal of the subtree of an unordered tree $T = (V, E)$ rooted at node $v \in V$, with the help of a stack of nodes, and builds a list L of the nodes of the subtree of T in the order in which they are visited during the traversal.

```
function preorder_subtree_list_traversal(T, r)
    let L be an empty list (of nodes)
    let S be an empty stack (of nodes)
    push r onto S
    while S is not empty do
        pop from S the top node v
        append v to L
        for all children w of node v in T in reverse order do
            push w onto S
    return L
```

Theorem 4.5 *Let $T_1 = (V_1, E_1)$ and $T_2 = (V_2, E_2)$ be unordered trees with, respectively, n_1 and n_2 nodes. The algorithm for bottom-up unordered maximum common subtree isomorphism runs in $O((n_1 + n_2)^2)$ time using $O(n_1 + n_2)$ additional space.*

Proof Let T_1 and T_2 be unordered trees with, respectively, n_1 and n_2 nodes, and let $k(v)$ denote the number of children of node v. The effort spent on a leaf node of T_1 or T_2 is $O(1)$, and the total effort spent on the leaves of T_1 and T_2 is bounded by $O(n_1 + n_2)$.

The effort spent on a non-leaf node v of T_1 or T_2, on the other hand, is dominated by bucket sorting a list of $k(v)$ integers, despite the expected $O(1)$ time taken to look up or insert the sorted list of integers in the dictionary. Since all these integers fall in the range $1, \ldots, n_1 + n_2$, the time taken for bucket sorting the children nodes is bounded by $O(k(v) + n_1 + n_2)$. The total effort spent on non-leaves of T_1 and T_2 is thus bounded by $O(\sum_{v \in V_1 \cup V_2} (k(v) + n_1 + n_2)) = O(\sum_{v \in V_1 \cup V_2} k(v)) + O((n_1 + n_2)^2) = O(n_1 + n_2) + O((n_1 + n_2)^2) = O((n_1 + n_2)^2)$. Therefore, the algorithm runs in $O((n_1 + n_2)^2)$ time using $O(n_1 + n_2)$ additional space. □

Lemma 4.14 *Let $T_1 = (V_1, E_1)$ and $T_2 = (V_2, E_2)$ be unordered trees of degree bounded by a constant k and with, respectively, n_1 and n_2 nodes. The algorithm for bottom-up unordered maximum common subtree isomorphism can be implemented to run in expected $O(n_1 + n_2)$ time using $O(n_1 + n_2)$ additional space.*

Proof For trees of degree bounded by a constant k, bucket sorting a list of at most k integers can be replaced with appropriate code performing at most k comparisons, and taking thus $O(1)$ time to sort a list of at most k integers. In this case, the effort spent on a non-leaf node v of T_1 or T_2 is dominated instead by the expected $O(1)$ time required to look up or insert the sorted list of k integers in the dictionary, and the algorithm runs in expected $O(n_1 + n_2)$ time, still using $O(n_1 + n_2)$ additional space. □

4.4 Applications

Isomorphism problems on trees find application whenever structures described by trees need to be identified or compared. In most application areas, tree isomorphism, subtree isomorphism, and maximal common subtree isomorphism can be seen as some form of pattern matching or information retrieval. Such an application will be further addressed in Chap. 7, within the broader context of isomorphism problems on graphs.

The application of the different isomorphism problems on trees to the comparison of RNA secondary structure in computational molecular biology is discussed next. See the bibliographic notes below for further applications of tree isomorphism and related problems.

An RNA (ribonucleic acid) molecule is composed of four types of nucleotides, also called *bases*. These are: adenine (A), cytosine (C), guanine (G), and uracil (U). Nucleotides A and U, as well as nucleotides C and G, form stable chemical bonds with each other and are thus called *complementary nucleotides*.

The structure of an RNA molecule can be described at different levels of abstraction. The *primary structure*, on the one hand, is the linear sequence of nucleotides in an RNA molecule. An RNA sequence is thus a finite sequence or string over the alphabet $\{A, C, G, U\}$. The *secondary structure*, on the other hand, is a simplified two-dimensional description of the three-dimensional structure of the RNA molecule in which some complementary nucleotides get paired, making the molecule fold in three-dimensional space. The secondary structure of an RNA molecule is thus the collection of all nucleotide pairs that occur in its three-dimensional structure. Almost all RNA molecules form secondary structure.

Now, the secondary structure of an RNA molecule can be represented by an undirected, connected, labeled graph, with one vertex for each nucleotide in the RNA sequence and where consecutive nucleotides in the sequence, as well as paired nucleotides, are joined by an edge. Vertices are labeled by the corresponding nucleotide, while edges are not labeled.

Example 4.21 Consider the following RNA sequence of 120 nucleotides from an Annelida species:

```
GUCUACGGCC AUACCACGUU GAAAGCACCG GUUCUCGUCC
GAUCACCGAA GUUAAGCAAC GUCGGGCCCG GUUAGUACUU
GGAUGGGUGA CCGCCUGGGA AUACCGGGUG CUGUAGACUU
```

and the following RNA sequence, also of 120 nucleotides, from an Hemichordata species:

```
GCCUACGGCC AUACCACGUA GAAUGCACCG GUUCUCGUCC
GAUCACCGAA GUUAAGCUGC GUCGGGCGUG GUUAGUACUU
GCAUGGGAGA CCGGCUGGGA AUACCACGUG CCGUAGGCUU
```

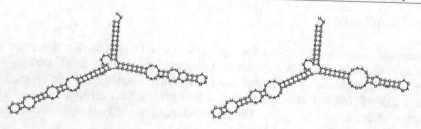

Fig. 4.19 Usual representation of the predicted secondary structure of RNA sequences from an Annelida species (left) and an Hemichordata species (right). The sequences are composed of 120 nucleotides each

Both sequences correspond to the same gene, the 5S rRNA. Their graphs of secondary structure are shown in Fig. 4.19.

The graph of an RNA secondary structure is much tree-like. As a matter of fact, it can be uniquely decomposed into structural elements called stacks, loops, and external nodes. Stacks are formed by paired nucleotides which are consecutive in the RNA sequence. Loops may differ in size (number of unpaired nucleotides) and branching degree (hairpin loops have degree one, and internal loops have degree two or more). External nodes are nucleotides that belong neither to a stack nor to a loop.

Now, the graph of an RNA secondary structure can also be represented by an ordered labeled tree, with non-leaves corresponding to nucleotide pairs and leaves corresponding to unpaired nucleotides. Since paired nucleotides are contracted to single nodes, stacks appear as chains of non-leaves, possibly ending at a leaf, while loops appear as bushes of leaves. An additional root node is added, as parent of the external nodes. Node labels are used in order to distinguish between leaves representing paired and unpaired nucleotides. In schematic diagrams, the label of a node standing for paired nucleotides is often represented by a black circle, while the label of a leaf standing for an unpaired nucleotide is represented by white circles. The tree representation of the predicted RNA secondary structures of Fig. 4.19 is shown in Fig. 4.20, where nodes standing for paired nucleotides are labeled by white squares instead of black circles.

Tree isomorphism, subtree isomorphism and, in general, maximum common subtree isomorphism find application in the comparison of RNA secondary structures. Maximum common subtrees of the RNA secondary structures of Annelida and Hemichordata of Fig. 4.20 are shown in Figs. 4.21, 4.22, and 4.23. While the trees representing these RNA secondary structures have, respectively, 81 and 82 nodes, a top-down ordered maximum common subtree has 37 nodes, a top-down unordered maximum common subtree has 62 nodes, and a bottom-up ordered maximum common subtree which, in this case, is also a bottom-up unordered maximum common subtree, has only 14 nodes.

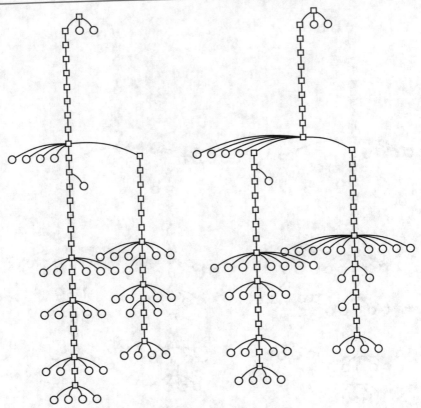

Fig. 4.20 Tree representation of the predicted RNA secondary structures of Annelida (left) and Hemichordata (right) of Fig. 4.19.

Summary

Several isomorphism problems on ordered and unordered trees were addressed in this chapter which form a hierarchy of pattern matching problems, where maximum common subtree isomorphism generalizes subtree isomorphism, which, in turn, generalizes tree isomorphism. Simple algorithms are given in detail for enumerating all solutions to these problems. Some of the algorithms for tree and subtree isomorphism are based on the methods of tree traversal discussed in Chap. 3, while some of the maximum common subtree algorithms are based on the divide-and-conquer technique reviewed in Sect. 2.4. References to more sophisticated algorithms are given in the bibliographic notes below. Computational molecular biology is also discussed as a prototypical application of isomorphism problems on trees.

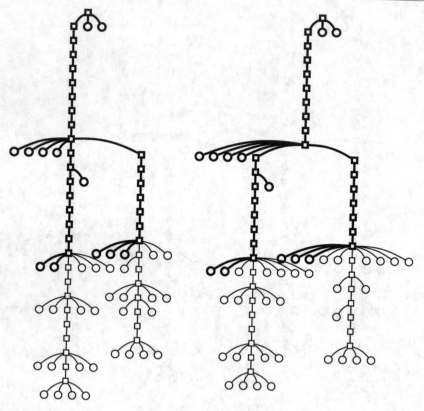

Fig. 4.21 A top-down ordered maximum common subtree isomorphism of the RNA secondary structures of Annelida (left) and Hemichordata (right) of Fig. 4.20

Bibliographic Notes

The algorithm for unordered tree isomorphism is based on [10, pp. 196–199]. Another, more efficient algorithm running in $O(n)$ time was proposed in [2, pp. 84–86]. Both algorithms for unordered tree isomorphism were actually found by J. Edmonds. See also Exercise 4.4.

Tree isomorphism is closely related to the problem of generating trees without repetitions. Algorithms for enumerating all non-isomorphic rooted ordered trees on n nodes were given in [29, 100], and for enumerating all non-isomorphic unordered trees on n nodes, either rooted or free, in [7, 56] and also in [96, Ch. 5]. See also Problem 4.2 and Exercise 4.1. Further, a certificate for free unordered trees was given in [53, Sect. 7.3.1]. See also Problem 4.3 and Exercise 4.2. Notice that enumerating non-isomorphic trees is different from enumerating *labeled* trees, that is, trees $T = (V, E)$ whose n nodes are labeled by a bijection $V \to \{1, 2, \ldots, n\}$. Algorithms for enumerating unordered labeled trees are often based on Prüfer sequences. See, for instance, [21] and [53, Sect. 3.3]. See also [11, 17, 50, 72, 93].

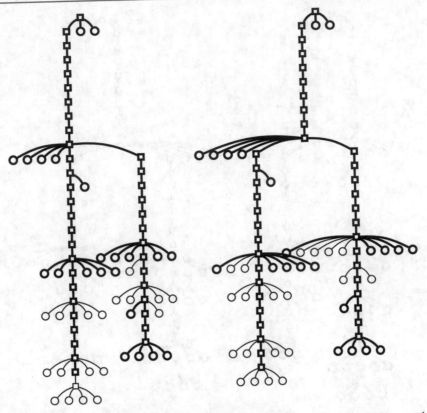

Fig. 4.22 A top-down unordered maximum common subtree isomorphism of the RNA secondary structures of Annelida (left) and Hemichordata (right) of Fig. 4.20

An algorithm for subtree isomorphism of binary trees was proposed in [6, Sect. 5d]. See also [86] and Exercise 4.5. The algorithm for top-down unordered subtree isomorphism is based on [64,68,90]. A more efficient algorithm was proposed in [18] which runs in $O(n_1\sqrt{n_1}n_2)$ time, and it was further improved in [73] to run in $O(n_1\sqrt{n_1}n_2/\log n_1)$ time. The algorithm for top-down unordered maximum common subtree isomorphism is also based on [64].

The simple algorithm for bottom-up ordered subtree isomorphism is based on [89]. Further algorithms were proposed in [26,34,35,66]. The algorithm for bottom-up unordered subtree isomorphism is based on [85], where the idea from [30] is exploited that a procedure for dynamically maintaining a global table of unique identifiers allows the compacted directed acyclic graph representation of a binary tree to be determined in expected $O(n)$ time. Further algorithms were proposed in [23,36]. The algorithms for bottom-up ordered and unordered maximum common subtree isomorphism are also based on [85]. See also [1,3,24,45,46,59,62,65,69,83].

Isomorphism problems on trees are also closely related to the problem of transforming or *editing* trees. The edit distance between two labeled trees is the cost of

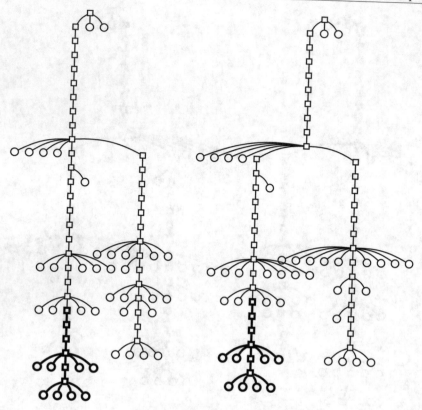

Fig. 4.23 A bottom-up ordered maximum common subtree isomorphism of the RNA secondary structures of Annelida (left) and Hemichordata (right) of Fig. 4.20.

a least-cost sequence of elementary edit operations, such as deletion and insertion of bottom-up subtrees (and, in particular, leaves) and modification of node or edge labels, that allows to transform a given tree into another given tree. Several such tree edit distances have been proposed [71,80,82,85,97]. Further, a hierarchy among several of these tree edit distances was given in [92,98]. See also [51].

The elementary edit operations often have a cost or weight associated to them, and under the assumption that the cost of modifying a node label is always less than the cost of deleting a node with the old label and inserting a node with the new label, the edit distance can be shown to correspond with a maximal common subtree, as shown in [9] in the more general context of elementary edit operations on graphs. Several algorithms were proposed for computing the edit distance between ordered trees [60,77,80,103]. Further measures of edit distance were proposed for ordered trees, either based on the elementary edit operation of edge rotation [20,101] or node splitting and merging [61], as well as measures of similarity based on the alignment of labeled trees [47–49,99].

Computing the edit distance between unordered trees is an NP-complete problem [104], even for trees of bounded degree $k \geqslant 2$, although under the constraint that deletion and insertion operations be made on leaves only [71] or that disjoint subtrees be mapped to disjoint subtrees [102], the distance can be computed in polynomial time. Some restricted edit distances between unrooted and unordered trees can also be computed in polynomial time [81]. Further algorithms were proposed for computing the edit distance between rooted unordered trees [76] and free unordered trees [105]. See also [22,31,43,84,87,88].

Tree isomorphism and related problems being a form of pattern matching, they should not be confused with the related problem of pattern matching in trees, where leaves in the smaller tree are labeled with variables (wildcards) and the pattern matching problem consists of finding all subtrees of the larger tree that are isomorphic to some extension of the smaller tree, which is obtained by replacing these variables by appropriate subtrees of the larger tree. Several algorithms were proposed for pattern matching in ordered trees [19,25,42,52].

The RNA sequences given in Example 4.21 for the 5S rRNA gene are taken from [78,79] for the Annelida species, and from [67] for the Hemichordata species. The usual representation of their predicted secondary structure shown in Fig. 4.19 was produced using the RNAfold program [40], which is based on the dynamic programming algorithm of [106] for computing minimum free energy secondary structures. See [5,94] for an introduction to computational molecular biology, see [41,95] for further details about RNA secondary structures and their mathematical properties, and [70] for an introduction to sequence comparison algorithms. See also [37] for dynamic programming algorithms on DNA and RNA sequences.

Beside the comparison of RNA secondary structures based on maximum common subtree isomorphism, discussed in this chapter, tree edit distance was used for comparing RNA secondary structures in [4,15,16,54,55,74,75], and tree alignment was used in [38,58]. See also [27,28,33,57,91]. Further applications of tree isomorphism and related problems to computational biology and chemistry include [32,39,44,63].

Another important application area of tree isomorphism and related problems is the comparison and retrieval of structured documents. Documents often display a structure, and the structural knowledge expressed in text documents marked up with languages such as SGML and XML can be exploited for more effective information retrieval, and detecting similarities and differences between structured documents is the basis of structured information retrieval. Tree editing was used in [97] to highlight differences between versions of a same computer program, and in [12–14] to find differences between structured documents. See also [8].

Problems

4.1 Determine whether the unordered trees T_1 and T_2 of Fig. 4.24 are isomorphic. Give also the isomorphism code for the root of each of them, as well as their canonical ordered trees.

4.2 Let $[v_1, v_2, \ldots, v_n]$ be the sequence of the n nodes of a tree T in the order in which they are visited during a preorder traversal. The *depth sequence* of T is the sequence of n integers $[depth[v_1], depth[v_2], \ldots, depth[v_n]]$. For instance, the depth sequences of the 14 non-isomorphic ordered trees on 5 nodes are shown in Fig. 4.25.

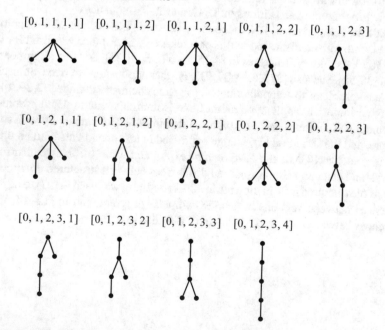

Fig. 4.24 Unordered trees for Problem 4.1

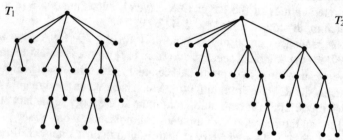

Fig. 4.25 Depth sequences of the 14 non-isomorphic ordered trees on 5 nodes, arranged in lexicographic order

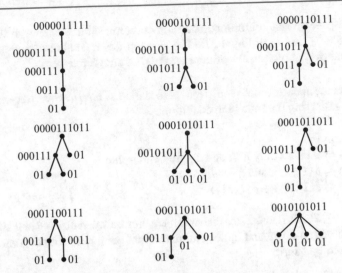

Fig. 4.26 Parenthesis strings of the 9 non-isomorphic unordered trees on 5 nodes, arranged in lexicographic order. The parenthesis string of each node is also shown next to the node

Fig. 4.27 Ordered trees for Problem 4.5

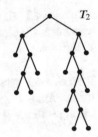

Prove that the depth sequence of an ordered tree is a certificate for ordered tree isomorphism.

4.3 Let the parenthesis string of an unordered tree on n nodes be the string of zeros and ones of length $2n$ defined as follows. The parenthesis string of a leaf node is 01, and the parenthesis string associated to a non-leaf node is obtained by concatenating the set of strings of the children of the node, sorted in lexicographic order, preceded by an additional 0 and followed by an additional 1. For instance, the parenthesis string of the 9 non-isomorphic unordered trees on 5 nodes are shown in Fig. 4.26. Prove that the parenthesis string of the root of an unordered tree is a certificate for unordered tree isomorphism.

4.4 Give upper and lower bounds on the number of subtree isomorphisms of an ordered tree on n_1 nodes into an ordered tree on n_2 nodes, where $n_1 \leqslant n_2$.

4.5 Find top-down and bottom-up maximum common subtree isomorphisms of the ordered trees T_1 and T_2 of Fig. 4.27. Find also top-down and bottom-up maximum common subtree isomorphisms of their underlying unordered trees.

4.6 A distance measure between trees, also called a *metric* over trees, is a real function δ satisfying the following conditions

- $\delta(T_i, T_j) \geqslant 0$,
- $\delta(T_i, T_j) = 0$ if and only if T_i and T_j are isomorphic,
- $\delta(T_i, T_j) = \delta(T_j, T_i)$, and
- $\delta(T_i, T_j) \leqslant \delta(T_i, T_k) + \delta(T_k, T_j)$

for all trees T_i, T_j, T_k. Define tree distance measures between ordered and unordered trees, based on top-down and bottom-up maximum common subtree isomorphism. Prove them to be metric.

Exercises

Exercise 4.1 Give an algorithm for ordered tree isomorphism based on the depth sequences discussed in Problem 4.2. Give also the time and space complexity of the algorithm.

Exercise 4.2 Give an algorithm for unordered tree isomorphism based on the paren-thesis strings discussed in Problem 4.3. Give also the time and space complexity of the algorithm.

Exercise 4.3 Extend the algorithm for unordered tree isomorphism, in order to test isomorphism of labeled trees. Assume that node labels are integers in the range $1, \ldots, n$, where n is the number of nodes in each of the trees. Give also a correctness proof, together with the time and space complexity of the extended algorithm.

Exercise 4.4 Consider the following procedure for testing isomorphism of unordered trees. Given two unordered trees, label first all the leaves in the two trees with the integer zero. If the number of leaves in the two trees differ, then the trees are not isomorphic.

Otherwise, determine for each tree the set of unlabeled nodes having all of their children already labeled. Tentatively, label each such node with the sequence of labels of its children, sorted in non-decreasing order. If the multisets of tentative labels for these nodes in the two trees differ, then the trees are not isomorphic. Otherwise, replace the tentative labels with *new* integer labels in a consistent way, that is, such that nodes with the same tentative label get the same new label. Repeat the procedure, until all nodes have been labeled. In such a case, the trees are isomorphic.

Give an algorithm for unordered tree isomorphism implementing the previous procedure. Give also the time and space complexity of the algorithm.

Exercise 4.5 All top-down subtree isomorphisms of an ordered tree $T_1 = (V_1, E_1)$ on n_1 nodes into an ordered tree $T_2 = (V_2, E_2)$ on n_2 nodes, where $n_1 \leqslant n_2$, can be obtained by repeated application of the top-down ordered subtree isomorphism algorithm, upon all bottom-up subtrees of T_2 of size and height at least as large as those of T_1. Consider, though, an alternative procedure based on the following remark.

For all nodes $v \in V_1$, let $Q(v) \subseteq V_2$ denote the set of all nodes $w \in V_2$ such that a path in T_2 originating at node w exactly replicates the path in T_1 from the root down to node v. Then, if some node $w \in V_2$ belongs to $Q(v)$ for *all the leaves* $v \in V_1$, there are paths in T_2 originating at node w that constitute a complete replica of T_1, that is, node w is the root of a top-down subtree of T_2 isomorphic to T_1. Therefore, the set of all nodes $w \in V_2$ that are roots of top-down subtrees of T_2 isomorphic to T_1 is just the intersection $\cap_{v \in V_1'} Q(v)$, where $V_1' \subseteq V_1$ is the set of the leaves of T_1.

Give an algorithm for finding all top-down ordered subtree isomorphisms of an ordered tree T_1 on n_1 nodes into an ordered tree T_2 on n_2 nodes, with $n_1 \leqslant n_2$, based on the previous remark. Give also the time and space complexity of the algorithm.

Exercise 4.6 Give algorithms for bottom-up subtree isomorphism on both ordered and unordered trees, based on the top-down subtree isomorphism algorithms. Give also a correctness proof, together with the time and space complexity of the algorithms.

Exercise 4.7 Extend the algorithm for bottom-up unordered subtree isomorphism, in order to test subtree isomorphism of labeled trees. Assume that node labels are integers in the range $1, \ldots, n$, where n is the number of nodes in each of the trees. Give also a correctness proof, together with the time and space complexity of the extended algorithm.

Exercise 4.8 Improve the algorithm for bottom-up ordered maximum common subtree isomorphism, replacing the use of node priority queues by bucket sorting. Give also a the time and space complexity of the improved algorithm.

Exercise 4.9 Give an efficient algorithm for computing tight upper and lower bounds on the size of a top-down maximum common subtree of two unordered trees. Assume the trees are unlabeled. Give also a correctness proof, together with the time and space complexity of the algorithm.

Exercise 4.10 Extend to labeled trees the algorithms for bottom-up ordered and unordered maximum common subtree isomorphism. Assume that node labels are integers in the range $1, \ldots, n$, where n is the total number of nodes in the trees. Give also a correctness proof, together with the time and space complexity of the extended algorithm.

References

1. Abboud A, Bačkurs A, Hansen TD, Williams VV, Zamir O (2018) Subtree isomorphism revisited. ACM Trans Algorithms 14(3):27:1–27:23
2. Aho AV, Hopcroft JE, Ullman JD (1974) The design and analysis of computer algorithms. Addison-Wesley, Reading MA
3. Akutsu T, Tamura T, Melkman AA, Takasu A (2015) On the complexity of finding a largest common subtree of bounded degree. Theor Comput Sci 590(1):2–16
4. Allali J, Sagot MF (2005) A new distance for high level RNA secondary structure comparison. IEEE/ACM Trans Comput Biol Bioinf 2(1):3–14
5. Attwood TK, Parry-Smith DG (1999) Introduction to bioinformatics. Prentice Hall, London, England
6. Berztiss AT (1975) Data structures: theory and practice, 2nd edn. Academic Press, New York NY
7. Beyer T, Hedetniemi SM (1980) Constant time generation of rooted trees. SIAM J Comput 9(4):706–712
8. Biswas P, Venkataramani G (2008) Comprehensive isomorphic subtree enumeration. In: Altman E (ed) Proceedings 2008 international conference on compilers, architectures and synthesis for embedded systems. ACM, pp 177–186
9. Bunke H (1997) On a relation between graph edit distance and maximum common subgraph. Pattern Recognit Lett 18(8):689–694
10. Busacker RG, Saaty TL (1965) Finite graphs and networks: an introduction with applications. McGraw-Hill, New York NY
11. Caminiti S, Finocchi I, Petreschi R (2007) On coding labeled trees. Theor Comput Sci 382(2):97–108
12. Chawathe SS (1999) Comparing hierarchical data in external memory. In: Malcolm MEO, Atkinson P, Valduriez P, Zdonik SB, Brodie ML (eds) Proceedings of 25th international conference on very large data bases. Morgan Kaufmann, New York NY, pp 90–101
13. Chawathe SS, García-Molina H (1997) Meaningful change detection in structured data. In: Peckman JM, Ram S, Franklin M (eds) Proceedings of ACM SIGMOD international conference on management of data. ACM, New York NY, pp 26–37
14. Chawathe SS, Rajaraman A, García-Molina H, Widom J (1996) Change detection in hierarchically structured information. In: Widom J (ed) Proceedings ACM SIGMOD international conference on management of data, ACM, New York NY, pp 493–504
15. Chen Q, Lan C, Chen B, Wang L, Li J, Zhang C (2017) Exploring consensus RNA substructural patterns using subgraph mining. IEEE/ACM Trans Comput Biol Bioinf 14(5):1134–1146
16. Chen S, Zhang K (2014) An improved algorithm for tree edit distance with applications for RNA secondary structure comparison. J Comb Optim 27(4):778–797
17. Chi Y, Nijsseny S, Muntz RR (2005) Canonical forms for labelled trees and their applications in frequent subtree mining. Knowl Inf Syst 8(2):203–234
18. Chung MJ (1987) $O(n^{5/2})$ time algorithms for the subgraph homeomorphism problem on trees. J Algorithms 8(1):106–112

19. Cole R, Hariharan R, Indyk P (1999) Tree pattern matching and subset matching in deterministic $O(nlog^3n)$-time. In: Tarjan RE, Warnow T (eds) Proceedings 10th annual ACM-SIAM symposium discrete algorithms. ACM, New York NY, pp 245–254

20. Culik K, Wood D (1982) A note on some tree similarity measures. Inf Process Lett 15(1):39–42

21. Deo NJ, Micikevicius P (2001) Prüfer-like codes for labeled trees. Congressus Numerantium 151(1):65–73

22. Dickinson PJ, Kraetzl M, Bunke H, Neuhaus M, Dadej A (2004) Similarity measures for hierarchical representations of graphs with unique node labels. Int J Pattern Recognit Artif Intell 3(2004), 425–442 (18)

23. Dinitz Y, Itai A, Rodeh M (1999) On an algorithm of Zemlyachenko for subtree isomorphism. Inf Process Lett 70(3):141–146

24. Droschinsky A, Kriege NM, Mutzel P (2016) Faster algorithms for the maximum common subtree isomorphism problem. In: Faliszewski P, Muscholl A, Niedermeier R (eds) Proceedings 41st international symposium mathematical foundations of computer science, leibniz international proceedings in informatics, vol 58 (2016), pp 33:1–33:14

25. Dubiner M, Galil Z, Magen E (1994) Faster tree pattern matching. J ACM 41(2):205–213

26. Dublish P (1990) Some comments on the subtree isomorphism problem for ordered trees. Inf Process Lett 36(5):273–275

27. Dulucq S, Tichit L (2003) RNA secondary structure comparison: Exact analysis of the Zhang-Shasha tree edit algorithm. Theor Comput Sci 306(1–3):471–484

28. Eliáš R, Hoksza D (2016) RNA secondary structure visualization using tree edit distance. Int J Biosci Biochem Bioinf 6(1):9–17

29. Er MC (1985) Enumerating ordered trees lexicographically. Comput J 28(5):538–542

30. Flajolet P, Sipala P, Steyaert JM (1990) Analytic variations on the common subexpression problem. In: Paterson MS (ed) Proceedings 17th international colloquium on automata, languages, and programming, Lecture notes in computer science, vol 443. Springer, Berlin Heidelberg, pp 220–234

31. Fukagawa D, Akutsu T (2006) Fast algorithms for comparison of similar unordered trees. Int J Found Comput Sci 17(3):703–729

32. Fukagawa D, Tamura T, Takasu A, Tomita E, Akutsu T (2011) A clique-based method for the edit distance between unordered trees and its application to analysis of glycan structures. BMC Bioinf 12(Suppl. 1):S13

33. Gao L, Xu C, Sun X, Song W, Xiao F, Zhang D (2019). In: Liu C (ed) Structural alignment of long, highly structured RNAs based on ordered tree method. VDE Verlag, pp 156–160

34. Grossi R (1991) Further comments on the subtree isomorphism for ordered trees. Inf Process Lett 40(5):255–256

35. Grossi R (1991) A note on the subtree isomorphism for ordered trees and related problems. Inf Process Lett 39(2):81–84

36. Grossi R (1993) On finding common subtrees. Theor Comput Sci 108(2):345–356

37. Gusfield D (1997) Algorithms on strings, trees, and sequences: computer science and computational biology. Cambridge University Press, Cambridge, England

38. Höchsmann M, Töller T, Giegerich R, Kurtz S (2003) Local similarity in RNA secondary structures. IEEE, pp 159–168

39. Heumann H, Wittum G (2009) The tree-edit-distance, a measure for quantifying neuronal morphology. Neuroinformatics 7(3):179–190

40. Hofacker IL, Fontana W, Stadler PF, höf fer SB, Tacker M, Schuster P (1994) Fast folding and comparison of RNA secondary structures. Monatshefte f. Chemie 125(2):167–188

41. Hofacker IL, Schuster P, Stadler PF (1998) Combinatorics of RNA secondary structures. Discret Appl Math 88(1–3):207–237

42. Hoffmann CM, O'Donnell MJ (1982) Pattern matching in trees. J ACM 29(1):68–95

43. Hokazono T, Kan T, Yamamoto Y, Hirata K (2012) An isolated-subtree inclusion for unordered trees. In: Hirokawa S, Hashimoto K (eds) Proceedings 2012 IIAI international conference on advanced applied informatics. IEEE, pp 345–350

44. Hufsky F, Dührkop K, Rasche F, Chimani M, Böker S (2012) Fast alignment of fragmentation trees. Bioinformatics 28(12):i265–i273
45. Čibej U, Mihelič J (2014) Search strategies for subgraph isomorphism algorithms. In: Gupta P, Zaroliagis C (eds) Proceedings 1st international conference applied algorithms, Lecture notes in computer science, vol 8321. Springer, pp 77–88
46. Čibej U, Mihelič J (2015) Improvements to Ullmann's algorithm for the subgraph isomorphism problem. Int J Pattern Recognit Artif Intell 29(7):1550025:1–1550025:26
47. Ishizaka Y, Yoshino T, Hirata K (2015) Anchored alignment problem for rooted labeled trees. In: Murata T, Mineshima K, Bekki D (eds) Proceeding 6th international symposium artificial intelligence, Lecture notes in computer science, vol 9067. Springer, pp 296–309
48. Jansson J, Lingas A (2001) A fast algorithm for optimal alignment between similar ordered trees. In: Amir A (ed) Proceedings 12th annual symposium combinatorial pattern matching, Lecture notes in computer science, vol 2089. Springer, Berlin Heidelberg, pp 232–243
49. Jiang T, Wang L, Zhang K (1995) Alignment of trees: An alternative to tree edit. Theor Comput Sci 143(1):137–148
50. Jovanovi A, Danilovi D (1984) A new algorithm for solving the tree isomorphism problem. Computing 32(3):187–198
51. Kan T, Higuchi S, Hirata K (2014) Segmental mapping and distance for rooted labeled ordered trees. Fundamenta Informaticae 132(4):461–483
52. Kosaraju SR (1989) Efficient tree pattern matching. In: Proceedings 30th annual symposium foundations of computer science. IEEE, Piscataway NJ, pp 178–183
53. Kreher DL, Stinson DR (1999) Combinatorial algorithms: generation, enumeration, and search. CRC Press, Boca Raton FL
54. Le SY, Nussinov R, Maizel JV (1989) Tree graphs of RNA secondary structures and their comparisons. Comput Biomed Res 22(5):461–473
55. Le SY, Owens J, Nussinov R, Chen JH, Shapiro B, Maizel JV (1989) RNA secondary structures: comparison and determination of frequently recurring substructures by consensus. Bioinformatics 5(3):205–210
56. Li G, Ruskey F (1999) The advantages of forward thinking in generating rooted and free trees. In Tarjan RE, Warnow T (eds) Proceedings 10th Annual ACM-SIAM symposium discrete algorithms. ACM, New York NY, pp 939–940
57. Liang Z, Zhang K (2014) Algorithms for local similarity between forests. J Comb Optim 27(1):14–31
58. Lozano A, Pinter RY, Rokhlenko O, Valiente G, Ziv-Ukelson M (2008) Seeded tree alignment. IEEE/ACM Trans Comput Biol Bioinf 5(4):503–513
59. Lozano A, Valiente G (2004) On the maximum common embedded subtree problem for ordered trees. In: Iliopoulos CS, Lecroq T (eds) String algorithmics, Texts in algorithmics, vol 2, Chap. 7. College Publications, pp 155–169
60. Lu SY (1979) A tree-to-tree distance and its applications to cluster analysis. IEEE Trans Pattern Anal Mach Intell 1(2):219–224
61. Lu SY (1984) A tree-matching algorithm based on node splitting and merging. IEEE Trans Pattern Anal Mach Intell 6(2):249–256
62. Luccio F, Pagli L, Enriquez AM, Rieumont PO (2007) Bottom-up subtree isomorphism for unordered labeled trees. Int J Pure Appl Math 38(3):325–343
63. Marín RM, Aguirre NF, Daza EE (2008) Graph theoretical similarity approach to compare molecular electrostatic potentials. J Chem Inf Model 48(1):109–118
64. Matula DW (1978) Subtree isomorphism in $O(n^{5/2})$. Ann Discret Math 2(1):91–106
65. Mihelič J, Čibej U, Fürst L (2021) A backtracking algorithmic toolbox for solving the subgraph isomorphism problem. In: Handbook of research on methodologies and applications of supercomputing, advances in systems analysis, software engineering, and high performance computing, Chap. 14. IGI Global, pp 208–246

66. Mäkinen E (1989) On the subtree isomorphism problem for ordered trees. Inf Process Lett 32(5):271–273
67. Ohama T, Kumazaki T, Hori H, Osawa S (1984) Evolution of multicellular animals as deduced from 5S rRNA sequences: A possible early emergence of the mesozoa. Nucleic Acids Res 12(12):5101–5108
68. Reyner SW (1977) An analysis of a good algorithm for the subtree problem. SIAM J Comput 6(4):730–732
69. Rosselló F, Valiente G (2006) An algebraic view of the relation between largest common subtrees and smallest common supertrees. Theor Comput Sci 362(1–3):33–53
70. Sankoff D, Kruskal JB (eds) (1999) Time warps, string edits, and macromolecules: the theory and practice of sequence comparison. Center for the study of language and information. Stanford CA
71. Selkow SM (1977) The tree-to-tree editing problem. Inf Process Lett 6(6):184–186
72. Seo S, Shin H (2007) A generalized enumeration of labeled trees and reverse Prüfer algorithm. J Comb Theory, Ser A 114(7):1357–1361
73. Shamir R, Tsur D (1999) Faster subtree isomorphism. J Algorithms 33(2):267–280
74. Shapiro BA (1988) An algorithm for comparing multiple RNA secondary structures. Bioinformatics 4(3):387–393
75. Shapiro BA, Zhang K (1990) Comparing multiple RNA secondary structures using tree comparisons. Comput Appl Biosci 6(4):309–318
76. Shasha D, Wang JTL, Zhang K, Shih FY (1994) Exact and approximate algorithms for unordered tree matching. IEEE Trans Syst Man Cybern 24(4):668–678
77. Shasha D, Zhang K (1990) Fast algorithms for the unit cost editing distance between trees. J Algorithms 11(4):581–621
78. Specht T, Ulbrich N, Erdmann V (1986) Nucleotide sequence of the 5S rRNA from the Annelida species Enchytraeus albidus. Nucleic Acids Res 14(10):4372
79. Specht T, Ulbrich N, Erdmann V (1987) The secondary structure of the 5S ribosomal ribonucleic acid from the Annelida Enchytraeus albidus. Endocytobiosis Cell Res 4(1):205–214
80. Tai KC (1979) The tree-to-tree correction problem. J ACM 26(3):422–433
81. Tanaka E (1994) A metric between unrooted and unordered trees and its bottom-up computing method. IEEE Trans Pattern Anal Mach Intell 16(12):1233–1238
82. Tanaka E, Tanaka K (1988) The tree-to-tree editing problem. Int J Pattern Recognit Artif Intell 2(2):221–240
83. Torsello A, Hidović D, Pelillo M (2005) Polynomial-time metrics for attributed trees. IEEE Trans Pattern Anal Mach Intell 27(7):1087–1099
84. Touzet H (2007) Comparing similar ordered trees in linear-time. J Discret Algorithms 5(4):696–705
85. Valiente G (2001) An efficient bottom-up distance between trees. Proceedings 8th international symposium string processing and information retrieval. IEEE Computer Science Press, Piscataway NJ, pp 212–219
86. Valiente G (2004) On the algorithm of Berztiss for tree pattern matching. In: Baeza-Yates R, Marroquí JL, Chávez E (eds) Proceedings 5th Mexican international conference computer science. IEEE, pp 43–49
87. Valiente G (2005) Constrained tree inclusion. J Discret Algorithms 3(2–4):431–447
88. Valiente G (2007) Efficient algorithms on trees and graphs with unique node labels. In: Kandel A, Bunke H, Last M (eds) Applied graph theory in computer vision and pattern recognition, Studies in computational intelligence, vol 52. Springer, Berlin, pp 137–149
89. Verma RM (1992) Strings, trees, and patterns. Inf Process Lett 41(3):157–161
90. Verma RM, Reyner SW (1989) An analysis of a good algorithm for the subtree problem, corrected. SIAM J Comput 18(5):906–908
91. Wang F, Akutsu T, Mori T (2020) Comparison of pseudoknotted RNA secondary structures by topological centroid identification and tree edit distance. J Comput Biol 27(9):1–9

92. Wang JTL, Zhang K (2001) Finding similar consensus between trees: An algorithm and a distance hierarchy. Pattern Recognit 34(1):127–137
93. Wang X, Wang L, Wu Y (2009) An optimal algorithm for Prüfer codes. J Softw Eng Appl 2(2):111–115
94. Waterman MS (1995) Introduction to computational biology: maps, sequences genomes. Chapman and Hall, London, England
95. Waterman MS, Smith TF (1978) RNA secondary structure: A complete mathematical analysis. Math Biosci 42(1):257–266
96. Wilf HS (1989) Combinatorial algorithms: an update. SIAM, Philadelphia PA
97. Yang W (1991) Identifying syntactic differences between two programs. Softw: Practice Exp 21(7):739–755
98. Yoshino T, Hirata K (2017) Tai mapping hierarchy for rooted labeled trees through common subforest. Theory Comput Syst 60(4):759–783
99. Yoshino T, Ishizaka Y, Hirata K (2017). In: Ganzha M, Maciaszek L, Paprzycki M (eds) Anchored alignment distance between rooted labeled unordered trees. IEEE, pp 433–440
100. Zaks S (1980) Lexicographic generation of ordered trees. Theor Comput Sci 10(1):63–82
101. Zelinka B (1991) Distances between rooted trees. Math Bohemica 116(1):101–107
102. Zhang K (1996) A constrained edit distance between unordered labeled trees. Algorithmica 15(3):205–222
103. Zhang K, Shasha D (1989) Simple fast algorithms for the editing distance between trees and related problems. SIAM J Comput 18(6):1245–1262
104. Zhang K, Statman R, Shasha D (1992) On the editing distance between unordered labeled trees. Inf Process Lett 42(3):133–139
105. Zhang K, Wang JTL, Shasha D (1996) On the editing distance between undirected acyclic graphs. Int J Found Comput Sci 7(1):43–57
106. Zucker M, Stiegler P (1981) Optimal computer folding of large RNA sequences using thermodynamics and auxiliary information. Nucleic Acids Res 9(1):133–148

Graph Traversal

5

5.1 Depth-First Traversal of a Graph

A traversal of a graph $G = (V, E)$ with n vertices is just a bijection $order : V \to \{1, \ldots, n\}$. In an operational view, a traversal of a graph consists of visiting first the vertex v with $order[v] = 1$, then the vertex w with $order[w] = 2$, and so on, until visiting last the vertex z with $order[z] = n$.

The generalization of tree traversal to graph traversal does not come for free, though. Graph traversal algorithms need to take into account the possibility that a vertex being visited had already been visited along a different path during the traversal, while in (rooted) tree traversal, a node can only be accessed along the unique path from the root to the node. Nevertheless, there is a close relationship between tree and graph traversal, which will be explored in detail in this chapter.

As with tree traversal, the order in which the edges going out of a given vertex are considered (and therefore, the order in which the vertices adjacent with a given vertex—the children of a node, in the case of trees—are considered) significant for an ordered graph, and it is also fixed by the representation adopted for both unordered and ordered graphs. Further, the order in which the vertices of the graph are considered, as well as the order in which the edges coming into and going out of a vertex are considered, are also significant.

The following assumption about the relative order of the vertices and edges of a graph will, without loss of generality, make it possible to give a precise notion of depth-first and breadth-first spanning forest of a graph.

Assumption 5.1 The order of the vertices of a graph is the order fixed by the representation of the graph, the order of the edges coming into a vertex is the order of their source vertices, and the order of the edges going out of a vertex is the order of their target vertices.

© The Author(s), under exclusive license to Springer Nature Switzerland AG 2021
G. Valiente, *Algorithms on Trees and Graphs*, Texts in Computer Science,
https://doi.org/10.1007/978-3-030-81885-2_5

The counterpart to the preorder traversal of a tree is the depth-first traversal of a graph, which is the preorder traversal of a depth-first spanning forest of a graph. Recall that a spanning forest of a graph is an ordered set of pairwise-disjoint subgraphs of the graph which are rooted trees and which, together, span all the vertices of the graph.

Definition 5.1 Let $G = (V, E)$ be a graph with n vertices. A spanning forest F of G is a **depth-first spanning forest** of G if the following conditions are satisfied:

- for all trees $T = (W, S)$ in F and for all edges $(v, w) \in S$, there is no edge $(x, y) \in E \cap (W \times W)$ such that $order[v] < order[x] < order[w] \leqslant order[y]$,
- for all trees $T_i = (W_i, S_i)$ and $T_j = (W_j, S_j)$ in F with $i < j$, there is no edge $(v, w) \in E$ with $v \in W_i$ and $w \in W_j$,

where $order : W \to \{1, \ldots, k\}$ is the preorder traversal of a rooted tree $T = (W, S)$ with $k \leqslant n$ nodes.

Example 5.1 The notion of depth-first spanning forest of a graph is illustrated in Fig. 5.1. Since all vertices of the graph are reachable from vertex v_1, every depth-first spanning forest of the graph whose initial vertex is v_1 must consist of a single tree, which is, therefore, a spanning tree of the graph. The spanning tree shown twice in the bottom row of Fig. 5.1 is not a depth-first spanning tree of the graph, because there is an edge (v, w) in the tree and an edge (x, y) not in the tree with $order[v] < order[x] < order[w] \leqslant order[y]$. In the picture to the left, $order[v] = 1$,

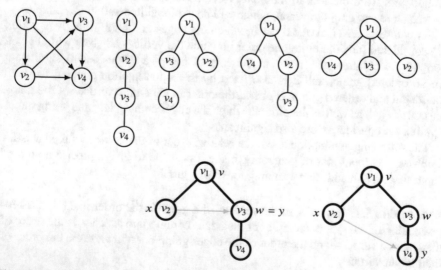

Fig. 5.1 Depth-first spanning forest of a graph. All the depth-first spanning trees rooted at vertex v_1 of the graph to the left of the top row are also shown in the top row to the right. The spanning tree shown twice in the bottom row is not a depth-first spanning tree

$order[x] = 2$, and $order[w] = order[y] = 3$, while in the picture to the right, $order[v] = 1$, $order[x] = 2$, $order[w] = 3$, and $order[y] = 4$.

Now, a depth-first traversal of a graph is just a preorder traversal of each of the trees in the depth-first spanning forest of the graph.

Definition 5.2 Let $G = (V, E)$ be a graph with n vertices, and let F be the depth-first spanning forest of G. A bijection $order : V \to \{1, \dots, n\}$ is a **depth-first traversal** of G if for all trees $T = (W, S)$ in F, the restriction of the bijection to $W \subseteq V$ is a preorder traversal of T.

In an operational view, a depth-first tree in a depth-first spanning forest of a graph is grown out of a single, initial vertex by adding edges leading to vertices not yet in the tree, together with their target vertices. The next edge to be added goes out of the latest-added vertex which is adjacent with some vertex not in the tree, and comes precisely into some vertex not in the tree. Then, the first condition in the previous definition ensures that the tree is grown in a depth-first fashion, that is, that edges going out of "deeper" vertices are added before edges going out of "shallower" vertices. The second condition guarantees that the tree is grown as much as possible, that is, that it spans all the vertices reachable in the graph from the initial vertex.

Example 5.2 In the spanning tree shown in the bottom row of Fig. 5.1, edge (v, w) cannot be added to the tree before having added edge (x, y), because vertex x is "deeper" than vertex v, that is, vertex x is visited after vertex v during a preorder traversal of the tree.

Now, the following result gives a constructive characterization of a depth-first tree in a particular depth-first spanning forest of a graph: the one generalizing to graphs the depth-first prefix leftmost traversal of a tree.

Lemma 5.1 *Let $G = (V, E)$ be a graph, let $u \in V$, let $W \subseteq V$ be the set of vertices reachable in G from vertex u, and let k be the number of vertices in W. Let also the subgraph $(W, S) = (W_k, S_k)$ of G be defined by induction as follows:*

- $(W_1, S_1) = (\{u\}, \emptyset)$,
- $(W_{i+1}, S_{i+1}) = (W_i \cup \{w\}, S_i \cup \{(v, w)\})$ *for* $1 \leqslant i < k$, *where*

  - *j is the largest integer between 1 and i such that there are vertices $v \in W_j$ and $w \in V \setminus W_i$ with $(v, w) \in E$*
  - *$\{v\} = W_j$ if $j = 1$, otherwise $\{v\} = W_j \setminus W_{j-1}$,*
  - *w is the smallest (according to the representation of the graph) such vertex in $V \setminus W_i$ with $(v, w) \in E$.*

Then, (W, S) is a depth-first spanning tree of the subgraph of G induced by W.

The following result, which will be used in the proof of Lemma 5.1, justifies the previous claim that the notion of depth-first spanning forest of a graph, indeed, generalizes the preorder traversal of a tree.

Lemma 5.2 *Let $G = (V, E)$ be a graph, and let $T = (W, S)$ be the depth-first spanning tree of the subgraph of G induced by W. Let also order $: W \to \{1, \ldots, k\}$, where k is the number of vertices in W, be the bijection defined by order$[v] = i$ if $\{v\} = W_i \setminus W_{i-1}$ for all vertices $v \in W$, where $W_0 = \emptyset$. Then, the bijection order $: W \to \{1, \ldots, k\}$ is the preorder traversal of $T = (W, S)$.*

Proof It has to be shown that in the inductive definition of depth-first spanning tree of Lemma 5.1, a tree is grown by adding vertices in the order given by the preorder traversal of the tree. The tree is grown by adding a vertex adjacent with the latest-added vertex (which is adjacent with some vertex not yet in the tree) and since by Assumption 5.1, the relative order among the vertices adjacent with a given vertex coincides with the relative order among the children of the vertex in the spanning tree, previous siblings are added before next siblings when growing the tree. Therefore, the tree is grown by adding vertices in preorder.   □

Proof (**Lemma** 5.1) Let $G = (V, E)$ be a graph with n vertices, let $u \in V$, let $W \subseteq V$ be the set of vertices reachable in G from vertex u, let k be the number of vertices in W, and let (W, S) be the subgraph of G given by the inductive definition in Lemma 5.1. Since $S_{i+1} \setminus S_i$ contains exactly one edge (v, w) going out of a vertex $v \in W_i$ and coming into a vertex $w \in W_{i+1} \setminus W_i$, for $1 \leqslant i < k$, it follows that (W, S) is a tree, rooted at vertex $u \in W \subseteq V$ and spanning the subgraph of G induced by W.

Now, suppose there are edges $(v, w) \in S$ and $(x, y) \in E \cap (W \times W)$ such that order$[v] <$ order$[x] <$ order$[w] \leqslant$ order$[y]$, and let $W_0 = \emptyset$. Let $j =$ order$[v]$, let $\ell =$ order$[x]$, and let $i =$ order$[w] - 1$. By Lemma 5.2, $\{v\} = W_j \setminus W_{j-1}$, $\{x\} = W_\ell \setminus W_{\ell-1}$, and $\{w\} = W_{i+1} \setminus W_i$. Furthermore, since order$[y] \geqslant$ order$[w]$ and $w \notin W_i$, it holds that $y \in V \setminus W_i$.

Then, $j < \ell \leqslant i$, $x \in W_\ell$, $y \in V \setminus W_i$, and $(x, y) \in E$, contradicting the hypothesis that j is the largest integer between 1 and i such that there is an edge in E from a vertex in W_j to a vertex in $V \setminus W_i$. Therefore, (W, S) is a depth-first spanning tree of the subgraph of G induced by W.   □

Example 5.3 The construction given in Lemma 5.1 for obtaining a depth-first spanning forest of a graph is illustrated in Fig. 5.2. Vertices are numbered according to the order in which they are added to the spanning forest. Since all of the vertices of the graph are reachable from the first-numbered vertex, the spanning forest consists of a single tree. Vertices not yet in the tree are shown in white, and vertices already in the tree are either shown in gray, if they are adjacent with some vertex which is not yet in the tree, otherwise they are shown in red.

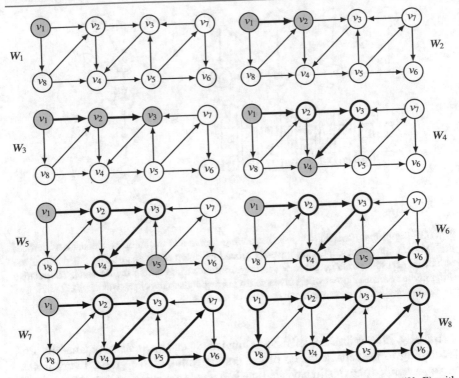

Fig. 5.2 Construction of a depth-first spanning forest $T = (V, S)$ of a graph $G = (V, E)$ with $n = 8$ vertices. Vertices are numbered according to the order in which they are added to the spanning forest. The relative order of the vertices adjacent with a given vertex corresponds to the counter-clockwise ordering of the outgoing edges of the vertex in the drawing of the graph

Now, a simple procedure for the depth-first traversal of a graph consists of performing a preorder traversal upon each of the depth-first trees in the depth-first spanning forest of the graph. The procedure differs from preorder tree traversal, though.

- During the depth-traversal of a graph, a vertex being visited may have already been visited along a different path in the graph. Vertices need to be marked as either unvisited or visited.
- Only those vertices which are reachable from an initial vertex are visited during the traversal of the depth-first tree rooted at the initial vertex. The procedure may need to be repeated upon still unvisited vertices, until all vertices have been visited.

As in the case of tree traversal, the stack implicit in the recursive algorithm for preorder tree traversal can be made explicit, yielding a simple iterative algorithm in which a stack of vertices holds those vertices which are still waiting to be traversed.

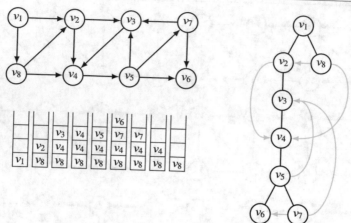

Fig. 5.3 Depth-first traversal, depth-first spanning forest, and evolution of the stack of vertices during execution of the depth-first traversal procedure, upon the graph of Fig. 5.2. Vertices are numbered according to the order in which they are first visited during the traversal. The relative order of the vertices adjacent with a given vertex corresponds to the counter-clockwise ordering of the outgoing edges of the vertex in the drawing of the graph

Initially, all vertices are marked as unvisited, and the stack contains the first (according to the representation of the graph) vertex of the graph. Every time a vertex is popped and (if still unvisited) visited, those unvisited vertices adjacent with the popped vertex are pushed, one after the other, starting with the last adjacent vertex, following with the previous adjacent vertex, and so on, until having pushed the first adjacent vertex of the popped vertex. When the stack has been emptied and no vertices remain to be pushed, the next (according to the representation of the graph) still unvisited vertex of the graph, if any, is pushed and the procedure is repeated, until no unvisited vertices remain in the graph.

Example 5.4 Consider the execution of the depth-first traversal procedure with the help of a stack of vertices, illustrated in Fig. 5.3. The evolution of the stack of vertices right before a vertex is popped shows that vertices are, as a matter of fact, popped in depth-first traversal order: $v_1, v_2, \ldots, v_8$.

The following algorithm performs a depth-first traversal of a graph $G = (V, E)$, using a stack of vertices S to implement the previous procedure. The order in which vertices are visited during the depth-first traversal is stored in the attribute $v.order$, for all vertices $v \in V$, where $v.order = -1$ if vertex v has not yet been visited.

```
procedure depth_first_traversal(G)
    for all vertices v of G do
        v.order ← −1
    num ← 0
    let S be an empty stack (of vertices)
    for all vertices v of G do
        if v.order = −1 then
            push v onto S
            while S is not empty do
                pop from S the top vertex v
                if v.order = −1 then
                    num ← num + 1
                    v.order ← num
                    for all w adjacent with v in G in reverse order do
                        if w.order = −1 then
                            push w onto S
```

Lemma 5.3 *The algorithm for depth-first traversal of a graph runs in $O(n + m)$ time using $O(m)$ additional space, where n is the number of vertices and m is the number of edges in the graph.*

Proof Let $G = (V, E)$ be a graph with n vertices and m edges. Since each vertex of the graph is pushed and popped at most once for each incoming edge, the loop is executed $n + m$ times and the algorithm runs in $O(n + m)$ time. Further, since each vertex is pushed at most once for each incoming edge, the stack cannot ever contain more than m vertices and the algorithm uses $O(m)$ additional space. □

The depth-first traversal procedure can also be applied to those vertices which are reachable from an initial vertex, in order to perform a traversal of the depth-first tree rooted at the initial vertex. The following function, which is based on the previous depth-first graph traversal algorithm, not only computes the order in which vertices are visited during the traversal, but also the depth-first spanning tree (W, S) of the subgraph of a graph $G = (V, E)$ induced by the set of vertices $W \subseteq V$ reachable from an initial vertex $v \in V$.

```
function depth_first_spanning_tree(G, v)
    let W be an empty set (of vertices)
    let S be an empty set (of edges)
    for all vertices w of G do
        w.order ← −1
    W ← W ∪ {v}
    num ← 1
    v.order ← num
    let Z be an empty stack (of edges)
    for all edges e coming out of vertex v in G in reverse order do
        push e onto Z
    while Z is not empty do
        pop from Z the top edge e
        let v be the target vertex of edge e in G
        if v.order = −1 then
            W ← W ∪ {v}
            S ← S ∪ {e}
            num ← num + 1
            v.order ← num
            for all edges e coming out of vertex v in G in reverse order do
                let w be the target vertex of edge e in G
                if w.order = −1 then
                    push e onto Z
    return (W, S)
```

Lemma 5.4 *The algorithm for computing the depth-first spanning tree of a graph from an initial vertex runs in $O(n + m)$ time using $O(m)$ additional space, where n is the number of vertices and m is the number of edges in the graph.*

Proof Let $G = (V, E)$ be a graph with n vertices and m edges. Since the edges coming out of each vertex of the graph are pushed and popped at most once, the loop is executed $n + m$ times and the algorithm runs in $O(n + m)$ time. Further, since each edge is pushed at most once, the stack cannot ever contain more than m edges and the algorithm uses $O(m)$ additional space. □

5.1.1 Leftmost Depth-First Traversal of a Graph

An interesting particular case of the depth-first traversal of an undirected graph consists of performing the traversal according to the ordered structure of the graph. Recall from Sect. 1.1 that an ordered graph is a graph that has been embedded in a certain surface, that is, such that the relative order of the adjacent vertices is fixed for each vertex. As a matter of fact, there is always a relative order of the vertices adjacent with each vertex, fixed by the graph representation.

The leftmost depth-first traversal of an undirected graph is related to the problem of finding an Euler trail through the graph. Recall from Sect. 1.1 that an undirected graph is Eulerian if it has an Euler cycle, that is, a closed trail containing all the vertices and edges of the graph, and that a non-trivial, connected undirected graph is Eulerian if and only if every vertex in the graph has even degree.

Now, an Euler cycle through a non-trivial, connected bidirected graph can be constructed by traversing each edge exactly once in each direction, what guarantees that the degree of each vertex is even, and therefore, the bidirected graph is indeed Eulerian.

Such a traversal is called a **leftmost depth-first traversal**, since the edges are explored in left-to-right order (if drawn downwards) for any vertex of the graph, and more generally, the whole graph is explored in a left-to-right fashion.

A simple method for the leftmost depth-first traversal of a bidirected graph is based on maze traversal methods. Starting with an edge traversed in one of its directions

- When an unvisited vertex is reached, the next (in the counter-clockwise ordering of the edges around the vertex) edge is traversed.
- When a visited vertex is reached along an unvisited edge, the same edge is traversed again but in the opposite direction.
- When a visited vertex is reached along a visited edge, the next (in the counter-clockwise ordering of the edges around the vertex) unvisited edge, if any, is traversed.

Example 5.5 The execution of the leftmost depth-first traversal procedure, starting, in turn, with each of the edges of the ordered graph of Fig. 5.4, is illustrated in Fig. 5.5. For instance, the closed trail $[v_1, v_2, v_8, v_1, v_8, v_7, v_6, v_5, v_4, v_3, v_2, v_3, v_7, v_3, v_6, v_3, v_4, v_6, v_4, v_5, v_6, v_7, v_2, v_7, v_8, v_2, v_1]$ corresponds to the leftmost depth-first traversal starting with edge (v_1, v_2).

The following recursive algorithm performs a leftmost depth-first traversal of a bidirected graph $G = (V, E)$, implementing the previous procedure. The edges of

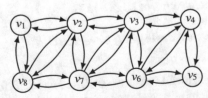

Fig. 5.4 A leftmost depth-first traversal of a bidirected, ordered graph. The relative order of the vertices adjacent with a given vertex reflects the counter-clockwise ordering of the outgoing edges of the vertex in the drawing. Vertices are numbered according to the order in which they are first visited during the traversal

$[v_1,v_2,v_8,v_1,v_8,v_7,v_6,v_5,v_4,v_3,v_2,v_3,v_7,v_3,v_6,v_3,v_4,v_6,v_4,v_5,v_6,v_7,v_2,v_7,v_8,v_2,v_1]$
$[v_1,v_8,v_7,v_6,v_5,v_4,v_3,v_2,v_1,v_2,v_8,v_2,v_7,v_2,v_3,v_7,v_3,v_6,v_3,v_4,v_6,v_4,v_5,v_6,v_7,v_8,v_1]$
$[v_2,v_1,v_8,v_7,v_6,v_5,v_4,v_3,v_2,v_3,v_7,v_3,v_6,v_3,v_4,v_6,v_4,v_5,v_6,v_7,v_2,v_7,v_8,v_2,v_8,v_1,v_2]$
$[v_2,v_8,v_1,v_2,v_1,v_8,v_7,v_6,v_5,v_4,v_3,v_2,v_3,v_7,v_3,v_6,v_3,v_4,v_6,v_4,v_5,v_6,v_7,v_2,v_7,v_8,v_2]$
$[v_2,v_7,v_8,v_2,v_8,v_1,v_2,v_1,v_8,v_7,v_6,v_5,v_4,v_3,v_2,v_3,v_7,v_3,v_6,v_3,v_4,v_6,v_4,v_5,v_6,v_7,v_2]$
$[v_2,v_3,v_7,v_2,v_7,v_8,v_2,v_8,v_1,v_2,v_1,v_8,v_7,v_6,v_5,v_4,v_3,v_4,v_6,v_4,v_5,v_6,v_3,v_6,v_7,v_3,v_2]$
$[v_3,v_2,v_1,v_8,v_7,v_6,v_5,v_4,v_3,v_4,v_6,v_4,v_5,v_6,v_3,v_6,v_7,v_3,v_7,v_2,v_7,v_8,v_2,v_8,v_1,v_2,v_3]$
$[v_3,v_7,v_2,v_3,v_2,v_1,v_8,v_7,v_8,v_2,v_8,v_1,v_2,v_7,v_6,v_5,v_4,v_3,v_4,v_6,v_4,v_5,v_6,v_3,v_6,v_7,v_3]$
$[v_3,v_6,v_7,v_3,v_7,v_2,v_3,v_2,v_1,v_8,v_7,v_8,v_2,v_8,v_1,v_2,v_7,v_6,v_5,v_4,v_3,v_4,v_6,v_4,v_5,v_6,v_3]$
$[v_3,v_4,v_6,v_3,v_6,v_7,v_3,v_7,v_2,v_3,v_2,v_1,v_8,v_7,v_8,v_2,v_8,v_1,v_2,v_7,v_6,v_5,v_4,v_5,v_6,v_4,v_3]$
$[v_4,v_3,v_2,v_1,v_8,v_7,v_6,v_5,v_4,v_5,v_6,v_4,v_6,v_3,v_6,v_7,v_3,v_7,v_2,v_7,v_8,v_2,v_8,v_1,v_2,v_3,v_4]$
$[v_4,v_6,v_3,v_4,v_3,v_2,v_1,v_8,v_7,v_6,v_7,v_3,v_7,v_2,v_7,v_8,v_2,v_8,v_1,v_2,v_3,v_6,v_5,v_4,v_5,v_6,v_4]$
$[v_4,v_5,v_6,v_4,v_6,v_3,v_4,v_3,v_2,v_1,v_8,v_7,v_6,v_7,v_3,v_7,v_2,v_7,v_8,v_2,v_8,v_1,v_2,v_3,v_6,v_5,v_4]$
$[v_5,v_4,v_3,v_2,v_1,v_8,v_7,v_6,v_5,v_6,v_4,v_6,v_3,v_6,v_7,v_3,v_7,v_2,v_7,v_8,v_2,v_8,v_1,v_2,v_3,v_4,v_5]$
$[v_5,v_6,v_4,v_5,v_4,v_3,v_2,v_1,v_8,v_7,v_6,v_7,v_3,v_7,v_2,v_7,v_8,v_2,v_8,v_1,v_2,v_3,v_6,v_3,v_4,v_6,v_5]$
$[v_6,v_5,v_4,v_3,v_2,v_1,v_8,v_7,v_6,v_7,v_3,v_7,v_2,v_7,v_8,v_2,v_8,v_1,v_2,v_3,v_6,v_3,v_4,v_6,v_4,v_5,v_6]$
$[v_6,v_4,v_5,v_6,v_5,v_4,v_3,v_2,v_1,v_8,v_7,v_6,v_7,v_3,v_7,v_2,v_7,v_8,v_2,v_8,v_1,v_2,v_3,v_6,v_3,v_4,v_6]$
$[v_6,v_3,v_4,v_6,v_4,v_5,v_6,v_5,v_4,v_3,v_2,v_1,v_8,v_7,v_6,v_7,v_3,v_7,v_2,v_7,v_8,v_2,v_8,v_1,v_2,v_3,v_6]$
$[v_6,v_7,v_3,v_6,v_3,v_4,v_6,v_4,v_5,v_6,v_5,v_4,v_3,v_2,v_1,v_8,v_7,v_8,v_2,v_8,v_1,v_2,v_7,v_2,v_3,v_7,v_6]$
$[v_7,v_6,v_5,v_4,v_3,v_2,v_1,v_8,v_7,v_8,v_2,v_8,v_1,v_2,v_7,v_2,v_3,v_7,v_3,v_6,v_3,v_4,v_6,v_4,v_5,v_6,v_7]$
$[v_7,v_3,v_6,v_7,v_6,v_5,v_4,v_3,v_4,v_6,v_4,v_5,v_6,v_3,v_2,v_1,v_8,v_7,v_8,v_2,v_8,v_1,v_2,v_7,v_2,v_3,v_7]$
$[v_7,v_2,v_3,v_7,v_3,v_6,v_7,v_6,v_5,v_4,v_3,v_4,v_6,v_4,v_5,v_6,v_3,v_2,v_1,v_8,v_7,v_8,v_2,v_8,v_1,v_2,v_7]$
$[v_7,v_8,v_2,v_7,v_2,v_3,v_7,v_3,v_6,v_7,v_6,v_5,v_4,v_3,v_4,v_6,v_4,v_5,v_6,v_3,v_2,v_1,v_8,v_1,v_2,v_8,v_7]$
$[v_8,v_1,v_2,v_8,v_2,v_7,v_8,v_7,v_6,v_5,v_4,v_3,v_2,v_3,v_7,v_3,v_6,v_3,v_4,v_6,v_4,v_5,v_6,v_7,v_2,v_1,v_8]$
$[v_8,v_7,v_6,v_5,v_4,v_3,v_2,v_1,v_8,v_1,v_2,v_8,v_2,v_7,v_2,v_3,v_7,v_3,v_6,v_3,v_4,v_6,v_4,v_5,v_6,v_7,v_8]$
$[v_8,v_2,v_7,v_8,v_7,v_6,v_5,v_4,v_3,v_2,v_3,v_7,v_3,v_6,v_3,v_4,v_6,v_4,v_5,v_6,v_7,v_2,v_1,v_8,v_1,v_2,v_8]$

Fig. 5.5 Execution of the leftmost depth-first traversal procedure, starting, in turn, with each edge of the ordered graph of Fig. 5.4

the graph are stored in the list of edges L in the leftmost depth-first order in which they are traversed.

```
function leftmost_depth_first_traversal(G, e)
    for all vertices v of G do
        v.visited ← false
    for all edges e' of G do
        e'.visited ← false
    let v be the source vertex of edge e in G
    v.visited ← true
    return rec_leftmost_depth_first_traversal(G, e)
```

The attributes $v.visited$, for all vertices $v \in V$, and $e.visited$, for all edges $e \in E$, are needed by the leftmost depth-first traversal procedure to keep track of traversed edges and visited vertices along the traversed edges. The following recursive algorithm implements the actual leftmost depth-first traversal procedure.

```
function rec_leftmost_depth_first_traversal(G, e)
    let L be an empty list (of edges)
    let v be the target vertex of edge e in G
    let e_r be the reverse of edge e in G
    append e to L
    if v.visited then
        if e_r.visited then
            e_p ← e_r
            repeat
                let e_p be the cyclic next edge after e_p coming out of v in G
            until e_p = e_r or not e_p.visited
            if e_p.visited then
                return L
        else
            e_p ← e_r
    else
        let e_p be the cyclic next edge after e_r coming out of vertex v in G
    e.visited ← true
    v.visited ← true
    concatenate rec_leftmost_depth_first_traversal(G, e_p) to L
    return L
```

5.2 Breadth-First Traversal of a Graph

The counterpart to the top-down traversal of a tree is the top-down traversal of a graph, which is the top-down traversal of a breadth-first spanning forest of a graph. Recall that a spanning forest of a graph is an ordered set of pairwise-disjoint subgraphs of the graph which are rooted trees and, which together, span all the vertices of the graph.

Definition 5.3 Let $G = (V, E)$ be a graph with n vertices. A spanning forest F of G is a **breadth-first spanning forest** of G if the following conditions are satisfied

- for all trees $T = (W, S)$ in F and for all edges $(v, w) \in S$, there is no edge $(x, y) \in E \cap (W \times W)$ such that $order[x] < order[v] < order[w] \leqslant order[y]$,
- for all trees $T_i = (W_i, S_i)$ and $T_j = (W_j, S_j)$ in F with $i < j$, there is no edge $(v, w) \in E$ with $v \in W_i$ and $w \in W_j$,

where $order : W \to \{1, \ldots, k\}$ is the top-down traversal of a rooted tree $T = (W, S)$ with $k \leqslant n$ nodes.

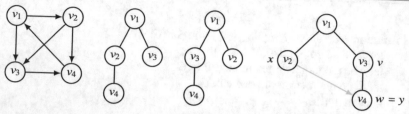

Fig. 5.6 Breadth-first spanning forest of a graph. All the breadth-first spanning trees rooted at vertex v_1 of the graph to the left of the figure are also shown in the middle of the figure. The spanning tree shown to the right of the figure is not a breadth-first spanning tree

Example 5.6 The notion of breadth-first spanning forest of a graph is illustrated in Fig. 5.6. Since all vertices of the graph are reachable from vertex v_1, every breadth-first spanning forest of the graph whose initial vertex is v_1 must consist of a single tree, which is, therefore, a spanning tree of the graph. The spanning tree shown to the right of Fig. 5.6 is not a breadth-first spanning tree of the graph, because there is an edge (v, w) in the tree and an edge (x, y) not in the tree with $order[x] = 2$, $order[v] = 3$, and $order[w] = order[y] = 4$, that is, with $order[x] < order[v] < order[w] \leqslant order[y]$.

Now, a breadth-first traversal of a graph is just a top-down traversal of each of the trees in the breadth-first spanning forest of the graph.

Definition 5.4 Let $G = (V, E)$ be a graph with n vertices, and let F be the breadth-first spanning tree of G. A bijection $order : V \rightarrow \{1, \ldots, n\}$ is a **breadth-first traversal** of G if for all trees $T = (W, S)$ in F, the restriction of the bijection to $W \subseteq V$ is a top-down traversal of T.

In an operational view, a breadth-first tree in a breadth-first spanning forest of a graph is grown out of a single, initial vertex by adding edges leading to vertices not yet in the tree, together with their target vertices. The next edge to be added goes out of the soonest-added vertex which is adjacent with some vertex not in the tree, and comes precisely into some vertex not in the tree. Then, the first condition in the previous definition ensures that the tree is grown in a breadth-first fashion, that is, that edges going out of "shallower" vertices are added before edges going out of "deeper" vertices. The second condition guarantees that the tree is grown as much as possible, that is, that it spans all the vertices reachable in the graph from the initial vertex.

Example 5.7 In the spanning tree shown to the right of Fig. 5.6, edge (v, w) cannot be added to the tree before having added arc (x, y), because vertex x is "shallower" than vertex v, that is, vertex x is visited before vertex v during a top-down traversal of the tree.

The following result gives a constructive characterization of a breadth-first tree in a particular breadth-first spanning forest of a graph: the one generalizing to graphs the breadth-first leftmost traversal of a tree.

Lemma 5.5 *Let $G = (V, E)$ be a graph, let $u \in V$, let $W \subseteq V$ be the set of vertices reachable in G from vertex u, and let k be the number of vertices in W. Let also the subgraph $(W, S) = (W_k, S_k)$ of G be defined by induction as follows:*

- $(W_1, S_1) = (\{u\}, \emptyset)$,
- $(W_{i+1}, S_{i+1}) = (W_i \cup \{w\}, S_i \cup \{(v, w)\})$ *for $1 \leqslant i < k$, where*

  - *j is the smallest integer between 1 and i such that there are vertices $v \in W_j$ and $w \in V \setminus W_i$ with $(v, w) \in E$,*
  - *$\{v\} = W_j$ if $j = 1$, otherwise $\{v\} = W_j \setminus W_{j-1}$,*
  - *w is the smallest (according to the representation of the graph) such vertex in $V \setminus W_i$ with $(v, w) \in E$.*

Then, (W, S) is a breadth-first spanning tree of the subgraph of G induced by W.

The following result, which will be used in the proof of Lemma 5.5, justifies the previous claim that the notion of breadth-first spanning forest of a graph, indeed, generalizes the top-down traversal of a tree.

Lemma 5.6 *Let $G = (V, E)$ be a graph, and let $T = (W, S)$ be the breadth-first spanning tree of the subgraph of G induced by W. Let also order$: W \rightarrow \{1, \ldots, k\}$, where k is the number of vertices in W, be the bijection defined by order$[v] = i$ if $\{v\} = W_i \setminus W_{i-1}$ for all vertices $v \in W$, where $W_0 = \emptyset$. Then, the bijection order$: W \rightarrow \{1, \ldots, k\}$ is the top-down traversal of $T = (W, S)$.*

Proof It has to be shown that in the inductive definition of breadth-first spanning tree of Lemma 5.5, a tree is grown by adding vertices in the order given by the top-down traversal of the tree. The tree is grown by adding a vertex adjacent with the soonest-added vertex (which is adjacent with some vertex not yet in the tree) and since by Assumption 5.1, the relative order among the vertices adjacent with a given vertex coincides with the relative order among the children of the vertex in the spanning tree, previous siblings are added before next siblings when growing the tree. Therefore, the tree is grown by adding vertices in top-down order. □

Proof (Lemma 5.5) Let $G = (V, E)$ be a graph with n vertices, let $u \in V$, let $W \subseteq V$ be the set of vertices reachable in G from vertex u, let k be the number of vertices in W, and let (W, S) be the subgraph of G given by the inductive definition in Lemma 5.5. Since $S_{i+1} \setminus S_i$ contains exactly one edge (v, w) going out of a vertex $v \in W_i$ and coming into a vertex $w \in W_{i+1} \setminus W_i$, for $1 \leqslant i < k$, it follows that (W, S) is a tree, rooted at vertex $u \in W \subseteq V$ and spanning the subgraph of G induced by W.

Now, suppose there are edges $(v, w) \in S$ and $(x, y) \in E \cap (W \times W)$ such that $order[x] < order[v] < order[w] \leqslant order[y]$, and let $W_0 = \emptyset$. Let $\ell = order[x]$, let $j = order[v]$, and let $i = order[w] - 1$. By Lemma 5.6, $\{x\} = W_\ell \setminus W_{\ell-1}$, $\{v\} = W_j \setminus W_{j-1}$, and $\{w\} = W_{i+1} \setminus W_i$. Furthermore, since $order[y] \geqslant order[w]$ and $w \notin W_i$, it holds that $y \in V \setminus W_i$.

Then, $1 \leqslant \ell < j \leqslant i$, $x \in W_\ell$, $y \in V \setminus W_i$, and $(x, y) \in E$, contradicting the hypothesis that j is the smallest integer between 1 and i such that there is an edge in E from a vertex in W_j to a vertex in $V \setminus W_i$. Therefore, (W, S) is a breadth-first spanning tree of the subgraph of G induced by W.   □

Example 5.8 The construction given in Lemma 5.5 above for obtaining a breadth-first spanning forest of a graph is illustrated in Fig. 5.7. Vertices are numbered according to the order in which they are added to the spanning forest. Since all of the vertices of the graph are reachable from the first-numbered vertex, the spanning forest consists of a single tree. Vertices not yet in the tree are shown in white, and

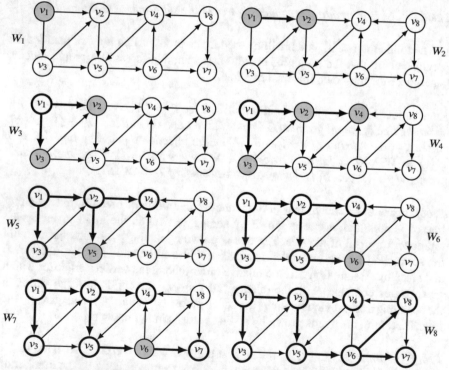

Fig. 5.7 Construction of a breadth-first spanning forest $T = (V, S)$ of a graph $G = (V, E)$ with $n = 8$ vertices. Vertices are numbered according to the order in which they are added to the spanning forest. The relative order of the vertices adjacent with a given vertex corresponds to the counter-clockwise ordering of the outgoing edges of the vertex in the drawing of the graph

vertices already in the tree are either shown in gray, if they are adjacent with some vertex which is not yet in the tree, otherwise they are shown in red.

Now, a simple procedure for the breadth-first traversal of a graph consists of performing a top-down traversal upon each of the breadth-first trees in the breadth-first spanning forest of the graph. The procedure differs from top-down tree traversal, though.

- During the breadth-traversal of a graph, a vertex being visited may have already been visited along a different path in the graph. Vertices need to be marked as either unvisited or visited.
- Only those vertices which are reachable from an initial vertex are visited during the traversal of the breadth-first tree rooted at the initial vertex. The procedure may need to be repeated upon still unvisited vertices, until all vertices have been visited.

As in the case of top-down tree traversal, a breadth-first traversal of a graph can be easily realized with the help of a queue of vertices, which holds those vertices which are still waiting to be traversed.

Initially, all vertices are marked as unvisited, and the queue contains the first (according to the representation of the graph) vertex of the graph. Every time a vertex is dequeued and (if still unvisited) visited, those unvisited vertices adjacent with the dequeued vertex are enqueued, one after the other, starting with the first adjacent vertex, following with the next adjacent vertex, and so on, until having pushed the last adjacent vertex of the popped vertex. When the queue has been emptied and no vertices remain to be enqueued, the next (according to the representation of the graph) still unvisited vertex of the graph, if any, is enqueued and the procedure is repeated, until no unvisited vertices remain in the graph.

Example 5.9 Consider the execution of the breadth-first traversal procedure with the help of a queue of vertices, illustrated in Fig. 5.8. The evolution of the queue of vertices right before a vertex is dequeued shows that vertices are, as a matter of fact, dequeued in breadth-first traversal order: $v_1, v_2, \ldots, v_8$.

The following algorithm performs a breadth-first traversal of a graph $G = (V, E)$, using a queue of vertices Q to implement the previous procedure. The order in which vertices are visited during the breadth-first traversal is stored in the attribute *order*, where $v.order = -1$ if vertex v has not yet been visited, for all vertices $v \in V$.

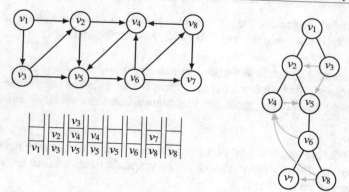

Fig. 5.8 Breadth-first traversal, breadth-first spanning forest, and evolution of the queue of vertices during execution of the breadth-first traversal procedure, upon the graph of Fig. 5.7. Vertices are numbered according to the order in which they are first visited during the traversal. The relative order of the vertices adjacent with a given vertex corresponds to the counter-clockwise ordering of the outgoing edges of the vertex in the drawing of the graph

procedure *breadth_first_traversal(G)*
　　for all vertices *v* of *G* **do**
　　　　v.order ← −1
　　num ← 1
　　let *Q* be an empty queue (of vertices)
　　for all vertices *v* of *G* **do**
　　　　if *v.order* = −1 **then**
　　　　　　enqueue *v* into *Q*
　　　　　　num ← *num* + 1
　　　　　　v.order ← *num*
　　　　　　while *Q* is not empty **do**
　　　　　　　　pop from *Q* the front vertex *v*
　　　　　　　　for all vertices *w* adjacent with vertex *v* in *G* **do**
　　　　　　　　　　if *w.order* = −1 **then**
　　　　　　　　　　　　enqueue *w* into *Q*
　　　　　　　　　　　　num ← *num* + 1
　　　　　　　　　　　　w.order ← *num*

Lemma 5.7 *The algorithm for breadth-first traversal of a graph runs in $O(n + m)$ time using $O(m)$ additional space, where n is the number of vertices and m is the number of edges in the graph.*

Proof Let $G = (V, E)$ be a graph with n vertices and m edges. Since each vertex of the graph is enqueued and dequeued at most once for each incoming edge, the loop is executed $n + m$ times and the algorithm runs in $O(n + m)$ time. Further, since each vertex is enqueued at most once for each incoming edge, the queue cannot ever contain more than m vertices and the algorithm uses $O(m)$ additional space. $\quad\square$

The breadth-first traversal procedure can also be applied to the those vertices which are reachable from an initial vertex, in order to perform a traversal of the breadth-first tree rooted at the initial vertex. The following function, which is based on the previous breadth-first graph traversal algorithm, not only computes the order in which vertices are visited during the traversal, but also the breadth-first spanning tree (W, S) of the subgraph of a graph $G = (V, E)$ induced by the set of vertices $W \subseteq V$ reachable from an initial vertex $v \in V$.

function *breadth_first_spanning_tree*(G, v)
    let W be an empty set (of vertices)
    let S be an empty set (of edges)
    for all vertices w of G **do**
        $w.order \leftarrow -1$
    $W \leftarrow W \cup \{v\}$
    $num \leftarrow 1$
    $v.order \leftarrow num$
    let Q be an empty queue (of edges)
    for all edges e coming out of vertex v in G **do**
        if v is not the target vertex of edge e in G **then**
            enqueue e into Q

    while Q is not empty **do**
        pop from Q the front edge e
        let w be the target vertex of edge e in G
        if $w \notin W$ **then**
            $num \leftarrow num + 1$
            $w.order \leftarrow num$
            $W \leftarrow W \cup \{w\}$
            $S \leftarrow S \cup \{e\}$
            for all edges e coming out of vertex w in G **do**
                let x be the target vertex of edge e in G
                if $x \notin W$ **then**
                    enqueue e into Q
    return (W, S)

Lemma 5.8 *The algorithm for computing the breadth-first spanning tree of a graph from an initial vertex runs in $O(n + m)$ time using $O(m)$ additional space, where n is the number of vertices and m is the number of edges in the graph.*

Proof Let $G = (V, E)$ be a graph with n vertices and m edges. Since the edges coming out of each vertex of the graph are enqueued and dequeued at most once, the loop is executed $n + m$ times and the algorithm runs in $O(n + m)$ time. Further, since each edge is enqueued at most once, the queue cannot ever contain more than m edges and the algorithm uses $O(m)$ additional space. □

5.3 Applications

Depth-first and breadth-first graph traversal are the basis of several fundamental graph algorithms which the reader may be familiar with, including algorithms for finding strong components in a graph and connected components in an undirected graph, finding cycles, topological sorting of an acyclic graph, and finding shortest paths and minimum spanning trees.

Let (W, S) be the breadth-first spanning tree of the subgraph of a graph $G = (V, E)$ reachable from a vertex $s \in V$. The shortest path $P \subseteq S$ from the source vertex s to a target vertex $t \in W$, can be reconstructed by tracing the edges of the spanning tree along the shortest path, from the edge coming into vertex t back to the edge going out of vertex s, as follows:

> **function** $acyclic\_shortest\_path(G, s, t, W, S)$
>   let D be an empty dictionary (of vertices to edges)
>   **for all** e in S **do**
>   let v be the target vertex of edge e in G
> $D[v] \leftarrow e$
>   let P be an empty list (of edges)
> $v \leftarrow t$
>   **while** $v \neq s$ **do**
> $e \leftarrow D[v]$
>   append e to P
>   let v be the source vertex of edge e in G
>   **return** P reversed

Lemma 5.9 *The algorithm for acyclic shortest path in a breadth-first spanning tree of a graph runs in $O(n)$ time using $O(n)$ additional space, where n is the number of vertices in the graph.*

Proof Let $G = (V, E)$ be a graph with n vertices, and let (W, S) be the breadth-first spanning tree of the subgraph of G induced by the set of vertices $W \subseteq V$ reachable from an initial vertex $s \in V$. Since each edge of the shortest path $P \subseteq S$ from the source vertex $s \in W \subseteq V$ to the target vertex $t \in W$ is considered only once, and the shortest path P cannot contain more than $n - 1$ edges, the loop is executed at most $n - 1$ times and the algorithm runs in $O(n)$ time using $O(n)$ additional space.   □

Another application of graph traversal is in maximum cardinality bipartite matching and maximum weight bipartite matching. Recall that a matching in a bipartite graph $G = (X \cup Y, E)$ with $E \subseteq X \times Y$ and $X \cap Y = \emptyset$ is a set $M \subseteq E$ of vertex-disjoint edges. Most known algorithms for maximum cardinality bipartite matching are based on the notion of augmenting path. An **augmenting path** in a bipartite graph $G = (X \cup Y, E)$ with respect to a matching $M \subseteq E$ is a path between unmatched vertices (those vertices which are not the source or the target of any edge in M) that alternates between edges in $E \setminus M$ and edges in M.

Example 5.10 The paths $[x_1, y_1], [x_2, y_2], [x_3, y_3], [x_4, y_5], [x_5, y_4]$ in the bipartite graph of Fig. 5.9 (left) are all augmenting paths with respect to the empty matching. The path $[x_6, y_2, x_2, y_6]$ in the bipartite graph of Fig. 5.9 (right) is an augmenting path with respect to the matching $\{(x_1, y_1), (x_2, y_2), (x_3, y_3), (x_4, y_5), (x_5, y_4)\}$.

Let $M \subseteq E$ be a matching in a bipartite graph $G = (X \cup Y, E)$, and let P be the set of edges along an augmenting path with respect to M. Notice that, since the edges of P are alternating in $E \setminus M$ and in M, they can be used to transform M into a larger matching $M' = M \oplus P$, the symmetric difference of M and P, which is also a matching and has one more edge than M, because P is a path between unmatched vertices.

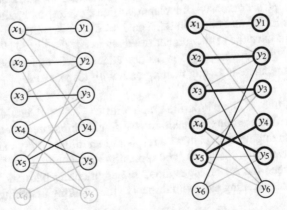

Fig. 5.9 Augmenting paths in the bipartite graph of Fig. 1.14, with respect to an empty matching (left) and with respect to matching $\{(x_1, y_1), (x_2, y_2), (x_3, y_3), (x_4, y_5), (x_5, y_4)\}$ (right). Matched vertices and edges in the matching are shown highlighted, while unmatched vertices and edges in an augmenting path but not in the matching are shown in black

The following theorem, first proved in [2], gives a simple algorithm for finding a maximum cardinality matching in a bipartite graph, by finding an augmenting path with respect to an initial matching (for example, the empty matching) and replacing the matching with the symmetric difference between the set of edges along the augmenting path and the matching, until no further augmenting path with respect to the resulting matching can be found.

Theorem 5.1 *A matching* $M \subseteq E$ *in a bipartite graph* $G = (X \cup Y, E)$ *with* $E \subseteq X \times Y$ *and* $X \cap Y = \emptyset$ *is a maximum cardinality bipartite matching if and only if there is no augmenting path with respect to* M *in* G.

Proof Let $M \subseteq E$ be a matching in a bipartite graph $G = (X \cup Y, E)$ with $E \subseteq X \times Y$ and $X \cap Y = \emptyset$, and let P be (the set of edges along) an augmenting path with respect to M. It has to be shown that M is not of maximum cardinality. But $M' = M \oplus P = (M \setminus P) \cup (P \setminus M) = (M \setminus (P \cap M)) \cup (P \setminus (P \cap M)) = (M \setminus (P \cap M)) \cup (P \cap (E \setminus M))$ is also a matching, and it has one more edge than M, because P begins and ends with edges in $E \setminus M$, alternating between edges in $E \setminus M$ and edges in M and thus, $P \cap (E \setminus M)$ has one more edge than $P \cap M$. Therefore, M is not a maximum cardinality matching.

Conversely, let $M \subseteq E$ be a matching in a bipartite graph $G = (X \cup Y, E)$ with $E \subseteq X \times Y$ and $X \cap Y = \emptyset$, and assume that M is not of maximum cardinality. It has to be shown that there is an augmenting path P with respect to M. Let M' be a maximum cardinality bipartite matching in G, and let $H = (X \cup Y, M \oplus M')$. This bipartite graph H has more edges in M' than in M, because $|M' \setminus M| = |M'| - |M' \cap M| > |M| - |M' \cap M| = |M \setminus M'|$. Since both M and M' are matchings in G, each vertex of H belongs to at most two edges of H, one edge in $M' \cap H$ and one edge in $M \cap H$. Then, H is composed of isolated vertices, cycles, and paths that alternate between edges in M' and edges in M. Since all cycles are alternating, they are of even length and have the same number of edges in M' than in M. Therefore, there is a path alternating between edges in M' and edges in M that has more edges in M' than in M, that is, there is an augmenting path with respect to M.   □

Let $G = (X \cup Y, E)$ be a bipartite graph with $E \subseteq X \times Y$ and $X \cap Y = \emptyset$. The following algorithm computes a maximum cardinality bipartite matching $M \subseteq E$ in G, by finding an augmenting path with respect to an initial empty matching, during a depth-first traversal of the graph, and replacing the matching with the symmetric difference between the set of edges along the augmenting path and the matching, until no further augmenting path with respect to the resulting matching can be found.

```
function maximum_cardinality_bipartite_matching(G)
    let M be an empty set (of edges)
    repeat
        P ← augmenting_path(G, M)
        M ← M ⊕ P
    until P is empty
    return M
```

Let $M \subseteq E$ be a (partial) matching in a bipartite graph $G = (X \cup Y, E)$ with $E \subseteq X \times Y$ and $X \cap Y = \emptyset$. An augmenting path in G, if it exists, can be obtained by graph traversal, for example, by a modified depth-first traversal of G starting with some unmatched vertex $x \in X$, and alternating between edges in $E \setminus M$ and edges in M, until reaching some unmatched vertex $y \in Y$.

In the following algorithm, the *depth_first_traversal* function gives the unmatched vertex $y \in Y$ of an augmenting path in the bipartite graph G with respect to matching M starting with an unmatched vertex $x \in X$, or *nil* if no such augmenting path can be found. Also, the predecessor of each vertex in the depth-first spanning tree of the subgraph of G induced by the set of vertices reachable from the initial vertex $x \in X$ is stored in the attribute $v.pred$, for all vertices $v \in X \cup Y$. The edges along the augmenting path P found are then recovered by following the sequence of predecessor vertices in the depth-first tree, from vertex $y \in Y$ back to vertex $x \in X$.

```
function augmenting_path(G, M)
    for all vertices v of G do
        v.visited ← false
        v.pred ← nil
    let P be an empty set (of edges)
    for all vertices x ∈ X do
        if x is not matched then
            y ← depth_first_traversal(G, M, x)
            if y ≠ nil then
                z ← y
                while z ≠ x do
                    add the edge of G between vertices z.pred and z to P
                    z ← z.pred
                break
    return P
```

The actual search for an augmenting path from a vertex in X to some unmatched vertex in Y is performed by the following recursive function, which is a modified depth-first traversal of the bipartite graph starting with a vertex $x \in X$ and alternating between edges in $E \setminus M$ and edges in M until an unmatched vertex $w \in Y$ is found.

```
function depth_first_traversal(G, M, x)
    x.visited ← true
    y ← nil
    for all edges e going out of vertex x in G do
        let w be the target vertex of edge e in G
        if not w.visited and e ∉ M then
            w.visited ← true
            w.pred ← x
            if w is not matched then
                return w
            else
                for all edges e' coming into vertex w in G do
                    let z be the source vertex of edge e' in G
                    if not z.visited and e' ∈ M then
                        z.visited ← true
                        z.pred ← w
                        y ← depth_first_traversal(G, M, z)
                        if y ≠ nil then
                            return y
    return y
```

Example 5.11 In the bipartite graph of Fig. 5.10, the empty matching yields the augmenting path $[x_1, y_1]$ (top left). Matching $\{(x_1, y_1)\}$ yields the augmenting path $[x_2, y_2]$ (top center). Matching $\{(x_1, y_1), (x_2, y_2)\}$ yields the augmenting path $[x_3, y_2, x_2, y_6]$ (top right). Matching $\{(x_1, y_1), (x_2, y_6), (x_3, y_2)\}$ yields the augmenting path $[x_4, y_3]$ (bottom left). Matching $\{(x_1, y_1), (x_2, y_6), (x_3, y_2), (x_4, y_3)\}$ yields the augmenting path $[x_5, y_3, x_4, y_5]$ (bottom center). Finally, matching $\{(x_1, y_1), (x_2, y_6), (x_3, y_2), (x_4, y_5), (x_5, y_3)\}$ yields the augmenting path $[x_6, y_2, x_3, y_3, x_5, y_4]$ (bottom right). The resulting maximum cardinality bipartite matching is $\{(x_1, y_1), (x_2, y_6), (x_3, y_3), (x_4, y_5), (x_5, y_4), (x_6, y_2)\}$.

Lemma 5.10 *The algorithm for maximum cardinality bipartite matching runs in* $O(nm)$ *time using* $O(m)$ *additional space, where n is the number of vertices and m is the number of edges in the bipartite graph, with* $n \leqslant m$.

Proof Let $G = (X \cup Y, E)$ be a bipartite graph with n vertices and m edges, and assume $n \leqslant m$. The augmenting path algorithm and the modified depth-first traversal algorithm both take $O(n + m)$ time using $O(n + m)$ additional space, and since the size of the matching is increased by one with each augmenting path, the augmenting path algorithm is executed at most n times. Thus, the algorithm takes $O(n(n + m)) = O(nm)$ time using $O(n + m) = O(m)$ additional space. $\square$

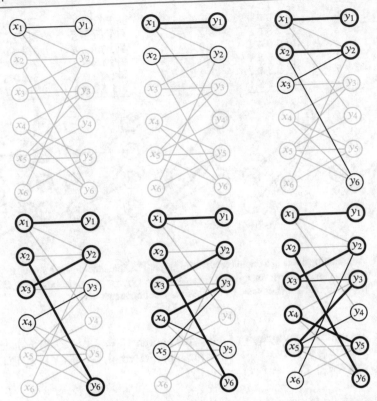

Fig. 5.10 Augmenting paths in the bipartite graph of Fig. 1.14, starting off with an empty matching. Matched vertices and edges in the matching are shown highlighted, while unmatched vertices and edges in an augmenting path but not in the matching are shown in black

Most known algorithms for maximum weight bipartite matching are also based on the notion of augmenting path, and they exploit a relationship between maximum weight matching in a bipartite graph and maximum cardinality matching in some subgraph of the bipartite graph. Such a subgraph of a weighted bipartite graph, known as *equality subgraph*, is based on the notion of *feasible labeling* for the vertices of the graph.

Let $G = (X \cup Y, E)$ be a bipartite graph with $E \subseteq X \times Y$ and $X \cap Y = \emptyset$, and let the weight of an edge $e \in E$ with source vertex $x \in X$ and target vertex $y \in Y$ be denoted by $w(x, y)$. A **feasible labeling** of G is a function $f : X \cup Y \to \mathbb{R}$ such that $f(x) + f(y) \geqslant w(x, y)$, for all edges $(x, y) \in E$. For example, the vertex labeling $f : X \cup Y \to \mathbb{R}$ given by $f(x) = \max\{w(x, y) \mid y \in Y, (x, y) \in E\}$ for all $x \in X$, and $f(y) = 0$ for all $y \in Y$, is a feasible labeling for the complete bipartite graph of Fig. 5.11 (left).

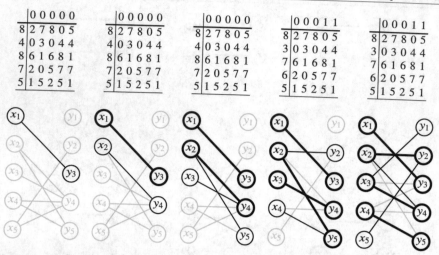

Fig. 5.11 Augmenting paths in a complete bipartite graph. Feasible labelings are shown to the left and on top of the edge weights, for x_1, x_2, x_3, x_4, x_5 and y_1, y_2, y_3, y_4, y_5, respectively. Only the edges of the equality subgraph induced by the feasible labeling are shown

Now, the **equality subgraph** of a bipartite graph $G = (X \cup Y, E)$ induced by a feasible labeling $f : X \cup Y \to \mathbb{R}$, is the bipartite graph $G' = (X \cup Y, E')$ with $E' = \{(x, y) \in E \mid f(x) + f(y) = w(x, y)\}$.

Example 5.12 In the complete bipartite graph of Fig. 5.11, where only the edges of the equality subgraph induced by the feasible labeling on the top are shown, the empty matching yields the augmenting path $[x_1, y_3]$ (left). Matching $\{(x_1, y_3)\}$ yields the augmenting path $[x_2, y_4]$. Matching $\{(x_1, y_3), (x_2, y_4)\}$ yields the augmenting path $[x_3, y_4, x_2, y_5]$. Matching $\{(x_1, y_3), (x_2, y_5), (x_3, y_4)\}$ yields the augmenting path $[x_4, y_5, x_2, y_2]$. Finally, matching $\{(x_1, y_3), (x_2, y_2), (x_3, y_4), (x_4, y_5)\}$ yields the augmenting path $[x_5, y_2, x_2, y_4, x_3, y_1]$ (right). The resulting maximum cardinality bipartite matching is $\{(x_1, y_3), (x_2, y_4), (x_3, y_1), (x_4, y_5), (x_5, y_2)\}$, with total edge weight 30, which is a maximum weight matching in the original complete bipartite graph.

The following theorem, first proved in [21,29], gives a simple algorithm for finding a maximum weight matching in a bipartite graph, by finding a maximum cardinality bipartite matching in the equality subgraph induced by an initial feasible labeling (for example, the largest weight of the outgoing edges for all $x \in X$, and zero for all $y \in Y$) and improving the feasible labeling until the maximum cardinality bipartite matching is a perfect matching.

Theorem 5.2 *Let $G' = (X \cup Y, E')$ be the equality subgraph of a bipartite graph $G = (X \cup Y, E)$ with $E \subseteq X \times Y$ and $X \cap Y = \emptyset$ induced by a feasible labeling $f : X \cup Y \rightarrow \mathbb{R}$. Then, a perfect matching $M \subseteq E'$ in G', if it exists, is also a maximum weight bipartite matching in G.*

Proof Let $G = (X \cup Y, E)$ be a bipartite graph with $E \subseteq X \times Y$ and $X \cap Y = \emptyset$, and let $M \subseteq E'$ be a perfect matching in the equality subgraph $G' = (X \cup Y, E')$ induced by a feasible labeling $f : X \cup Y \rightarrow \mathbb{R}$. Assume, without loss of generality, that $|X| = |Y|$, that $f(v) \geqslant 0$ for all vertices $v \in X \cup Y$, and that $w(x, y) \geqslant 0$ for all edges $(x, y) \in E$. Suppose there is a matching $M' \subseteq E$ in G with $\sum_{(x,y)\in M'} w(x, y) > \sum_{(x,y)\in M} w(x, y)$. But $\sum_{(x,y)\in M} w(x, y) = \sum_{(x,y)\in M}(f(x) + f(y)) = \sum_{x\in X} f(x) + \sum_{y\in Y} f(y)$, because M is a matching in the equality subgraph G' and because M is a perfect matching. Then, $\sum_{(x,y)\in M'}(f(x) + f(y)) \geqslant \sum_{(x,y)\in M'} w(x, y) > \sum_{x\in X} f(x) + \sum_{y\in Y} f(y)$, because f is a feasible labeling of G, contradicting the assumption that M' is a matching in G. □

Let $G = (X \cup Y, E)$ be a complete bipartite graph with $E \subseteq X \times Y$ and $X \cap Y = \emptyset$, and let the weight of an edge $e \in E$ with source vertex $x \in X$ and target vertex $y \in Y$ be denoted by $w(x, y)$. The following algorithm computes a maximum weight bipartite matching $M \subseteq E$ in G, by finding a maximum cardinality bipartite matching in the equality subgraph $G' = (X \cup Y, E')$ induced by the initial feasible labeling $f : X \cup Y \rightarrow \mathbb{R}$ given by $f(x) = \max\{w(x, y) \mid y \in Y\}$ for all $x \in X$, and $f(y) = 0$ for all $y \in Y$, and improving the feasible labeling until $|M| = \min(|X|, |Y|)$, that is, until the maximum cardinality bipartite matching in G' is a perfect matching in G.

The sets S and T contain, respectively, the vertices of X and Y in the current alternating path. At each step of the algorithm, either an augmenting path is found and the current matching M is augmented, or the current feasible labeling f is improved by computing the value $\epsilon = \min\{f(x) + f(y) - w(x, y) \mid x \in S, y \in Y \setminus T\}$, which is subtracted from $f(x)$ for all $x \in S$, and added to $f(y)$ for all $y \in T$. In this way, f remains a feasible labeling for the edges in the alternating path, while some new edges in $S \times (Y \setminus T)$, $(X \setminus S) \times T$, and $(X \setminus S) \times (Y \setminus T)$ are added to the equality subgraph.

Instead of computing such a value $\epsilon = \min\{f(x) + f(y) - w(x, y) \mid x \in S, y \in Y \setminus T\}$ from scratch, though, additional variables $\bar{y} = \min\{f(x) + f(y) - w(x, y) \mid x \in S\}$ are maintained for all $y \in Y \setminus T$, and they are updated whenever any of S, T, f changes. These additional variables, which are stored in the attribute $y.slack$, for all vertices $y \in Y$, allow for a more efficient implementation of the algorithm.

```
function maximum_weight_bipartite_matching(G)
    for all x ∈ X do
        x.f ← max{w(x, y) | y ∈ Y}
    for all y ∈ Y do
        y.f ← 0
    let M be an empty set (of edges)
    while |M| ≠ min(|X|, |Y|) do
        let x be an unmatched vertex in X
        let S, T be empty sets (of vertices)
        S ← S ∪ {x}
        for all y' ∈ Y \ T do
            y'.slack ← x.f + y'.f − w(x, y')
        while true do
            slack ← min{y'.slack | y' ∈ Y \ T}
            if slack ≠ 0 then
                ε ← slack
                for all x' ∈ S do
                    x'.f ← x'.f − ε
                for all y' ∈ T do
                    y'.f ← y'.f + ε
                for all y' ∈ Y \ T do
                    y'.slack ← y'.slack − ε
                    slack ← min(slack, y'.slack)
            if slack = 0 then
                for all y ∈ Y \ T do
                    if y.slack = slack then
                        break
                if y is not matched then
                    E' ← {(x', y') ∈ E | x'.f + y'.f = w(x, y)}
                    P ← augmenting_path((X ∪ Y, E'), M, x, y)
                    M ← M ⊕ P
                    break
                else
                    let z be the vertex in X such that (z, y) ∈ M
                    S ← S ∪ {z}
                    T ← T ∪ {y}
                    for all y' ∈ Y \ T do
                        y'.slack ← min(y'.slack, z.f + y'.f − w(z, y'))
    return M
```

Let $M \subseteq E$ be a (partial) matching in a weighted complete bipartite graph $G = (X \cup Y, E)$. An actual augmenting path between an unmatched vertex $x \in X$ and an unmatched vertex $y \in Y$ can also be obtained by graph traversal, for example, by a modified depth-first traversal of G starting with vertex x, and alternating between edges in $E \setminus M$ and edges in M, until reaching vertex y.

In the following algorithm, the predecessor of each vertex in the depth-first spanning tree of the subgraph of G induced by the set of vertices visited during the search for the final vertex $y \in Y$ from the initial vertex $x \in X$, is stored in the attribute $v.pred$, for all vertices $v \in X \cup Y$, during execution of the *depth_first_traversal* procedure. The edges along the augmenting path P found are then recovered by following the sequence of predecessor vertices in the depth-first tree, from vertex $y \in Y$ back to vertex $x \in X$.

```
function augmenting_path(G, M, x, y)
    for all vertices v of G do
        v.visited ← false
        v.pred ← nil
    depth_first_traversal(G, M, x, y)
    let P be an empty set (of edges)
    z ← y
    while z ≠ x do
        add the edge of G between vertices z.pred and z to P
        z ← z.pred
    return P
```

The actual search for an augmenting path from a vertex in X to a vertex in Y is performed by the following recursive procedure, which is a modified depth-first traversal of the bipartite graph starting with a given vertex $x \in X$, alternating between edges in $E \setminus M$ and edges in M, and finishing with another given vertex $y \in Y$.

```
procedure depth_first_traversal(G, M, x, y)
    x.visited ← true
    for all edges e going out of vertex x in G do
        let w be the target vertex of edge e in G
        if not w.visited and e ∉ M then
            w.visited ← true
            w.pred ← x
            if w = y then
                return
            else
                for all edges e' coming into vertex w in G do
                    let z be the source vertex of edge e' in G
                    if not z.visited and e' ∈ M then
                        z.visited ← true
                        z.pred ← w
                        depth_first_traversal(G, M, z, y)
```

Lemma 5.11 *The algorithm for maximum weight bipartite matching runs in $O(n^3)$ time using $O(n^2)$ additional space, where n is the number of vertices in the complete bipartite graph.*

Proof Let $G = (X \cup Y, E)$ be a complete bipartite graph with n vertices. The augmenting path algorithm takes $O(n^2)$ time using $O(n^2)$ additional space, and the outer loop is executed at most n times, since the size of the matching is increased by one with each augmenting path. When no augmenting path is found, computing the value ϵ from the additional variables, improving the labeling, and updating the additional variables all take $O(n)$ time, and this is done at most n times for each outer loop because, when $N(S) = T$, the set of equality edges E' is changed to enforce $N(S) \neq T$ and then, at most n matched vertices $z \in X$ can be inserted into S for each outer loop, contributing thus $O(n^2)$ to the time taken by the outer loop. Thus, the algorithm takes $O(n^3)$ time using $O(n^2)$ additional space. $\qquad\square$

A quite different application of graph traversal will be addressed now, related to the isomorphism problems on trees discussed in Chap. 4, as an appetizer to Chap. 7. Recall that two ordered trees are isomorphic if there is a bijective correspondence between their node sets which preserves and reflects the structure of the ordered trees. In the same sense, two ordered graphs $G_1 = (V_1, E_1)$ and $G_2 = (V_2, E_2)$ are isomorphic if there is a bijection $M \subseteq V_1 \times V_2$ which preserves and reflects the ordered structure of the graphs—that is, such that the ordered sequence of vertices adjacent with vertex $v \in V_1$ coincides (up to a cyclic rotation) with the ordered sequence of vertices adjacent with vertex $M[v] \in V_2$, for all vertices $v \in V_1$. In such a case, M is an **ordered graph isomorphism** of G_1 to G_2.

Example 5.13 The ordered graphs of Fig. 5.12 are isomorphic, and $M = \{(v_1, w_1), (v_2, w_2), (v_3, w_3), (v_4, w_4), (v_5, w_5), (v_6, w_6), (v_7, w_7), (v_8, w_8)\}$ is an ordered graph isomorphism of G_1 to G_2. As a matter of fact, the ordered sequence of vertices $[v_1, v_8, v_7, v_3]$ adjacent with vertex $v_2 \in V_1$ coincides, up to a cyclic rota-

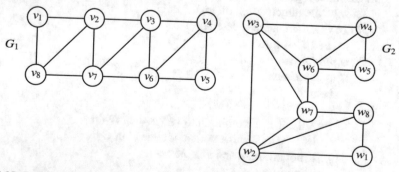

Fig. 5.12 Isomorphic ordered graphs. The relative order of the vertices adjacent with a given vertex reflects the counter-clockwise ordering of the outgoing edges of the vertex in the drawing of each of the graphs

tion, with the ordered sequence $[w_7, w_3, w_1, w_8]$ of vertices adjacent with vertex $M[v_2] = w_2 \in V_2$, and the same holds for all of the vertices in the graphs.

A straightforward procedure for testing isomorphism of two connected, ordered undirected graphs consists of performing a leftmost depth-traversal of the bidirected graphs underlying both undirected graphs, and then testing whether the vertex mapping induced by the traversal is an isomorphism between the ordered undirected graphs. Since a leftmost depth-first traversal of a bidirected graph is only unique for a given initial edge, though, the leftmost depth-first traversal starting in turn with each of the edges of one graph, has to be tested against the leftmost depth-first traversal of the other graph starting with some (fixed) edge.

Let $G_1 = (V_1, E_1)$ and $G_2 = (V_2, E_2)$ be ordered bidirected graphs with n vertices, let $order_1 : V_1 \to \{1, \dots, n\}$ be a leftmost depth-first traversal of G_1, and let $order_2 : V_2 \to \{1, \dots, n\}$ be a leftmost depth-first traversal of G_2. The vertex mapping $M \subseteq V_1 \times V_2$ induced by $order_1$ and $order_2$ is given by $M[v] = w$ if and only if $order_1[v] = order_2[w]$, for all vertices $v \in V_1$ and $w \in V_2$. Since both $order_1$ and $order_2$ are bijections, M is also a bijection.

The following algorithm implements the previous procedure for testing isomorphism of two ordered undirected graphs, based on a series of leftmost depth-first traversals of the underlying bidirected graphs. The bijection $M \subseteq V_1 \times V_2$ computed by the function upon two isomorphic ordered undirected graphs $G_1 = (V_1, E_1)$ and $G_2 = (V_2, E_2)$ is stored in the attribute $v.M$, for all vertices v of G_1.

```
function ordered_graph_isomorphism(G₁, G₂)
    if G₁ and G₂ do not have the same number of vertices and edges then
        return false
    if G₁ has no edges then
        for all vertices v of G₁ and w of G₂ do
            v.M ← w
        return true
    let e₁ be the first edge of G₁
    L₁ ← leftmost_depth_first_traversal(G₁, e₁)
    obtain_order_in_which_vertices_were_visited(G₁, L₁)
    for all edges e₂ of G₂ do
        L₂ = leftmost_depth_first_traversal(G₂, e₂)
        obtain_order_in_which_vertices_were_visited(G₂, L₂)
        build_ordered_graph_isomorphism_mapping(G₁, G₂)
        if test_mapping_for_ordered_graph_isomorphism(G₁, G₂) then
            return true
    return false
```

An actual ordered graph isomorphism mapping $M \subseteq V_1 \times V_2$ of graph $G_1 = (V_1, E_1)$ to graph $G_2 = (V_2, E_2)$ can be constructed by mapping the ith-visited vertex $v \in V_1$ during the leftmost depth-first traversal of G_1 to the ith-visited vertex $w \in V_2$ during the leftmost depth-first traversal of G_2, for $1 \leqslant i \leqslant n$.

procedure *build_ordered_graph_isomorphism_mapping*(G_1, G_2)
    let n be the number of vertices of G_1
    let *disorder*$_1[1..n]$ be a new array (of vertices)
    for all vertices v of G_1 **do**
        disorder$_1[v.order] \leftarrow v$
    let *disorder*$_2[1..n]$ be a new array (of vertices)
    for all vertices v of G_2 **do**
        disorder$_2[v.order] \leftarrow v$
    for all i from 1 to n **do**
        disorder$_1[i].M \leftarrow$ *disorder*$_2[i]$

Testing a candidate vertex mapping $M \subseteq V_1 \times V_2$ for ordered graph isomorphism of $G_1 = (V_1, E_1)$ to $G_2 = (V_2, E_2)$ accounts to testing whether the ordered sequence L_1 of vertices adjacent with vertex $v \in V_1$ coincides, up to a cyclic rotation, with the ordered sequence L_2 of vertices adjacent with vertex $M[v] \in V_2$, for all vertices $v \in V_1$.

function *test_mapping_for_ordered_graph_isomorphism*(G_1, G_2)
    for all vertices v of G_1 **do**
        let L_1 be an empty list (of vertices)
        for all edges e coming out of vertex v in G_1 **do**
            append $G_1.target(e)$ to L_1
        let L_2 be an empty list (of vertices)
        for all edges e coming out of vertex $v.M$ in G_2 **do**
            append $G_2.target(e)$ to L_2
        if L_1 and L_2 do not have the same size **then**
            return false
        if L_1 and L_2 are not empty **then**
            let v_1 be the vertex at the front of L_1
            let v_2 be the vertex at the front of L_2
            let z be the vertex at the back of L_2
            while $v_2 \neq z$ and $v_1.M \neq v_2$ **do**
                move to the back of L_2 the vertex at the front of L_2
                let v_2 be the vertex at the front of L_2
        for all vertices v_1 in L_1 and v_2 in L_2 **do**
        if $v_1.M \neq v_2$ **then**
            return false
    return true

The order in which the vertices of G_1 and G_2 are visited is just the order in which they are first reached during the leftmost depth-first traversal of G_1 and G_2, respectively.

> **procedure** *obtain_order_in_which_vertices_were_visited*(G, L)
>   **for all** vertices v of G **do**
> $v.order \leftarrow -1$
>   let e be the first edge in L
>   let v be the source vertex of edge e in G
> $num \leftarrow 1$
> $v.order \leftarrow num$
>   **for all** edges e in L **do**
>   let w be the target vertex of edge e in G
>   **if** $w.order = -1$ **then**
> $num \leftarrow num + 1$
> $w.order \leftarrow num$

Lemma 5.12 *The algorithm for ordered graph isomorphism runs in $O((n + m)m)$ time using $O(n + m)$ additional space, where n is the number of vertices and m is the number of edges in the ordered graphs.*

Proof Let $G_1 = (V_1, E_1)$ and $G_2 = (V_2, E_2)$ be ordered graphs with n vertices and m edges. Since one leftmost depth-first traversal of G_1 and m leftmost depth-first traversals of G_2 are performed, and each leftmost depth-first traversal, as well as the test of a mapping for ordered graph isomorphism, takes $O(n+m)$ time, the algorithm runs in $O((n+m)m)$ time. Further, since only the leftmost depth-first traversal of G_1 and one leftmost depth-first traversal of G_2 are kept at the same time, the algorithm uses $O(n + m)$ additional space. □

Summary

The most common methods of exploring a graph were addressed in this chapter. Simple algorithms are given in detail for two different methods of graph traversal: depth-first traversal and breadth-first traversal. While the former generalizes the depth-first prefix leftmost (preorder) traversal of a rooted tree, the latter is a generalization of breadth-first leftmost (top-down) tree traversal. A particular case of the depth-first traversal of an undirected graph, called leftmost depth-first traversal, is also discussed and detailed algorithms tailored to this particular case are also given. The application of leftmost depth-first graph traversal to the isomorphism of ordered graphs is also discussed in detail.

Bibliographic Notes

Depth-first graph traversal was first described in [37], and is the basis of many fundamental graph algorithms. Breadth-first graph traversal was first described in [23, 28]. See also [7,38]. The recognition of depth-first and breadth-first spanning trees was addressed, respectively, in [20,31,33,34] and in [25].

The algorithm for leftmost depth-first traversal of a bidirected graph is based on the maze traversal algorithm found by Trémaux and recalled in [41]. An algorithm for the leftmost depth-first traversal of an undirected graph is included in LEDA [27, Sect. 8.7.2]. The algorithm for isomorphism of ordered graphs is based on [17]. See also [16,35,40].

The relationship between maximum cardinality bipartite matchings and augmenting paths was first proved in [2]. See also [3, Chap. 7], [24, Chap. 1], and [32]. The maximum cardinality bipartite matching algorithm was first proposed in [15] and further studied in [11]. See [1] for a faster implementation, running in $O(n\sqrt{n}\sqrt{m}/\log n)$ time on a bipartite graph with n vertices and m edges. Maximum cardinality bipartite matching can also be stated as a network flow problem, see [14]. See also [5, Chap. 3], [8, Problem 26-6], and [30, Sect. 10.2].

The method for solving the maximum weight bipartite matching problem based on augmenting paths is known as the *Hungarian method for the assignment problem*, since it is based on the work of two Hungarian mathematicians [13,19]. See also [26]. The method itself was first published in [21,29] in terms of matrices, and in [22] in terms of bipartite graphs. See [27, Sect. 7.8] for a faster implementation, running in $O(n(m + n\log n))$ time on a bipartite graph with n vertices and m edges. The presentation of the algorithm is based on [9]. See also [4,6,10,12,18,36,39,42], [5, Chap. 4], and [30, Sect. 11.2].

Problems

5.1 Give the depth-first tree rooted in turn at each of the vertices of the graph in Fig. 5.13. Explain why the depth-first tree rooted at vertex v_2 differs from the tree rooted at the same vertex v_2 in the depth-first forest of the graph.

5.2 In the depth-first traversal of a graph $G = (V, E)$ with n vertices, the set of edges E can be partitioned into four classes, according to $order : V \to \{1, \ldots, n\}$, the order in which their source and target vertices are first visited and also according to $comp : V \to \{1, \ldots, n\}$, the order in which processing of their source and target vertices is completed or finished, where a vertex is completed as soon as all its adjacent vertices have been visited. An edge (v, w) is called a *tree edge* if it belongs to the depth-first forest of the graph, and it is called a *forward edge* if it is parallel to a path of tree edges. In both cases, it holds that $order[v] < order[w]$ and $comp[v] > comp[w]$. An edge (v, w) is called a *backward edge* if it is anti-parallel to a path of tree edges and in this case, $order[v] \geqslant order[w]$ and $comp[v] \leqslant comp[w]$. Finally, an edge (v, w) is called a *cross edge* if it is neither a tree, nor a forward, nor

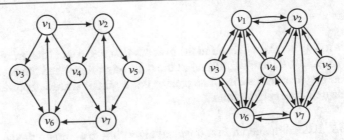

Fig. 5.13 Graph (left) and bidirected graph (right) for problems 5.1–5.5. The relative order of the vertices adjacent with a given vertex corresponds to the counter-clockwise ordering of the outgoing edges of the vertex in the drawing of the graph

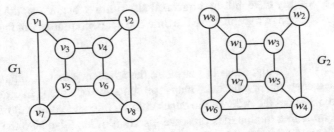

Fig. 5.14 Ordered undirected graphs for problem 5.6. The relative order of the vertices adjacent with a given vertex corresponds to the counter-clockwise ordering of the outgoing edges of the vertex in each of the graphs

a backward edge, and in such a case, $order[v] > order[w]$ and $comp[v] > comp[w]$. Give the classification of the edges in the depth-first traversal of the graph in Fig. 5.13.

5.3 Give the leftmost depth-first traversal of the bidirected graph in Fig. 5.13, starting in turn at each of the edges of the graph. Explain whether there is any relationship holding among all the leftmost depth-first traversals of a bidirected graph.

5.4 Give the breadth-first tree rooted in turn at each of the vertices of the graph in Fig. 5.13. Explain why the breadth-first tree rooted at vertex v_2 differs from the tree rooted at the same vertex v_2 in the breadth-first forest of the graph.

5.5 The same classification of the edges of a graph explained in Problem 5.2 applies to breadth-first traversal, but there are no forward edges in the breadth-first traversal of a graph. Give the classification of the edges in the breadth-first traversal of the graph in Fig. 5.13.

5.6 Determine whether the ordered undirected graphs in Fig. 5.14 are isomorphic and, in such a case, give an ordered graph isomorphism of G_1 to G_2.

Exercises

Exercise 5.1 The depth-first tree and the breadth-first tree rooted at vertex v_2 of the graph in Fig. 5.13 are identical, although the depth-first forest and the breadth-first forest of the graph are different. Characterize those graphs whose depth-first forest and breadth-first forest are identical.

Exercise 5.2 Extend the depth-first traversal algorithm to compute the completion order of all the vertices in a graph, as well as to distinguish among tree, forward, backward, and cross edges.

Exercise 5.3 Modify the depth-first traversal algorithm to operate on the adjacency matrix representation of a graph. Give the time and space complexity of the modified algorithm.

Exercise 5.4 Give an algorithm to rearrange the adjacency matrix, adjacency list, extended adjacency list, or adjacency map representation of a graph in order to meet Assumption 5.1 about the order of the edges coming into and going out of a vertex. You may assume that dictionaries preserve insertion order. Give also the time and space complexity of the algorithm for each graph representation.

Exercise 5.5 Give an algorithm to compute the bidirected graph underlying a graph, in such a way that the ordered structure of the given graph is preserved. For instance, for the graph in Fig. 5.13, the cyclic sequence of vertices adjacent with vertex v_1 is $[v_2, v_3, v_4]$ in the graph but $[v_2, v_3, v_6, v_4]$ in the underlying bidirected graph.

Exercise 5.6 Modify the breadth-first traversal algorithm to find a shortest path in a graph from a given vertex to every other vertex reachable in the graph from the given vertex.

Exercise 5.7 Extend the breadth-first traversal algorithm to compute the completion order of all the vertices in a graph, as well as to distinguish among tree, backward, and cross edges.

Exercise 5.8 Modify the breadth-first traversal algorithm to operate on the adjacency matrix representation of a graph. Give the time and space complexity of the modified algorithm.

Exercise 5.9 Modify the ordered graph isomorphism algorithm given in Sect. 5.3 to compare isomorphism codes, where the isomorphism code of a connected, ordered undirected graph $G = (V, E)$ with m edges is a sequence of $2m$ vertices in the order in which they are visited during a leftmost depth-first traversal of the underlying bidirected graph, and equality of isomorphism codes holds up to a cyclic rotation of the sequences. Give also the time and space complexity of the modified algorithm.

Exercise 5.10 The ordered graph isomorphism algorithm given in Sect. 5.3 is based on performing a leftmost depth-first traversal of one of the graphs and a series of leftmost depth-first traversals of the other graph, testing in each case, the induced vertex mapping for ordered graph isomorphism. Non-isomorphism of two connected, ordered undirected graphs could be detected earlier, though, by performing a series of simultaneous leftmost depth-first traversals of the two graphs, in the same way as done in Chap. 4 for solving isomorphism problems on ordered trees. Give an algorithm for ordered graph isomorphism based on simultaneous leftmost depth-first traversals of the graphs. Give also the time and space complexity of the algorithm.

References

1. Alt H, Blum N, Mehlhorn K, Paul M (1991) Computing a maximum cardinality matching in a bipartite graph in time $O(n^{1.5}sqrtm/logn)$. Inf Process Lett 37(4):237–240
2. Berge C (1957) Two theorems in graph theory. Proc Natl Acad Sci U S A 43(9):842–844
3. Berge C (1985) Graphs and hypergraphs, 2nd edn. North-Holland, Amsterdam
4. Bourgeois F, Lassalle JC (1971) An extension of the Munkres algorithm for the assignment problem to rectangular matrices. Commun ACM 14(12):802–804
5. Burkard R, Dell'Amico M, Martello S (2009) Assignment problems. Society for Industrial and Applied Mathematics, Philadelphia PA
6. Carpaneto G, Martello S, Toth P (1988) Algorithms and codes for the assignment problem. Ann Oper Res 13(1):191–223
7. Chakraborty S, Mukherjee A, Satti SR (2019) Space efficient algorithms for breadth-depth search. In: Gąsieniec LA, Jansson J, Levcopoulos C (eds) Proceedings 22nd international symposium fundamentals of computation theory, lecture notes in computer science, vol 11651. Springer, Berlin, pp 201–212
8. Cormen TH, Leiserson CE, Rivest RL, Stein C (2009) Introduction to algorithms, 3rd edn. MIT Press, Cambridge MA
9. Cui H, Zhang J, Cui C, Chen Q (2016) Solving large-scale assignment problems by Kuhn-Munkres algorithm. In: Xu M, Zhang K (eds) Proceedings, vol 73. 2nd international conference on advances in mechanical engineering and industrial informatics, Advances in engineering research. Atlantis Press, Paris, France, pp 822–827
10. Dell'Amico M, Toth P (2000) Algorithms and codes for dense assignment problems: the state of the art. Discret Appl Math 100(1–2):17–48
11. Duff IS, Wiberg T (1988) Remarks on implementation of $O(n^{1/2}\tau)$ assignment algorithms. ACM Trans Math Softw 14(3):267–287
12. Edmonds J, Karp RM (1972) Theoretical improvements in algorithmic efficiency for network flow problems. J ACM 19(2):248–264
13. Egerváry J (1931) Matrixok kombinatorius tulajdonságairól. Matematikai és Fizikai Lapok 38(1):16–28
14. Ford LR, Fulkerson DR (1962) Flows in networks. Princeton University Press, Princeton NJ
15. Hopcroft JE, Karp RM (1973) An $n^{5/2}$ algorithm for maximum matchings in bipartite graphs. SIAM J Comput 2(4):225–231
16. Hopcroft JE, Tarjan RE (1973) An $O(n \log n)$ algorithm for isomorphism of triconnected planar graphs. J Comput Syst Sci 7(3):323–331
17. Jiang XY, Bunke H (1999) Optimal quadratic-time isomorphism of ordered graphs. Pattern Recognit 32(7):1273–1283

18. Karp RM (1980) An algorithm to solve the $m \times n$ assignment problem in expected time $O(mn \log n)$. Networks 10(2):143–152

19. Kőnig D (1931) Graphok és matrixok. Matematikai és Fizikai Lapok 38(1):116–119

20. Korach E, Ostfeld Z (1989) DFS tree construction: algorithms and characterization. In: van Leeuwen J (ed) Proceedings 14th international workshop graph-theoretic concepts in computer science, vol 344. Lecture notes in computer science. Springer, Berlin Heidelberg, pp 87–106

21. Kuhn HW (1955) The Hungarian method for the assignment problem. Naval Res Logistics Q 2(1–2):83–97

22. Kuhn HW (1956) Variants of the Hungarian method for assignment problems. Naval Res Logistics Q 3(4):253–258

23. Lee CY (1961) An algorithm for path connection and its applications. IRE Trans Electron Comput 10(3):346–365

24. Lovász L, Plummer MD (1986) Matching theory, North-Holland Mathematics Studies, vol 121. North-Holland, Leipzig, Germany

25. Manber U (1990) Recognizing breadth-first search trees in linear time. Inf Process Lett 34(4):167–171

26. Martello S (2010) Jenő Egerváry: From the origins of the Hungarian algorithm to satellite communication. Central Euro J Oper Res 18(1):47–58

27. Mehlhorn K, Näher S (1999) The LEDA platform of combinatorial and geometric computing. Cambridge University Press, Cambridge, England

28. Moore EF (1959) The shortest path through a maze. In: Proceedings of international symposium theory of switching. Harvard University Press, Cambridge MA, pp 285–292

29. Munkres J (1957) Algorithms for the assignment and transportation problems. J Soc Ind Appl Math 5(1):32–38

30. Papadimitriou CH, Steiglitz K (1998) Combinatorial optimization: algorithms and complexity. Dover, Mineola NY

31. Peng CH, Wang BF, Wang JS (2000) Recognizing unordered depth-search trees of an undirected graph in parallel. IEEE Trans Parallel Distrib Syst 11(6):559–570

32. Petersen J (1891) Die Theorie der regulären Graphs. Acta Math 15(1):193–220

33. Reif JH (1985) Depth-first search is inherently sequential. Inf Process Lett 20(5):229–234

34. Schevon CA, Vitter JS (1988) A parallel algorithm for recognizing unordered depth-first search. Inf Process Lett 28(2):105–110

35. Schlieder C (1998) Diagrammatic transformation processes on two-dimensional relational maps. J Vis Lang Comput 9(1):45–59

36. Silver R (1960) An algorithm for the assignment problem. Commun ACM 3(11):605–606

37. Tarjan RE (1972) Depth-first search and linear graph algorithms. SIAM J Comput 1(2):146–160

38. Tarjan RE (1983) Space-efficient implementations of graph search methods. ACM Trans Math Softw 9(3):326–339

39. Tomizawa N (1971) On some techniques useful for solution of transportation network problems. Networks 1(2):173–194

40. Valiente G (2001) A general method for graph isomorphism. In: Freivalds R (ed) Proceedings 13th international symposium fundamentals of computation theory, vol 2138. Lecture notes in computer science. Springer, Berlin Heidelberg, pp 428–431

41. Weinberg L (1966) A simple and efficient algorithm for determining isomorphism of planar triply connected graphs. IEEE Trans Circuit Theory 13(2):142–148

42. Wong JK (1979) A new implementation of an algorithm for the optimal assignment problem: An improved version of Munkres' algorithm. BIT Numer Math 19(3):418–424

Clique, Independent Set, and Vertex Cover

6.1 Cliques, Maximal Cliques, and Maximum Cliques

A clique of an undirected graph is a complete subgraph of the graph. A maximal clique of a graph is a clique of the graph that is not properly included in any other clique of the graph, and a maximum clique of a graph is a (maximal) clique of the graph with the largest number of vertices.

Recall from Sect. 1.1 that an undirected graph is a complete graph if every pair of distinct vertices in the graph is joined by an edge.

Definition 6.1 A **clique** of an undirected graph $G = (V, E)$ is a set of vertices $C \subseteq V$ such that the subgraph of G induced by C is complete, that is, such that $\{v, w\} \in E$ for all distinct vertices $v, w \in C$. A clique C of an undirected graph $G = (V, E)$ is **maximal** if there is no clique D of G such that $C \subseteq D$ and $C \neq D$, and it is **maximum** if there is no clique of G with more vertices than C. The **clique number** of G, denoted by $\omega(G)$, is the cardinality of a maximum clique of G.

It follows from the previous definition that maximum cliques are also maximal.

Lemma 6.1 *Every maximum clique $C \subseteq V$ of an undirected graph $G = (V, E)$ is maximal.*

Proof Let $C \subseteq V$ be a maximum clique of an undirected graph $G = (V, E)$, and suppose C is not maximal. There is, by Definition 6.1, a clique $D \subseteq V$ such that $C \subseteq D$ and $C \neq D$. Then, $D \setminus C \neq \emptyset$, that is, clique D has more vertices than clique C, contradicting the hypothesis that C is a maximum clique. Therefore, C is a maximal clique. $\qquad \square$

© The Author(s), under exclusive license to Springer Nature Switzerland AG 2021
G. Valiente, *Algorithms on Trees and Graphs*, Texts in Computer Science,
https://doi.org/10.1007/978-3-030-81885-2_6

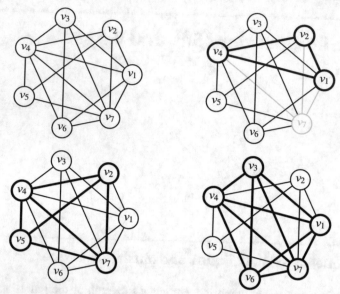

Fig. 6.1 Maximal and maximum cliques of an undirected graph. The clique $\{v_1, v_2, v_4\}$ is not maximal, because it is included in clique $\{v_1, v_2, v_4, v_7\}$. The cliques $\{v_1, v_2, v_4, v_7\}$, $\{v_1, v_3, v_4, v_6, v_7\}$, and $\{v_2, v_4, v_5, v_7\}$ are all maximal. Clique $\{v_1, v_3, v_4, v_6, v_7\}$ is also a maximum clique

Example 6.1 The undirected graph shown in Fig. 6.1 has three maximal cliques: $\{v_1, v_2, v_4, v_7\}$, $\{v_2, v_4, v_5, v_7\}$, and $\{v_1, v_3, v_4, v_6, v_7\}$. The latter is also the only maximum clique of the graph.

Notice that being a clique of an undirected graph is a hereditary graph property, that is, a property shared by all those subgraphs of the given graph which are induced by some subset of the given clique.

Lemma 6.2 *Let $C \subseteq V$ be a clique of an undirected graph $G = (V, E)$, and let $B \subseteq C$. Then, B is also a clique of G.*

Proof Let $C \subseteq V$ be a clique of an undirected graph $G = (V, E)$, and let $B \subseteq C$. Then, $\{v, w\} \in E$ for all distinct vertices $v, w \in C$ by Definition 6.1 and, in particular, $\{v, w\} \in E$ for all distinct vertices $v, w \in B$. Therefore, $B \subseteq V$ is also a clique. $\square$

Maximal Cliques of a Graph

Finding all maximal cliques of an undirected graph is an NP-hard problem, because the maximum number of maximal cliques of an undirected graph is exponential in the number of vertices of the graph.

Consider first the problem of finding all cliques of an undirected graph. A clique of an undirected graph can be extended to a larger clique if there is a vertex not in the clique which is adjacent with all of the vertices in the clique.

A simple procedure to extend a clique $C \subseteq V$ of an undirected graph $G = (V, E)$ in all possible ways is the following. Let $P \subseteq V \setminus C$ be the set of *candidate* vertices which can be used to extend C to a larger clique, that is, $P = \{w \in V \setminus C \mid \{v, w\} \in E$ for all $v \in C\}$. Then, all possible extensions of C with vertices from P can be obtained by taking each candidate vertex of P in turn and removing it from P, adding it to C, creating a new set P from the old set by removing those vertices which are not adjacent with the selected candidate vertex, recursively performing the procedure upon the new sets C and P, and then removing the selected candidate vertex from C.

In order to avoid obtaining duplicate cliques, that is, to avoid obtaining all permutations of the vertices of a clique C, though, the set of candidate vertices $P \subseteq V \setminus C$ is considered to include only those vertices which are adjacent with all of the vertices in C and which are greater than the vertices in C, according to the order on the vertices fixed by the representation of the graph. That is, $P = \{w \in V \setminus C \mid order[v] < order[w], \{v, w\} \in E$ for all $v \in C\}$, where $order : V \rightarrow \{1, \ldots, n\}$ be the order on V fixed by the representation of an undirected graph $G = (V, E)$ with n vertices. This is achieved by just iterating over the vertices of the graph, or over the vertices adjacent with a given vertex, and enforced by most of the data structures used for representing graphs. For example, it is enforced by the Python code for the adjacency map representation of a graph given in Appendix A, because Python dictionaries preserve insertion order.

The extension of a clique C to a larger clique by adding a candidate vertex $v \in P$ is illustrated in Fig. 6.2. The new set P of candidate vertices, shown encircled with dashed lines, contains those vertices $w \in P$ which are adjacent with vertex v.

Now, the following backtracking algorithm for enumerating cliques extends an (initially empty) clique C in all possible ways, in order to enumerate all cliques $C \subseteq V$ of an undirected graph $G = (V, E)$. The cliques found are collected in a list L of sets of vertices.

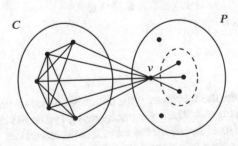

Fig. 6.2 Extending a clique C to a larger clique by adding a candidate vertex $v \in P$. The new set of candidate vertices contains those vertices $w \in P$ which are adjacent with vertex v. Edges joining vertices in C with vertices in $P \setminus \{v\}$ are omitted, for clarity

```
function all_cliques(G)
    let C, P be empty sets (of vertices)
    for all vertices v of G do
        insert v in P
    return next_clique(G, C, P)
```

The actual extension of a clique is done by the following recursive function, which extends a clique $C \subseteq V$ of the undirected graph $G = (V, E)$ by adding to C each vertex $w \in V \setminus C$ in turn which is adjacent with all those vertices which are already in C.

For each extension of clique C by adding a vertex $v \in P$, a new set P' of candidate vertices is created containing only those vertices $w \in P$ which are adjacent with vertex v, instead of removing from a copy of set P those vertices which are not adjacent with vertex v.

Notice that every vertex and every pair of adjacent vertices constitute a (trivial) clique of an undirected graph. All non-trivial cliques are collected in a list L of sets of vertices.

```
function next_clique(G, C, P)
    let L be an empty list (of sets of vertices)
    if C has more than two vertices then          ignore trivial cliques
        append C to L
    for all vertices v in P do
        P ← P \ {v}
        let P' be an empty set (of vertices)
        for all vertices w adjacent with vertex v in G do
            if w ∈ P then
                P' ← P' ∪ {w}
        C ← C ∪ {v}
        concatenate next_clique(G, C, P') to L
        C ← C \ {v}
    return L
```

Example 6.2 Consider the execution of the clique enumeration procedure illustrated in Fig. 6.3 and also in Fig. 6.4. The given undirected graph has 48 cliques, 24 of which are non-trivial, and also 24 for which the set of candidate vertices P with which they can be extended is empty. Only three out of these 24 cliques are maximal, namely: $\{v_1, v_2, v_4, v_7\}$, $\{v_1, v_3, v_4, v_6, v_7\}$, and $\{v_2, v_4, v_5, v_7\}$.

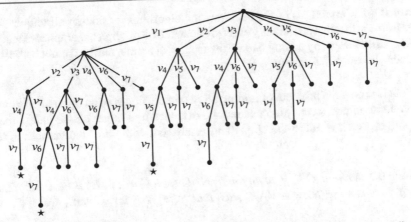

Fig. 6.3 Search tree for the execution of the clique enumeration procedure, upon the undirected graph of Fig. 6.1. Every edge of the tree represents an extension of the current clique by adding one vertex, and the edge is labeled by the vertex added to the current clique. Every path in the tree from the root to some node represents a clique of the graph. Those leaves of the tree representing a maximal clique are marked by a star

C	P
	$v_1 v_2 v_3 v_4 v_5 v_6 v_7$
v_1	$v_2 v_3 v_4 v_6 v_7$
$v_1 v_2$	$v_4 v_7$
$v_1 v_2 v_4$	v_7
$v_1 v_2 v_4 v_7$	
$v_1 v_2 v_7$	
$v_1 v_3$	$v_4 v_6 v_7$
$v_1 v_3 v_4$	$v_6 v_7$
$v_1 v_3 v_4 v_6$	v_7
$v_1 v_3 v_4 v_6 v_7$	
$v_1 v_3 v_4 v_7$	
$v_1 v_3 v_6$	v_7
$v_1 v_3 v_6 v_7$	
$v_1 v_3 v_7$	
$v_1 v_4$	$v_6 v_7$
$v_1 v_4 v_6$	v_7
$v_1 v_4 v_6 v_7$	
$v_1 v_4 v_7$	
$v_1 v_6$	v_7
$v_1 v_6 v_7$	
$v_1 v_7$	
v_2	$v_4 v_5 v_7$
$v_2 v_4$	$v_5 v_7$
$v_2 v_4 v_5$	v_7

C	P
$v_2 v_4 v_5 v_7$	
$v_2 v_4 v_7$	
$v_2 v_5$	v_7
$v_2 v_5 v_7$	
$v_2 v_7$	
v_3	$v_4 v_6 v_7$
$v_3 v_4$	$v_6 v_7$
$v_3 v_4 v_6$	v_7
$v_3 v_4 v_6 v_7$	
$v_3 v_4 v_7$	
$v_3 v_6$	v_7
$v_3 v_6 v_7$	
$v_3 v_7$	
v_4	$v_5 v_6 v_7$
$v_4 v_5$	v_7
$v_4 v_5 v_7$	
$v_4 v_6$	v_7
$v_4 v_6 v_7$	
$v_4 v_7$	
v_5	v_7
$v_5 v_7$	
v_6	v_7
$v_6 v_7$	
v_7	

Fig. 6.4 Execution of the clique enumeration procedure, upon the undirected graph of Fig. 6.1. For each clique C, the corresponding set P contains those vertices which can be used to extend C to a larger clique

Remark 6.1 Correctness of the backtracking algorithm for finding all cliques follows from the fact that an (initially empty) clique C of an undirected graph is extended in all possible ways, by adding one vertex from P at a time, until it cannot be further extended because P is empty.

Consider now the problem of finding all maximal cliques of an undirected graph. A clique of an undirected graph cannot be extended to a larger clique unless there is a vertex which is not in the clique and is adjacent with all of the vertices in the clique.

Lemma 6.3 *A clique $C \subseteq V$ of an undirected graph $G = (V, E)$ is maximal if and only if there is no vertex $w \in V \setminus C$ such that $\{v, w\} \in E$ for all vertices $v \in C$.*

Proof Let $C \subseteq V$ be a maximal clique of an undirected graph $G = (V, E)$. Then, by Definition 6.1, $\{u, v\} \in E$ for all distinct vertices $u, v \in C$. Suppose now that there is a vertex $w \in V \setminus C$ such that $\{v, w\} \in E$, for all vertices $v \in C$. Then, $\{u, v\} \in E$ for all distinct vertices $u, v \in C \cup \{w\}$, that is, $C \cup \{w\}$ is a clique, contradicting the hypothesis that C is maximal. Therefore, there is no vertex $w \in V \setminus C$ such that $\{v, w\} \in E$ for all vertices $v \in C$.

Let now $C \subseteq V$ be a clique of an undirected graph $G = (V, E)$ such that there is no vertex $w \in V \setminus C$ with $\{v, w\} \in E$ for all vertices $v \in C$, and let $D \subseteq V \setminus C$ be a set of vertices such that $C \cup D$ is a maximal clique. Then, by Definition 6.1, $\{v, w\} \in E$ for all distinct vertices $v, w \in C \cup D$ and, in particular, for all vertices $v \in C$ and $w \in D$, contradicting the hypothesis that there is no vertex $w \in V \setminus C$ with $\{v, w\} \in E$ for all vertices $v \in C$. Therefore, C is maximal. $\square$

In the backtracking algorithm for enumerating cliques, a necessary condition for a clique C to be maximal is that P be empty, for otherwise C can still be extended to a larger clique by adding a vertex from P. Such a necessary condition is not sufficient for C to be a maximal clique, though, because a clique C may be contained in some already obtained clique and still P be empty.

Therefore, in order to extend the backtracking algorithm to the enumeration of maximal cliques, the already obtained cliques need to be recognized. The set S of those vertices such that all extensions of clique C with vertices from S were already obtained, allows the efficient recognition of already obtained cliques.

Let $P \subseteq V \setminus C$ be, again, the set of candidate vertices which can be used to extend C to a larger clique, and let now $S \subseteq V \setminus C \setminus P$ be the set of vertices which cannot be used to extend C to a larger clique, because all cliques D with $C \subseteq D \subseteq C \cup S$ were already obtained.

Then, all possible extensions of C with vertices from P and without vertices from S can be obtained by taking each candidate vertex of P in turn and removing it from P, adding it to C, creating new sets P and S from the old sets by removing those vertices which are not adjacent with the selected candidate vertex, recursively performing the procedure upon the new sets C, P, and S, and then removing the selected candidate vertex from C and adding it to S. Throughout this procedure, a clique C is maximal if both P and S are empty.

Lemma 6.4 *A clique C is maximal if and only if both P and S are empty.*

Proof Let C be a maximal clique, and suppose $P \neq \emptyset$. Then, there is a vertex $w \in P$ such that $\{v, w\} \in E$, for all vertices $v \in C$, that is, $C \cup \{w\}$ is a clique. Further, since $P \subseteq V \setminus C$, it follows that $C \neq C \cup \{w\}$, contradicting the hypothesis that clique C is maximal. Therefore, $P = \emptyset$.

Suppose now $S \neq \emptyset$. Then, there is a vertex $w \in S$ such that clique $C \cup \{w\}$ was already obtained, contradicting again the hypothesis that C is maximal. Therefore, $S = \emptyset$.

Let now C be a clique such that both $P = \emptyset$ and $S = \emptyset$, and suppose C is not maximal. Then, there is a clique D such that $C \subseteq D$ and $C \neq D$. Let $w \in D \setminus C$. Since D is a clique, $\{v, w\} \in E$ for all vertices $v \in D$ with $v \neq w$ and, in particular, $\{v, w\} \in E$ for all vertices $v \in C$, that is, $w \in P$, contradicting the hypothesis that $P = \emptyset$. Therefore, C is maximal. $\square$

Now, the following simple backtracking algorithm for enumerating maximal cliques extends an (initially empty) clique C in all possible ways, in order to enumerate all maximal cliques $C \subseteq V$ of an undirected graph $G = (V, E)$. The maximal cliques found are collected in a list L of sets of vertices.

```
function simple_all_maximal_cliques(G)
    let C, P, S be empty sets (of vertices)
    for all vertices v of G do
        P ← P ∪ {v}
    return simple_next_maximal_clique(G, C, P, S)
```

The actual extension of a clique is done by the following recursive function, which extends a clique $C \subseteq V$ of the undirected graph $G = (V, E)$ by adding to C each vertex $w \in V \setminus C$ in turn which is adjacent with all those vertices which are already in C.

For each extension of clique C by adding a vertex $v \in P$, a new set P' of candidate vertices is created containing only those vertices $w \in P$ which are adjacent with vertex v, and a new set S' of non-candidate vertices is created containing only those vertices $w \in S$ which are adjacent with vertex v.

```
function simple_next_maximal_clique(G, C, P, S)
    let L be an empty list (of sets of vertices)
    if P is empty and S is empty then
        append C to L
    else
        for all vertices v in P do
            P ← P \ {v}
            let P′, S′ be empty sets (of vertices)
            for all vertices w adjacent with vertex v do
                if w ∈ P then
                    P′ ← P′ ∪ {w}
                if w ∈ S then
                    S′ ← S′ ∪ {w}
            C ← C ∪ {v}
            concatenate simple_next_maximal_clique(G, C, P′, S′) to L
            C ← C \ {v}
            S ← S ∪ {v}
    return L
```

Example 6.3 Consider the execution of the simple maximal clique enumeration procedure illustrated in Fig. 6.3 and also in Fig. 6.5. The maximal cliques of the given undirected graph are those cliques C for which the set of candidate vertices P with which they can be extended is empty and the set of vertices S with which all extensions of C were already obtained is also empty. For instance, clique $\{v_1, v_3, v_4\}$ is not maximal, because it can be extended with vertices from $P = \{v_6, v_7\}$ in order to obtain the larger cliques $\{v_1, v_3, v_4, v_6\}$, $\{v_1, v_3, v_4, v_6, v_7\}$, and $\{v_1, v_3, v_4, v_7\}$. Neither is clique $\{v_1, v_3, v_7\}$ maximal, although $P = \emptyset$, because its extensions with vertices from $S = \{v_4, v_6\}$, namely: $\{v_1, v_3, v_4, v_7\}$ and $\{v_1, v_3, v_6, v_7\}$, were already obtained. The only maximal cliques are $\{v_1, v_2, v_4, v_7\}$, $\{v_1, v_3, v_4, v_6, v_7\}$, and $\{v_2, v_4, v_5, v_7\}$.

The following simple result yields an effective improvement of the maximal clique enumeration procedure.

Lemma 6.5 *No extension of a clique $C \cup \{v\}$ to a larger clique can include vertex u if $\{u, v\} \notin E$, for all vertices $u, v \in P$.*

Proof Let $u, v \in P$, and let $C \cup \{v\}$ and D be cliques such that $C \cup \{u, v\} \subseteq D$. Suppose $\{u, v\} \notin E$. Then, $C \cup \{u, v\}$ is not a clique, contradicting thus by Lemma 6.2 the hypothesis that D is a clique. □

C	P	S
	$v_1 v_2 v_3 v_4 v_5 v_6 v_7$	
v_1	$v_2 v_3 v_4 v_6 v_7$	
$v_1 v_2$	$v_4 v_7$	
$v_1 v_2 v_4$	v_7	
$v_1 v_2 v_4 v_7$		
$v_1 v_2 v_7$		v_4
$v_1 v_3$	$v_4 v_6 v_7$	
$v_1 v_3 v_4$	$v_6 v_7$	
$v_1 v_3 v_4 v_6$	v_7	
$v_1 v_3 v_4 v_6 v_7$		
$v_1 v_3 v_4 v_7$		v_6
$v_1 v_3 v_6$	v_7	v_4
$v_1 v_3 v_6 v_7$		v_4
$v_1 v_3 v_7$		$v_4 v_6$
$v_1 v_4$	$v_6 v_7$	$v_2 v_3$
$v_1 v_4 v_6$	v_7	v_3
$v_1 v_4 v_6 v_7$		v_3
$v_1 v_4 v_7$		$v_2 v_3 v_6$
$v_1 v_6$	v_7	$v_3 v_4$
$v_1 v_6 v_7$		$v_3 v_4$
$v_1 v_7$		$v_2 v_3 v_4 v_6$
v_2	$v_4 v_5 v_7$	v_1
$v_2 v_4$	$v_5 v_7$	v_1
$v_2 v_4 v_5$	v_7	

C	P	S
$v_2 v_4 v_5 v_7$		
$v_2 v_4 v_7$		$v_1 v_5$
$v_2 v_5$	v_7	v_4
$v_2 v_5 v_7$		v_4
$v_2 v_7$		$v_1 v_4 v_5$
v_3	$v_4 v_6 v_7$	v_1
$v_3 v_4$	$v_6 v_7$	v_1
$v_3 v_4 v_6$	v_7	v_1
$v_3 v_4 v_6 v_7$		v_1
$v_3 v_4 v_7$		$v_1 v_6$
$v_3 v_6$	v_7	$v_1 v_4$
$v_3 v_6 v_7$		$v_1 v_4$
$v_3 v_7$		$v_1 v_4 v_6$
v_4	$v_5 v_6 v_7$	$v_1 v_2 v_3$
$v_4 v_5$	v_7	v_2
$v_4 v_5 v_7$		v_2
$v_4 v_6$	v_7	$v_1 v_3$
$v_4 v_6 v_7$		$v_1 v_3$
$v_4 v_7$		$v_1 v_2 v_3 v_5 v_6$
v_5	v_7	$v_2 v_4$
$v_5 v_7$		$v_2 v_4$
v_6	v_7	$v_1 v_3 v_4$
$v_6 v_7$		$v_1 v_3 v_4$
v_7		$v_1 v_2 v_3 v_4 v_5 v_6$

Fig. 6.5 Execution of the simple maximal clique enumeration procedure, upon the undirected graph of Fig. 6.1. For each clique C, the corresponding set P contains those vertices which can be used to extend C to a larger clique, and the corresponding set S contains those vertices which cannot be used to further extend C to a larger clique, because all extensions of clique C with vertices from S were already obtained

The improvement of the maximal clique enumeration procedure consists of considering only those candidate vertices $v \in P$ which are not adjacent with some vertex $u \in P$, in order to extend C to a larger clique.

Again, let $P \subseteq V \setminus C$ be the set of candidate vertices which can be used to extend C to a larger clique, and let $S \subseteq V \setminus C \setminus P$ be the set of vertices which cannot be used to extend C to a larger clique, because all cliques D with $C \subseteq D \subseteq C \cup S$ were already obtained. All possible extensions of C with vertices from P and without vertices from S can also be obtained by taking first some candidate vertex $u \in P$ and then, taking each candidate vertex $v \in P$ in turn which is not adjacent with vertex u and removing vertex v from P, adding vertex v to C, creating new sets P and S from the old sets by removing those vertices which are not adjacent with vertex v, recursively performing the procedure upon the new sets C, P, and S, and then removing vertex v from C and adding vertex v to S. Throughout this procedure, again, a clique C is maximal if both P and S are empty.

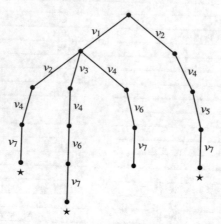

Fig. 6.6 Search tree for the execution of the maximal clique enumeration procedure, upon the undirected graph of Fig. 6.1. Every edge of the tree represents an extension of the current clique by adding one vertex, and the edge is labeled by the vertex added to the current clique. Every path in the tree from the root to a leaf node represents a clique of the graph. Those leaves of the tree representing a maximal clique are marked by a star

The effectiveness of the simple improvement to the maximal clique enumeration procedure is reflected in the reduction of the search tree for the execution of the recursive procedure.

Example 6.4 Consider the execution illustrated in Fig. 6.6 and also in Fig. 6.7 of the improved maximal clique enumeration procedure. The maximal cliques of the given undirected graph are those cliques C for which the set of candidate vertices P with which they can be extended is empty and the set of vertices S with which all extensions of C were already obtained is also empty. For instance, clique $\{v_1, v_4, v_6\}$ is not maximal, because it can be extended with a vertex from $P = \{v_7\}$ in order to obtain the larger clique $\{v_1, v_4, v_6, v_7\}$ which is also not maximal, although in this case $P = \emptyset$, because its extension with a vertex from $S = \{v_3\}$, namely: $\{v_1, v_3, v_4, v_6, v_7\}$, was already obtained. The only maximal cliques are $\{v_1, v_2, v_4, v_7\}$, $\{v_1, v_3, v_4, v_6, v_7\}$, and $\{v_2, v_4, v_5, v_7\}$.

The effectiveness of the simple improvement to the maximal clique enumeration procedure is reflected in the reduction of the search tree for the execution of the recursive procedure shown in Fig 6.3, to the smaller search tree shown in Fig 6.6. As a matter of fact, the search tree for the improved procedure has 16 nodes, corresponding to 15 non-trivial cliques obtained, and only 4 leaves, three of which correspond to maximal cliques. The search tree for the simple procedure, on the other hand, had 48 nodes, corresponding to 47 non-trivial cliques obtained, and 24 leaves, only three of which correspond to maximal cliques.

	C	P	S
		$v_1 v_2 v_3 v_4 v_5 v_6 v_7$	
	v_1	$v_2 v_3 v_4 v_6 v_7$	
	$v_1 v_2$	$v_4 v_7$	
	$v_1 v_2 v_4$	v_7	
maximal clique	$v_1 v_2 v_4 v_7$		
	$v_1 v_3$	$v_4 v_6 v_7$	
	$v_1 v_3 v_4$	$v_6 v_7$	
	$v_1 v_3 v_4 v_6$	v_7	
maximum clique	$v_1 v_3 v_4 v_6 v_7$		
	$v_1 v_6$	$v_4 v_7$	v_3
	$v_1 v_4 v_6$	v_7	v_3
	$v_1 v_4 v_6 v_7$		v_3
	v_5	$v_2 v_4 v_7$	
	$v_2 v_5$	$v_4 v_7$	
	$v_2 v_4 v_5$	v_7	
maximal clique	$v_2 v_4 v_5 v_7$		

Fig. 6.7 Execution of the maximal clique enumeration procedure upon the undirected graph of Fig. 6.1. For each clique C, the corresponding set P contains those vertices which can be used to extend C to a larger clique, and the corresponding set S contains those vertices which cannot be used to further extend C to a larger clique, because all extensions of clique C with vertices from S were already obtained

The following improved backtracking algorithm for enumerating maximal cliques extends an (initially empty) clique C in all possible ways, in order to enumerate all maximal cliques $C \subseteq V$ of an undirected graph $G = (V, E)$, implementing the improved maximal clique enumeration procedure. The maximal cliques found are collected in a list L of sets of vertices.

> **function** *all_maximal_cliques*(G)
>   let C, P, S be empty sets (of vertices)
>   **for all** vertices v of G **do**
> $P \leftarrow P \cup \{v\}$
>   **return** *next_maximal_clique*(G, C, P, S)

The actual extension of a clique is done by the following recursive function, which extends a clique $C \subseteq V$ of the undirected graph $G = (V, E)$ by adding to C each vertex $w \in V \setminus C$ in turn which is adjacent with all those vertices which are already in C.

Again, for each extension of clique C by adding a vertex $v \in P$, a new set P' of candidate vertices is created containing only those vertices $w \in P$ which are adjacent with vertex v, and a new set S' of non-candidate vertices is created containing only those vertices $w \in S$ which are adjacent with vertex v.

```
function next_maximal_clique(G, C, P, S)
    let L be an empty list (of sets of vertices)
    if P is empty then
        if S is empty then
            append C to L
    else
        let u be an arbitrary vertex from P
        for all vertices v in P do
            if vertices u and v are not adjacent in G then
                P ← P \ {v}
                let P', S' be empty sets (of vertices)
                for all vertices w adjacent with vertex v do
                    if w ∈ P then
                        P' ← P' ∪ {w}
                    if w ∈ S then
                        S' ← S' ∪ {w}
                C ← C ∪ {v}
                concatenate next_maximal_clique(G, C, P', S') to L
                C ← C \ {v}
                S ← S ∪ {v}
    return L
```

Remark 6.2 Correctness of the improved backtracking algorithm for finding all maximal cliques follows from the fact that an (initially empty) clique C of an undirected graph is extended in all possible ways, by adding one vertex from P at a time, until it cannot be further extended because P is empty. In such a case, Lemma 6.4 ensures that clique C is maximal if and only if S is also empty.

Maximum Clique of a Graph

The problem of determining whether an undirected graph $G = (V, E)$ with n vertices has a clique with k or more vertices, for a fixed integer k with $0 \leqslant k \leqslant n$, belongs to the class of NP-complete problems. This means that all known algorithms for solving the maximum clique problem upon general graphs (that is, graphs without any restriction on any graph parameter) take time exponential in the size of the graph, and that it is highly unlikely that such an algorithm will be found which takes time polynomial in the size of the graph.

All maximum cliques of an undirected graph can be obtained by first finding all maximal cliques of the graph, and then selecting the largest among all these maximal cliques. The improved backtracking algorithm for finding all maximal cliques, however, can be easily turned into a practical branch-and-bound algorithm for finding a maximum clique of an undirected graph.

Notice first that the maximum vertex degree in an undirected graph gives an upper bound on the size of a maximum clique of the graph.

Lemma 6.6 *Let $C \subseteq V$ be a maximum clique of an undirected graph $G = (V, E)$. Then, $size[C] \leqslant \max_{v \in V} \deg(v) + 1$.*

Proof Let $C \subseteq V$ be a maximum clique of an undirected graph $G = (V, E)$, let $maxdeg = \max_{v \in V} \deg(v)$, and suppose $size[C] > maxdeg + 1$. Then, by Definition 6.1, $\{v, w\} \in E$ for all distinct vertices $v, w \in C$ and then, $\deg(v) \geqslant size[C] - 1 > maxdeg$ for all vertices $v \in C$, contradicting the hypothesis that $maxdeg = \max_{v \in V} \deg(v)$. Therefore, $size[C] \leqslant \max_{v \in V} \deg(v) + 1$. □

Now, in the improved backtracking algorithm for finding all maximal cliques, an upper bound on the size of an extension of a clique $C \subseteq V$ of an undirected graph $G = (V, E)$ is given by the total size of the clique C being extended and the set P of candidate vertices with which clique C can be extended.

Lemma 6.7 *Let $D \subseteq V$ be a maximum clique of an undirected graph $G = (V, E)$ and let $C \subseteq D$. Then, $size[D] \leqslant size[C] + size[P]$.*

Proof Let $C \subseteq D$ be a clique contained in a maximum clique $D \subseteq V$ of an undirected graph $G = (V, E)$. Since clique C can only be extended by adding vertices from P, it follows that $size[E] \leqslant size[C] + size[P]$ for any extension E of clique C and, in particular, $size[D] \leqslant size[C] + size[P]$. □

Both the upper bound on the size of a maximum clique and the upper bound on the size of an extension of a clique can be used to turn the backtracking algorithm into a practical branch-and-bound algorithm. The procedure is identical to the backtracking procedure for finding all maximal cliques of an undirected graph, except that a clique C is recursively extended if and only if the size of clique C is still below the upper bound on the size of an extension of the clique, and moreover, the size of largest clique found so far is below the upper bound given by the maximum vertex degree.

Example 6.5 Consider the execution of the procedure for finding a maximum clique, illustrated in Figs. 6.8 and 6.9. The maximal cliques of the given undirected graph are those cliques C for which the set of candidate vertices P with which they can be extended is empty and the set of vertices S with which all extensions of C were already obtained is also empty. A clique C is actually extended only if there are enough candidate vertices in P to eventually obtain a clique larger than the current largest clique, and unless the current largest clique is already as large as one plus the maximum vertex degree.

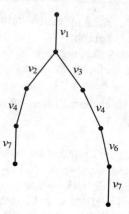

Fig. 6.8 Search tree for the execution of the maximum clique procedure, upon the undirected graph of Fig. 6.1. Every edge of the tree represents an extension of the current clique by adding one vertex, and the edge is labeled by the vertex added to the current clique. Every path in the tree from the root to a leaf node represents a clique of the graph. Both leaves of the tree represent a maximal clique, and the rightmost leaf node also stands for a maximum clique

	C	P	S
		$v_1\, v_2\, v_3\, v_4\, v_5\, v_6\, v_7$	
	v_1	$v_2\, v_3\, v_4\, v_6\, v_7$	
	$v_1\, v_2$	$v_4\, v_7$	
	$v_1\, v_2\, v_4$	v_7	
maximal clique	$v_1\, v_2\, v_4\, v_7$		
	$v_1\, v_3$	$v_4\, v_6\, v_7$	
	$v_1\, v_3\, v_4$	$v_6\, v_7$	
	$v_1\, v_3\, v_4\, v_6$	v_7	
maximum clique	$v_1\, v_3\, v_4\, v_6\, v_7$		

Fig. 6.9 Execution of the procedure for finding a maximum clique, upon the undirected graph of Fig. 6.1. For each clique C, the corresponding set P contains those vertices which can be used to extend C to a larger clique, and the corresponding set S contains those vertices which cannot be used to further extend C to a larger clique, because all extensions of clique C with vertices from S were already obtained. However, a clique C is actually extended only if there are enough candidate vertices in P to eventually obtain a clique larger than the current largest clique

For instance, clique $C = \{v_1, v_3, v_4\}$ is further extended, because the current largest clique is $|MAX| = \{v_1, v_2, v_4, v_7\}$, $P = \{v_6, v_7\}$, $size[MAX] = 4 \leqslant 3 + 2 = size[C] + size[P]$, and also $size[MAX] \leqslant 7$, one plus the maximum vertex degree in the graph.

The following branch-and-bound algorithm for finding a maximum clique extends an (initially empty) clique C in all possible ways, in order to enumerate all maximal cliques $C \subseteq V$ of an undirected graph $G = (V, E)$, implementing the previous procedure.

```
function maximum_clique(G)
    let C, P, S be empty sets (of vertices)
    d ← 0
    for all vertices v of G do
        P ← P ∪ {v}
        d ← max(d, v.outdeg)
    let M be an empty set (of vertices)
    return next_maximum_clique(G, C, P, S, d, M)
```

The actual extension of a clique is done by the following recursive function, which extends a clique $C \subseteq V$ of the undirected graph $G = (V, E)$ by adding to C each vertex $w \in V \setminus C$ in turn which is adjacent with all those vertices which are already in C.

Again, for each extension of clique C by adding a vertex $v \in P$, a new set P' of candidate vertices is created containing only those vertices $w \in P$ which are adjacent with vertex v, and a new set S' of non-candidate vertices is created containing only those vertices $w \in S$ which are adjacent with vertex v.

Further, an extension is only computed if it may eventually lead to a clique larger than the largest clique found so far, that is, if the size of the current largest clique M is less than the size of clique C plus the size of the new set P' of candidate vertices with which clique C can be extended, and unless M is already as large as one plus the maximum vertex degree.

```
function next_maximum_clique(G, C, P, S, d, M)
    if M is smaller than C then
        M ← C
    if P is not empty then
        let u be an arbitrary vertex from P
        for all vertices v in P do
            if vertices u and v are not adjacent in G then
                P ← P \ {v}
                let P', S' be empty sets (of vertices)
                for all vertices w adjacent with vertex v in G do
                    if w ∈ P then
                        P' ← P' ∪ {w}
                    if w ∈ S then
                        S' ← S' ∪ {w}
                C ← C ∪ {v}
                if M is smaller than the size of C plus the size of P' then
                    if the size of M is less than d + 1 then
                        M ← next_maximum_clique(G, C, P', S', d, M)
                C ← C \ {v}
                S ← S ∪ {v}
    return M
```

Remark 6.3 Correctness of the branch-and-bound algorithm for finding a maximum clique follows from the fact that an (initially empty) clique C of an undirected graph is extended in all possible ways by adding one vertex from P at a time, until it cannot be further extended because P is empty. In such a case, Lemma 6.4 ensures that clique C is maximal if and only if S is also empty, and clique C being maximal is a necessary condition for C to be a maximum clique by Lemma 6.1.

Further, clique C is actually extended only if there are enough candidate vertices in P to eventually obtain a clique larger than the current largest clique, according to Lemma 6.7, and unless the current largest clique is already as large as one plus the maximum vertex degree, according to Lemma 6.6.

6.2 Maximal and Maximum Independent Sets

An independent set of an undirected graph is an induced subgraph of the graph which has no edges. A maximal independent set of a graph is an independent set of the graph that is not properly included in any other independent set of the graph, and a maximum independent set of a graph is a (maximal) independent set of the graph with the largest number of vertices.

Definition 6.2 An **independent set** of an undirected graph $G = (V, E)$ is a set of vertices $I \subseteq V$ such that the subgraph of G induced by I has no edges, that is, such that $\{v, w\} \notin E$ for all distinct vertices $v, w \in I$. An independent set I of an undirected graph $G = (V, E)$ is **maximal** if there is no independent set J of G such that $I \subseteq J$ and $I \neq J$, and it is **maximum** if there is no independent set of G with more vertices than I. The **independence number** of G, denoted by $\beta(G)$, is the cardinality of a maximum independent set of G.

It follows from the previous definition that maximum independent sets are also maximal.

Lemma 6.8 *Every maximum independent set* $I \subseteq V$ *of an undirected graph* $G = (V, E)$ *is maximal.*

Proof Let $I \subseteq V$ be a maximum independent set of an undirected graph $G = (V, E)$, and suppose I is not maximal. There is, by Definition 6.2, an independent set $J \subseteq V$ such that $I \subseteq J$ and $I \neq J$. Then, $J \setminus I \neq \emptyset$, that is, independent set J has more vertices than I, contradicting the hypothesis that I is a maximum independent set. Therefore, I is also a maximal independent set. $\square$

Fig. 6.10 Maximal and maximum independent sets of an undirected graph. Independent set $\{v_1, v_6\}$ is not maximal, because it is included in independent set $\{v_1, v_5, v_6\}$, which is a maximal and also a maximum independent set

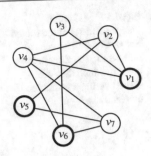

Example 6.6 The undirected graph shown in Fig. 6.10 has five maximal independent sets: $\{v_1, v_7\}$, $\{v_2, v_6\}$, $\{v_1, v_5, v_6\}$, $\{v_2, v_3, v_7\}$, and $\{v_3, v_4, v_5\}$. The latter three are also maximum independent sets of the graph.

Notice that being an independent set of an undirected graph is another hereditary graph property.

Lemma 6.9 *Let $I \subseteq V$ be an independent set of an undirected graph $G = (V, E)$, and let $H \subseteq I$. Then, H is also an independent set of G.*

Proof Let $I \subseteq V$ be an independent set of an undirected graph $G = (V, E)$, and let $H \subseteq I$. Then, $\{v, w\} \notin E$ for all distinct vertices $v, w \in I$ by Definition 6.2 and, in particular, $\{v, w\} \notin E$ for all distinct vertices $v, w \in H$. Therefore, $H \subseteq V$ is also an independent set. $\qquad\square$

Maximum Independent Set of a Tree

An independent set of a tree is a set of nodes of the tree which does not include any of the children of the nodes in the independent set or, in an equivalent formulation, a set of nodes of the tree which does not include the parent of any of the nodes in the independent set. A maximal independent set of a tree is an independent set of the tree that is not properly included in any other independent set of the tree, and a maximum independent set of a tree is a (maximal) independent set of the tree with the largest number of nodes.

Definition 6.3 An **independent set** of a rooted tree $T = (V, E)$ is a set of nodes $I \subseteq V$ such that $parent[v] \notin I$ for all non-root nodes $v \in I$. An independent set I of a rooted tree $T = (V, E)$ is **maximal** if there is no independent set J of T such that $I \subseteq J$ and $I \neq J$, and it is **maximum** if there is no independent set of T with more nodes than I. The **independence number** of T, denoted by $\beta(T)$, is the cardinality of a maximum independent set of T.

Fig. 6.11 Maximum
independent set of the rooted
tree of Fig. 3.3. Nodes are
numbered according to the
order in which they are
visited during a postorder
traversal

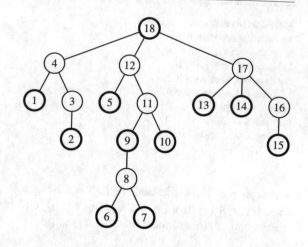

Example 6.7 A maximum independent set of a rooted tree is illustrated in Fig. 6.11. All the leaves of the tree belong to the maximum independent set, together with all those non-leaves, none of whose children but at least one of whose grandchildren belong to the maximum independent set.

In a rooted tree, the children of a node constitute an independent set, because otherwise the undirected graph underlying the tree would not be acyclic anymore. For the same reason, all the leaves of a tree also constitute an independent set of the tree.

A particular maximum independent set of a tree can be obtained as follows. All the leaves of the tree belong to the maximum independent set, together with all those non-leaves none of whose children, but at least one of whose grandchildren, belong to the maximum independent set.

Lemma 6.10 *Let $T = (V, E)$ be a tree with n nodes, and let $I = I_n \subseteq V$ be the set of nodes defined by induction as follows:*

- $I_1 = \{v \in V \mid (v, w) \notin E \text{ for all } w \in V\}$,
- $I_{i+1} = I_i \cup \{u \in V \setminus I_i \mid v \notin I_i \text{ for all } (u, v) \in E\} \cap \{u \in V \setminus I_i \mid (u, v), (v, w) \in E \text{ for some } w \in I_i\}$, for $1 \leqslant i < n$.

Then, I is a maximum independent set of T.

Proof Suppose $I \subseteq V$ is not an independent set of $T = (V, E)$. Then, there are nodes $v, w \in I$ such that $(v, w) \in E$. Now, it must be $v \notin I_1$, because $(v, w) \in E$, and it must also be $v \notin I_i$ for $2 \leqslant i \leqslant n$, because $(v, w) \in E$ and $w \in I$, contradicting the hypothesis that $v \in I$. Therefore, I is an independent set of T.

Suppose now that I is not maximal. Then, there is a node $v \in V \setminus I$ which is neither the parent nor a child of any node $w \in I$, that is, whose parent and all of whose children belong to $V \setminus I$. Then, for some i, with $1 \leqslant i < n$, it holds that $v \in \{x \in V \setminus I_i \mid y \notin I_i \text{ for all } (x, y) \in E\}$. Further, for all children w of node v, there must be $(w, z) \in E$ for some node $z \in I$, because otherwise it would be $w \in I$, and then $v \in \{x \in V \setminus I_i \mid (x, y), (y, z) \in E \text{ for some } z \in I_i\}$. Therefore, $v \in I_i$ for some i, with $1 \leqslant i \leqslant n$, contradicting the hypothesis that $v \in V \setminus I$. Then, I is a maximal independent set of T.

Finally, notice that the set I_1 contains all the nodes of least height of T and for $1 \leqslant i < n$, the set I_{i+1} contains all those nodes of least height which are not the parent of any node in I_i. Since the number of nodes of height $i + 1$ is less than or equal to the number of nodes of height i, it follows that I is a maximum independent set of T. □

Now, a maximum independent set of a tree can be obtained by collecting *independent* nodes during a postorder traversal of the tree. All the leaves of the tree are independent, and non-leaves are independent if none of their children belong to the maximum independent set.

The following algorithm finds a maximum independent set $I \subseteq V$ of a tree $T = (V, E)$, implementing the previous procedure.

```
function maximum_independent_set(T)
    let I be an empty set (of nodes)
    for all nodes v of T do
        v.independent ← true
    for all nodes v in T in postorder do
        if v is not a leaf node of T then
            for all children w of node v in T do
                if w.independent then
                    v.independent ← false
                    break
        if v.independent then
            I ← I ∪ {v}
    return I
```

Lemma 6.11 *The algorithm for finding a maximum independent set of a rooted tree runs in $O(n \log n)$ time using $O(n)$ additional space, where n is the number of nodes in the tree.*

Proof Let $T = (V, E)$ be a rooted tree with n nodes. A postorder traversal is made which, by Lemma 3.4, takes $O(n)$ time using $O(n)$ additional space. Further, $O(n)$ insertions in an initially empty set of nodes are made, corresponding to the nodes in $I \subseteq V$. Therefore, the algorithm runs in $O(n \log n)$ time using $O(n)$ additional space. □

Remark 6.4 The tree maximum independent set algorithm can also be implemented to run in $O(n)$ time, where n is the number of nodes in the tree, while still using $O(n)$ additional space. See Exercise 6.4.

Maximum Independent Set of a Graph

The problem of determining whether an undirected graph $G = (V, E)$ with n vertices has an independent set with k or more vertices, for a fixed integer k with $0 \leqslant k \leqslant n$, also belongs to the class of NP-complete problems. This means that all known algorithms for solving the maximum independent set problem upon general graphs (that is, graphs without any restriction on any graph parameter) take time exponential in the size of the graphs, and that it is highly unlikely that such an algorithm will be found which takes time polynomial in the size of the graph.

Finding a maximum independent set of an undirected graph can be reduced to the problem of finding a maximum clique in the complement of the graph. Recall from Sect. 1.1 that the complement of an undirected graph (V, E) is the undirected graph (V, F), where $\{v, w\} \in F$ if and only if $\{v, w\} \notin E$, for all distinct vertices $v, w \in V$.

Lemma 6.12 *Let $C \subseteq V$ be a maximum clique of the complement of a graph $G = (V, E)$. Then, C is a maximum independent set of graph G.*

Proof It suffices to show that a clique $C \subseteq V$ of the complement of a graph $G = (V, E)$ is an independent set of graph G. Let (V, F) be the complement of graph G. Then, by Definition 6.1, $\{v, w\} \in F$ for all distinct vertices $v, w \in C$, that is, $\{v, w\} \notin E$ for all distinct vertices $v, w \in C$. Therefore, C is, by Definition 6.2, an independent set of graph G. $\qquad\square$

Example 6.8 The relationship between maximum clique and maximum independent set is illustrated in Fig. 6.12. The set of vertices $\{v_1, v_5, v_6\}$ is shown as a maximum

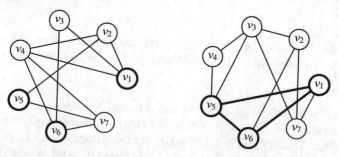

Fig. 6.12 Complement (right) of the undirected graph (left) of Fig. 6.10. A maximum independent set in the graph corresponds to a maximum clique in the complement of the graph

independent set of the graph of Fig. 6.10, and also as a maximum clique of the complement of the graph.

The following algorithm implements the construction of the complement H of an undirected graph G. The correspondence among the vertices of G and H is stored in the attribute $v.M$, for all vertices v of G and H.

> **function** *graph_complement*(G)
>   **let** H be an empty graph
>   **for all** vertices v of G **do**
>   add a new vertex w to H
> $v.M \leftarrow w$
> $w.M \leftarrow v$
>   **for all** vertices v of G **do**
>   **for all** vertices w of G **do**
>   **if** $v \neq w$ and vertices v and w are not adjacent in G **then**
>   add a new edge $(v.M, w.M)$ to H
>   **return** H

Now, the following algorithm implements the reduction of maximum independent set to maximum clique. The maximum independent set found is represented by a set I of vertices.

> **function** *maximum_independent_set*(G)
> $H \leftarrow$ *graph_complement*(G)
> $C \leftarrow$ *maximum_clique*(H)
>   **let** I be an empty set (of vertices)
>   **for all** vertices v in C **do**
> $I \leftarrow I \cup \{v.M\}$
>   **return** I

6.3   Minimal and Minimum Vertex Covers

A vertex cover of an undirected graph is a set of vertices covering all of the edges of the graph, that is, such that every edge of the graph is incident with some vertex of the vertex cover. A minimal vertex cover of a graph is a vertex cover of the graph that does not properly include any other vertex cover of the graph, and a minimum vertex cover of a graph is a (minimal) vertex cover of the graph with the smallest number of vertices.

Definition 6.4 A **vertex cover** of an undirected graph $G = (V, E)$ is a set of vertices $C \subseteq V$ such that $v \in C$ or $w \in C$ for all edges $\{v, w\} \in E$. A vertex cover C of an undirected graph $G = (V, E)$ is **minimal** if there is no vertex cover D of G such that $D \subseteq C$ and $C \neq D$, and it is **minimum** if there is no vertex cover of G with less vertices than C. The **vertex cover number** of G, denoted by $\alpha(G)$, is the cardinality of a minimum vertex cover of G.

It follows from the previous definition that minimum vertex covers are also minimal.

Lemma 6.13 *Every minimum vertex cover $C \subseteq V$ of an undirected graph $G = (V, E)$ is minimal.*

Proof Let $C \subseteq V$ be a minimum vertex cover of an undirected graph $G = (V, E)$, and suppose C is not minimal. There is, by Definition 6.4, a vertex cover $D \subseteq V$ such that $D \subseteq C$ and $C \neq D$. Then, $C \setminus D \neq \emptyset$, that is, vertex cover D has less vertices than C, contradicting the hypothesis that C is a minimum vertex cover. Therefore, C is also a minimal vertex cover. $\qquad\square$

Example 6.9 The undirected graph shown in Fig. 6.13 has five minimal vertex covers: $\{v_1, v_2, v_6, v_7\}$, $\{v_1, v_4, v_5, v_6\}$, $\{v_2, v_3, v_4, v_7\}$, $\{v_1, v_3, v_4, v_5, v_7\}$, and $\{v_2, v_3, v_4, v_5, v_6\}$. The first three are also minimum vertex covers of the graph.

The notions of vertex cover and independent set are dual to each other, although in a different sense than for cliques and independent sets. Given an independent set of an undirected graph, those vertices of the graph which do not belong to the independent set constitute a vertex cover of the graph.

Lemma 6.14 *A set of vertices $W \subseteq V$ is an independent set of an undirected graph $G = (V, E)$ if and only if the set of vertices $V \setminus W$ is a vertex cover of G.*

Proof Let $W \subseteq V$ be an independent set of an undirected graph $G = (V, E)$, and suppose $V \setminus W$ is not a vertex cover of G. Then, $v, w \notin V \setminus W$ for some edge

Fig. 6.13 Minimal and minimum vertex covers of an undirected graph. Vertex cover $\{v_1, v_2, v_3, v_4, v_7\}$ is not minimal, because it includes vertex cover $\{v_2, v_3, v_4, v_7\}$, which is a minimal and also a minimum vertex cover

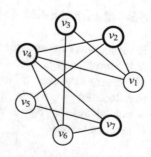

$\{v, w\} \in E$, that is, there are vertices $v, w \in W$ with $\{v, w\} \in E$, contradicting the hypothesis that W is an independent set of G. Therefore, $V \setminus W$ is a vertex cover of G.

Let now $V \setminus W$ be a vertex cover of an undirected graph $G = (V, E)$, where $W \subseteq V$, and suppose W is not an independent set of G. Then, $\{v, w\} \in E$ for some vertices $v, w \in W$, that is, there is an edge $\{v, w\} \in E$ such that neither $v \in V \setminus W$ nor $w \in V \setminus W$, contradicting the hypothesis that $V \setminus W$ is a vertex cover of G. Therefore, W is an independent set of G. $\square$

There is also a close relationship between the cardinalities of a maximum independent set and a minimum vertex cover of an undirected graph.

Theorem 6.1 *Let $G = (V, E)$ be an undirected graph with n vertices. Then, $\alpha(G) + \beta(G) = n$.*

Proof Let $G = (V, E)$ be an undirected graph with n vertices, and let $W \subseteq V$ be a maximum independent set of G, that is, an independent set of G with $\beta(G)$ vertices. Since, by Lemma 6.14, $V \setminus W$ is a vertex cover of G, it follows that $\alpha(G) \leqslant n - \beta(G)$. Let now $W \subseteq V$ be a minimum vertex cover of G, that is, a vertex cover of G with $\alpha(G)$ vertices. Since, by Lemma 6.14, $V \setminus W$ is an independent set of G, it follows that $\beta(G) \geqslant n - \alpha(G)$. Therefore, $\alpha(G) + \beta(G) = n$. $\square$

Remark 6.5 Notice that the existence of isolated vertices in an undirected graph does not alter the previous relationship between the independence number and the vertex cover number of the graph. As a matter of fact, isolated vertices must belong to every maximum independent set, but no minimum vertex cover can have an isolated vertex.

Now, given a maximum independent set of an undirected graph, those vertices of the graph which do not belong to the maximum independent set constitute a minimum vertex cover of the graph.

Corollary 6.1 *A set of vertices $W \subseteq V$ is a maximum independent set of an undirected graph $G = (V, E)$ if and only if the set of vertices $V \setminus W$ is a minimum vertex cover of G.*

Proof Let $W \subseteq V$ be a maximum independent set of an undirected graph $G = (V, E)$, that is, an independent set of G with $\beta(G)$ vertices. Then, by Lemma 6.14, $V \setminus W$ is a vertex cover by G, and it has $n - \beta(G)$ vertices. Now, it follows from Theorem 6.1 that $V \setminus W$ has $\alpha(G)$ vertices, and therefore, $V \setminus W$ is a minimum vertex cover of G.

Let now $W \subseteq V$ be a minimum vertex cover of an undirected graph $G = (V, E)$, that is, a vertex cover of G with $\alpha(G)$ vertices. Then, by Lemma 6.14, $V \setminus W$ is an independent set of G, and it has $n - \alpha(G)$ vertices. Now, it follows from Theorem 6.1 that $V \setminus W$ has $\beta(G)$ vertices, and therefore, $V \setminus W$ is a maximum independent set of G.

□

Minimum Vertex Cover of a Tree

A vertex cover of a tree is a set of nodes covering all of the edges of the tree, that is, such that for all nodes in the tree, the node or the parent of the node belong to the vertex cover. A minimal vertex cover of a tree is a vertex cover of the tree that does not properly include any other vertex cover of the tree, and a minimum vertex cover of a tree is a (minimal) vertex cover of the tree with the smallest number of nodes.

Definition 6.5 A **vertex cover** of a rooted tree $T = (V, E)$ is a set of nodes $C \subseteq V$ such that $v \in C$ or $parent[v] \in C$ for all non-root nodes $v \in V$. A vertex cover C of a rooted tree $T = (V, E)$ is **minimal** if there is no vertex cover D of T such that $D \subseteq C$ and $C \neq D$, and it is **minimum** if there is no vertex cover of T with less nodes than C. The **vertex cover number** of T, denoted by $\alpha(T)$, is the cardinality of a minimum vertex cover of T.

Example 6.10 The set of nodes $\{v_3, v_4, v_8, v_{11}, v_{12}, v_{16}, v_{17}\}$ is a minimum vertex cover $C \subseteq V$ of the rooted tree $T = (V, E)$ shown in Fig. 6.14. The parent of every non-root node of the tree which does not belong to the minimum vertex cover, does indeed belong to the minimum vertex cover. In fact, $parent[v_1] = v_4 \in C$, $parent[v_2] = v_3 \in C$, $parent[v_5] = v_{12} \in C$, $parent[v_6] = parent[v_7] = v_8 \in C$, $parent[v_9] = parent[v_{10}] = v_{11} \in C$, $parent[v_{13}] = parent[v_{14}] = v_{17} \in C$, and $parent[v_{15}] = v_{16} \in C$. Further, root node $v_{18} \notin C$, because all of children nodes $v_4, v_{12}, v_{17} \in C$.

Fig. 6.14 Minimum vertex cover of the rooted tree of Fig. 3.3. Nodes are numbered according to the order in which they are visited during a postorder traversal

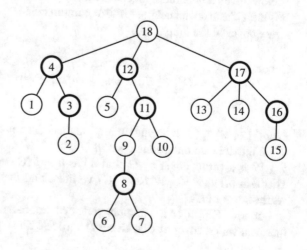

A minimum vertex cover of a tree can be obtained by first finding a maximum independent set of the tree, and then collecting all those nodes of the tree which do not belong to the maximum independent set.

The following algorithm finds a minimum vertex cover $C \subseteq V$ of a tree $T = (V, E)$, implementing the previous procedure.

> **function** *minimum_vertex_cover*(T)
>   let C be an empty set (of nodes)
> $I \leftarrow$ *maximum_independent_set*(T)
>   **for all** nodes v of T **do**
>   **if** $v \notin I$ **then**
> $C \leftarrow C \cup \{v\}$
>   **return** C

Lemma 6.15 *The algorithm for finding a minimum vertex cover of a rooted tree runs in $O(n \log n)$ time using $O(n)$ additional space, where n is the number of nodes in the tree.*

Proof Let $T = (V, E)$ be a rooted tree with n nodes. Finding a maximum independent set $I \subseteq V$ takes $O(n \log n)$ time using $O(n)$ additional space, by Lemma 6.11. Further, $O(n)$ insertions in an initially empty set of nodes are made, corresponding to the nodes in $C = V \setminus I \subseteq V$. Therefore, the algorithm runs in $O(n \log n)$ time using $O(n)$ additional space.   □

Remark 6.6 The tree minimum vertex cover algorithm can also be implemented to run in $O(n)$ time, where n is the number of nodes in the tree, while still using $O(n)$ additional space. See Exercise 6.6.

Minimum Vertex Cover of a Graph

The problem of determining whether an undirected graph $G = (V, E)$ with n vertices has a vertex cover with k or less vertices, for a fixed integer k with $0 \leqslant k \leqslant n$, also belongs to the class of NP-complete problems, meaning that all known algorithms for solving the minimum vertex cover problem upon general graphs (that is, graphs without any restriction on any graph parameter) take time exponential in the size of the graphs, and that it is highly unlikely that such an algorithm will be found which takes time polynomial in the size of the graph.

Finding a minimum vertex cover of an undirected graph can be reduced to the problem of finding a maximum independent set. As a matter of fact, a minimum vertex cover of an undirected graph can be obtained by first finding a maximum independent set of the graph, and then collecting all those vertices of the graph which do not belong to the maximum independent set.

The following algorithm finds a minimum vertex cover $C \subseteq V$ of an undirected graph $G = (V, E)$, implementing the previous procedure.

```
function minimum_vertex_cover(G)
    let C be an empty set (of vertices)
    I ← maximum_independent_set(G)
    for all vertices v of G do
        if v ∉ I then
            C ← C ∪ {v}
    return C
```

6.4 Applications

Clique, independent set, and vertex cover algorithms find application in the comparison of structures described by trees or graphs. In particular, finding maximal and maximum cliques is the basis of the algorithms discussed in Chap. 7 for finding maximal and maximum common subgraph isomorphisms between two unordered graphs.

The application of maximum clique algorithms to the multiple alignment of nucleic acid or amino acid (protein) sequences in computational molecular biology is discussed next. See the bibliographic notes below for further applications of clique and independent set algorithms.

Recall from Sect. 4.4 that a ribonucleic acid molecule is composed of four types of nucleotides. Protein molecules in animals are instead composed of 20 types of amino acids, also called *residues*. These are: alanine (A), cysteine (C), aspartic acid (D), glutamic acid (E), phenylalanine (F), glycine (G), histidine (H), isoleucine (I), lysine (K), leucine (L), methionine (M), asparagine (N), proline (P), glutamine (Q), arginine (R), serine (S), threonine (T), valine (V), tryptophan (W), and tyrosine (Y). Protein molecules in plants are composed of about 100 types of residues.

Protein molecules are composed of one or more sequences of residues in a specific order, determined by the sequence of nucleotides in the gene encoding the protein.

Example 6.11 Six sequences of 60 residues each, extracted from the Protein Data Bank [10] and corresponding to the hemoglobin protein of different animal species, are shown in Fig. 6.15. The hemoglobin protein consists of four sequences of about 140 residues each.

The edit distance between two protein sequences is the cost of a least-cost sequence of elementary edit operations, such as deletion, substitution, and insertion of residues, that allows to transform one protein sequence into the other. An alignment of two

```
MVHLTPEEKSAVTALWGKVNVDEVGGEALGRLLVVYPWTQRFFESFGDLSTPDAVMGNPK
VQLSGEEKAAVLALWDKVNEEEVGGEALGRLLVVYPWTQRFFDSFGDLSNPGAVMGNPKV
MVLSPADKTNVKAAWGKVGAHAGEYGAEALERMFLSFPTTKTYFPHFDLSHGSAQVKGHG
MVLSAADKTNVKAAWSKVGGHAGEYGAEALERMFLGFPTTKTYFPHFDLSHGSAQVKAHG
VLSEGEWQLVLHVWAKVEADVAGHGQDILIRLFKSHPETLEKFDRFKHLKTEAEMKASED
PIVDTGSVAPLSAAEKTKIRSAWAPVYSTYETSGVDILVKFFTSTPAAQEFFPKFKGLTT
```

Fig. 6.15 Sample protein sequences

```
VQLSGEEKAAVLALWDKVNEE--EVGGEALGRLLVVYPWTQRFFDSFGDLSNPGAVMGNPKV
MVLSPADKTNVKAAWGKVGAHAGEYGAEALERMFLSFPTTKTYFPHFDLSHGSAQVKGHG--
```

Fig. 6.16 An alignment of two protein sequences

protein sequences is an alternative representation of the transformation of one protein sequence into the other.

An alignment of two protein sequences is obtained by first inserting dashes, either into or at the end of the sequences, in such a way that the resulting sequences have the same length, and then placing the resulting sequences one on top of the other, in such a way that every residue or dash in either sequence is opposite to a unique residue or dash in the other sequence. Then, a dash in the sequence at the top represents the insertion of a residue at the same position in the sequence at the bottom, and a dash in the sequence at the bottom represents the deletion of a residue at the same position from the sequence at the top. The absence of dashes at a particular position in both sequences indicates a substitution of the residues, if the residues at that position differ.

Example 6.12 An alignment of the second and the third protein sequences of Fig. 6.15, obtained with the CLUSTAL W program [51], is shown in Fig. 6.16. The dashes in the second sequence (shown on top of the third sequence) represent the insertion of residues A and G into the third sequence, and the dashes at the end of the third sequence represent the deletion of residues K and V from the second sequence. Further, 38 out of the other 58 positions represent substitutions of different residues and only 20 positions represent identical substitutions.

Alignment is an important form of protein sequence comparison. The biological significance of sequence alignment lies in the fact that, in deoxiribonucleic acid, ribonucleic acid, or amino acid sequences, high sequence similarity usually implies functional or structural similarity.

The alignment of multiple protein sequences, although being a natural generalization of protein sequence alignment, has an even more profound biological significance. Multiple sequence alignment may reveal evolutionary history, reveal common two-dimensional and three-dimensional molecular structure, and suggest common biological function.

A multiple alignment of several protein sequences is obtained by first inserting dashes, either into or at the end of the sequences, in such a way that the resulting

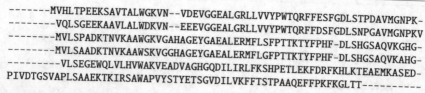

```
-------MVHLTPEEKSAVTALWGKVN--VDEVGGEALGRLLVVYPWTQRFFESFGDLSTPDAVMGNPK-
--------VQLSGEEKAAVLALWDKVN--EEEVGGEALGRLLVVYPWTQRFFDSFGDLSNPGAVMGNPKV
--------MVLSPADKTNVKAAWGKVGAHAGEYGAEALERMFLSFPTTKTYFPHF-DLSHGSAQVKGHG-
--------MVLSAADKTNVKAAWSKVGGHAGEYGAEALERMFLGFPTTKTYFPHF-DLSHGSAQVKAHG-
---------VLSEGEWQLVLHVWAKVEADVAGHGQDILIRLFKSHPETLEKFDRFKHLKTEAEMKASED-
PIVDTGSVAPLSAAEKTKIRSAWAPVYSTYETSGVDILVKFFTSTPAAQEFFPKFKGLTT----------
```

Fig. 6.17 A multiple alignment of six protein sequences

sequences have the same length, and then placing the resulting sequences one on top of the other, in such a way that every residue or dash in either sequence is opposite to a unique residue or dash in every other sequence.

Example 6.13 A multiple alignment of the protein sequences of Fig. 6.15, also obtained with the CLUSTAL W program [51], is shown in Fig. 6.17. Only 9 out of the 60 positions represent identical substitutions in all of the protein sequences, that is, residues common to all six sequences.

Most protein sequence alignment algorithms run in $O(n^2)$ time, where n is the number of residues in each of the sequences. By extension to several sequences, the multiple alignment of $k \geqslant 2$ protein sequences takes $O(n^k)$ time. That is, the time taken to compute an alignment grows exponentially with the number of sequences to be aligned.

However, since there are $\binom{k}{2} = k(k-1)/2$ pairwise alignments of $k \geqslant 2$ protein sequences, and it holds that $O(k^2n^2) \subseteq O(n^k)$ if $k \leqslant \log n$, it follows that an approach to multiple sequence alignment based on combining the results of all pairwise alignments could improve the $O(n^k)$ time bound for all practical purposes, that is, unless the number of protein sequences to be aligned exceeds the logarithm of the number of residues in each of the sequences.

A simple and natural way to combine alignments of protein sequences is the following. Given the $k(k-1)/2$ pairwise alignments of $k \geqslant 2$ protein sequences with n residues each, their *alignment graph* is a k-partite undirected graph with nk vertices and at most $nk(k-1)/2$ edges. There is a vertex in the alignment graph for each residue in each sequence, and for each pair of aligned residues in each of the pairwise sequence alignments, there is an edge in the alignment graph between the corresponding vertices.

Now, a multiple sequence alignment is given by the maximum cliques in the alignment graph of the sequences.

Example 6.14 The alignment graph of the protein sequences given in Fig. 6.15 is shown in Fig. 6.16. All pairwise alignments of the protein sequences were obtained with the CLUSTAL W multiple alignment program [51]. The alignment graph has $360 = 60 \cdot 6$ vertices and $841 \leqslant 900 = 60 \cdot 6 \cdot 5/2$ edges. There are 149 maximal cliques in the alignment graph, only 36 of which are also maximum cliques.

The multiple sequence alignment based on maximum cliques in the alignment graph of the sequences, complements multiple sequence alignments obtained by other methods by emphasizing those residues which are aligned in a consistent way in all of the pairwise alignments of the sequences.

Example 6.15 The multiple sequence alignment based on the maximum cliques in the alignment graph of Fig. 6.18, corresponding to the protein sequences of Fig. 6.15, is shown in Fig. 6.19. Vertical bars between residues in consecutive protein sequences represent maximum clique edges in the alignment graph, that is, residues that are aligned in a consistent way in all of the pairwise alignments of the sequences. The maximum clique edges in the alignment graph induce a unique padding of the sequences with initial and final dashes, but a non-unique padding of the sequences with inner dashes. The padding shown in Fig. 6.19 is identical to the multiple alignment shown in Fig. 6.17.

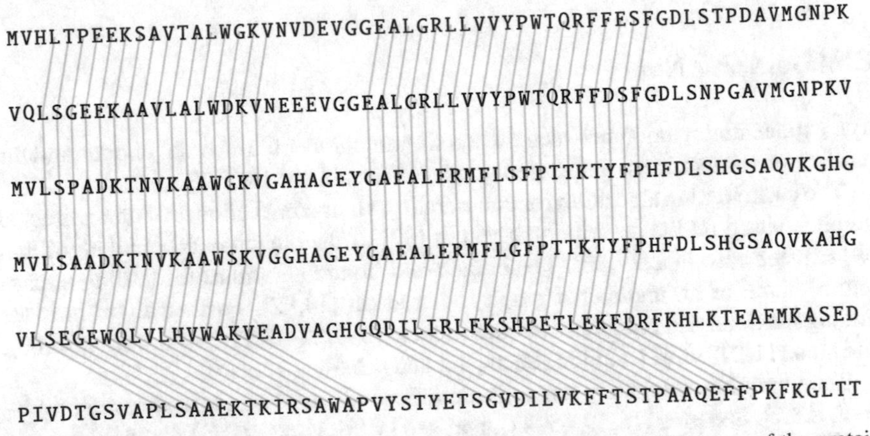

```
MVHLTPEEKSAVTALWGKVNVDEVGGEALGRLLVVYPWTQRFFESFGDLSTPDAVMGNPK

VQLSGEEKAAVLALWDKVNEEEVGGEALGRLLVVYPWTQRFFDSFGDLSNPGAVMGNPKV

MVLSPADKTNVKAAWGKVGAHAGEYGAEALERMFLSFPTTKTYFPHFDLSHGSAQVKGHG

MVLSAADKTNVKAAWSKVGGHAGEYGAEALERMFLGFPTTKTYFPHFDLSHGSAQVKAHG

VLSEGEWQLVLHVWAKVEADVAGHGQDILIRLFKSHPETLEKFDRFKHLKTEAEMKASED

PIVDTGSVAPLSAAEKTKIRSAWAPVYSTYETSGVDILVKFFTSTPAAQEFFPKFKGLTT
```

Fig. 6.18 Maximum cliques in the alignment graph for the pairwise alignments of the protein sequences of Fig. 6.15. Only maximum clique edges between consecutive layers in the alignment graph are shown for clarity

```
-------MVHLTPEEKSAVTALWGKVN--VDEVGGEALGRLLVVYPWTQRFFESFGDLSTPDAVMGNPK-
       ||||||||||||||||        |||||||||||||||||||||||
--------VQLSGEEKAAVLALWDKVN--EEEVGGEALGRLLVVYPWTQRFFDSFGDLSNPGAVMGNPKV
        ||||||||||||||        |||||||||||||||||||||||
--------MVLSPADKTNVKAAWGKVGAHAGEYGAEALERMFLSFPTTKTYFPHF-DLSHGSAQVKGHG-
        ||||||||||||||        |||||||||||||||||||||||
--------MVLSAADKTNVKAAWSKVGGHAGEYGAEALERMFLGFPTTKTYFPHF-DLSHGSAQVKAHG-
        ||||||||||||||        |||||||||||||||||||||||
---------VLSEGEWQLVLHVWAKVEADVAGHGQDILIRLFKSHPETLEKFDRFKHLKTEAEMKASED-
         ||||||||||||||        |||||||||||||||||||||||
PIVDTGSVAPLSAAEKTKIRSAWAPVYSTYETSGVDILVKFFTSTPAAQEFFPKFKGLTT----------
```

Fig. 6.19 Multiple sequence alignment for the protein sequences of Fig. 6.15, based on the maximum cliques in the alignment graph of Fig. 6.18

Summary

The related problems of finding maximal and maximum cliques, independent sets, and vertex covers of an undirected graph were addressed in this chapter. Simple algorithms are given in detail for enumerating all maximal cliques and finding a maximum clique in an undirected graph. The dual problem of finding a maximum independent set in an undirected graph is solved by finding a maximum clique in the complement of the graph, and a simpler algorithm is given for finding a maximum independent set in a tree. Most of these algorithms are based on the backtracking and branch-and-bound techniques reviewed in Sect. 2.4, while the algorithm for finding a maximum independent set in a tree is based on the methods of tree traversal discussed in Chap. 3. The algorithms for finding a maximum independent set are also used for solving the related problem of finding a minimum vertex cover, in both trees and undirected graphs. Further, the application of maximum cliques to the multiple alignment of protein sequences in computational molecular biology is also discussed in detail.

Bibliographic Notes

The exponential relation between the maximum number of maximal cliques and the number of vertices in an undirected graph was established in [37].

The backtracking algorithm for enumerating all maximal cliques of an undirected graph is based on the description given in [33] of the branch-and-bound algorithm of [12]. See also [1,2,29,38]. The branch-and-bound algorithm for finding a maximum clique of an undirected graph is based on [14,42]. Further algorithms for finding maximal and maximum cliques include [3–7,22,27,32,34,40,41,54,56,60]. See also [11,21,28,61] and [35, Sects. 4.3 and 4.6.3].

The algorithm for finding a maximum independent set of a tree is based on [49, Sect. 8.5.2]. See also [16,30,45–47], and see [9,26,31,43,50,55] for maximal and maximum independent set enumeration algorithms upon general graphs. Algorithms for finding a minimum vertex cover include [8,13,17,20,39]. See also [15,18,19].

The application of maximal cliques to multiple sequence alignment is based on [49, Sect. 8.7.8]. See also [23,44,48,57–59]. A comprehensive comparison of different approaches and programs for multiple sequence alignment can be found in [25,36,52,53]. See also [24, Chap. 14].

Problems

6.1 Find all non-trivial cliques, all maximal cliques, and all maximum cliques of the graph in Fig. 6.20.

6.2 Find all maximal independent sets and all maximum independent sets of the graph in Fig. 6.20.

Fig. 6.20 Undirected graph
for problems 6.1–6.4

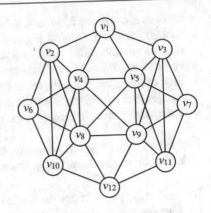

6.3 Give the complement of the graph in Fig. 6.20, and find all maximal cliques and all maximum cliques of the graph complement.

6.4 Find all minimal vertex covers and all minimum vertex covers of the graph in Fig. 6.20.

6.5 Give the number of maximal independent sets of the complete bipartite graph $K_{p,q}$ on $p + q$ vertices, as well as the number of vertices in each of them.

6.6 Recall from Chap. 2 that a vertex coloring of an undirected graph is an assignment of a color to each vertex of the graph, such that adjacent vertices are assigned different colors. Recall also that the chromatic number $\chi(G)$ of an undirected graph G is the minimum number of different colors required for a vertex coloring of G. Show that $\chi(G) \geqslant \omega(G)$, and that $\chi(G) \geqslant \lceil n/\beta(G) \rceil$, for an undirected graph G with n vertices.

Exercises

Exercise 6.1 Modify the maximal clique enumeration algorithm to collect the maximal cliques in a rooted tree, as shown in Fig. 6.6, instead of a list of sets of vertices. Use the simple tree layout algorithm of Sect. 3.5 for the visualization of the search tree of maximal cliques.

Exercise 6.2 In the maximal clique enumeration algorithm, the generation of duplicate cliques is avoided by only considering those candidate vertices which are greater than the vertices in the current clique, according to the order on the vertices fixed by the representation of the graph. However, the particular representation of the graph may have a significant influence on the performance of the algorithm. Some natural orderings of the vertices of an undirected graph include the following:

- Vertices are ordered from lowest to highest degree,
- Each of the vertices has the highest degree in the subgraph induced by the previous vertices and the vertex of highest degree,
- Vertices are ordered from highest to lowest degree,
- Each of the vertices has the lowest degree in the subgraph induced by the previous vertices and the vertex of lowest degree,
- A random order is imposed on the vertices of the graph.

Perform experiments on random graphs to determine the effect of the previous vertex orderings on the performance of the maximal clique enumeration algorithm.

Exercise 6.3 A benchmark for maximum cliques was set up for the Second DIMACS Implementation Challenge [28] and is available from

ftp://dimacs.rutgers.edu/pub/challenge/graph/

Apply the branch-and-bound maximum clique algorithm to the benchmark graphs.

Exercise 6.4 The $O(n \log n)$ time taken by the tree maximum independent set algorithm upon a tree with n nodes, comes from the $O(\log n)$ time taken to insert a node in a set of $O(n)$ nodes. However, the fact that the potential elements of the set are known beforehand, allows using a more efficient set data structure. Give an alternative implementation of the tree maximum independent set algorithm running in $O(n)$ time using $O(n)$ additional space, where n is the number of nodes in the tree.

Exercise 6.5 If the root of a tree belongs to an independent set of the tree, then none of the children of the root may belong to the independent set. Otherwise, if the root does not belong to the independent set, any independent sets of the subtrees rooted at each of the children of the root of the tree, may be combined into an independent set of the tree. The size of a maximum independent set $I \subseteq V$ of a rooted tree $T = (V, E)$ is thus described by the recurrence relations

$$\begin{cases} \beta(T[v]) = \max\left(\beta'(T[v]), 1 + \sum_{(v,w) \in E} \beta'(T[w])\right) \\ \beta'(T[v]) = \sum_{(v,w) \in E} \beta(T[w]) \end{cases}$$

for all non-leaves $v \in V$ and, for all leaves $w \in V$,

$$\begin{cases} \beta(T[w]) = 1 \\ \beta'(T[w]) = 0 \end{cases}$$

where $T[v]$ denotes the subtree of T rooted at node $v \in V$. Give an algorithm for computing the independence number $\beta(T) = \beta(T[root[T]])$ of a rooted tree T. Give also the time and space complexity of the algorithm.

Exercise 6.6 The $O(n \log n)$ time taken by the tree minimum vertex cover algorithm upon a tree with n nodes, comes from the $O(\log n)$ time taken to insert a node in a set of $O(n)$ nodes. However, the fact that the potential elements of the set are known beforehand, allows using a more efficient set data structure. Give an alternative implementation of the tree minimum vertex cover algorithm running in $O(n)$ time using $O(n)$ additional space, where n is the number of nodes in the tree.

Exercise 6.7 Adapt the branch-and-bound maximum clique algorithm, using the relationship between the chromatic number and the clique number of an undirected graph of Problem 6.6 as an alternative upper bound on the size of a maximum clique.

Exercise 6.8 Perform experiments on random graphs to compare the efficiency of the branch-and-bound maximum clique algorithm, using the upper bound on the size of a maximum clique based on the size of the clique being extended and the number of candidate vertices with which it can be extended, on the one hand and using, on the other hand, the upper bound of Problem 6.6 and Exercise 6.7.

Exercise 6.9 Recall from Sect. 6.4 that the alignment graph of k sequences with n residues each is an undirected graph with nk vertices and at most $nk(k-1)/2$ edges, where there is a vertex in the alignment graph for each residue in each sequence, and for each pair of aligned residues in each of the pairwise sequence alignments, there is an edge in the alignment graph between the corresponding vertices. Give an algorithm for building the alignment graph of $k \geqslant 2$ sequences with n residues each, given their $k(k-1)/2$ pairwise alignments.

Exercise 6.10 Both the backtracking maximal clique enumeration algorithm and the branch-and-bound maximum clique algorithm take general graphs as input. In the particular case of an alignment graph of $k \geqslant 2$ sequences, though, the maximal cliques of interest are those having exactly k vertices, which are thus (as long as there is such a k-vertex clique in the alignment graph) maximum cliques. Adapt the branch-and-bound maximum clique algorithm in order to find all k-cliques in an alignment graph of $k \geqslant 2$ sequences.

References

1. Akkoyunlu EA (1973) The enumeration of maximal cliques of large graphs. SIAM J Comput 2(1):1–6
2. Augustson JG, Minker J (1970) An analysis of some graph theoretical cluster techniques. J ACM 17(4):571–588
3. Babel L, Tinhofer G (1990) A branch and bound algorithm for the maximum clique problem. Zeitschrift für Operations Res 34(3):207–217
4. Balas E, Xue J (1991) Minimum weighted coloring of triangulated graphs, with application to maximum weight vertex packing and clique finding in arbitrary graphs. SIAM J Comput 20(2):209–221

5. Balas E, Xue J (1992) Addendum: Minimum weighted coloring of triangulated graphs, with application to maximum weight vertex packing and clique finding in arbitrary graphs. SIAM J Compu 21(5):1000

6. Balas E, Xue J (1996) Weighted and unweighted maximum clique algorithms with upper bounds from fractional coloring. Algorithmica 15(5):397–412

7. Balas E, Yu CS (1986) Finding a maximum clique in an arbitrary graph. SIAM J Comput 15(4):1054–1068

8. Balasubramanian R, Fellows MR, Raman V (1998) An improved fixed-parameter algorithm for vertex cover. Inf Process Lett 65(3):163–168

9. Beigel R (1999) Finding maximum independent sets in sparse and general graphs. In: Robert TW, Tarjan E (eds) Proceedings 10th annual ACM-SIAM Symp. Discrete algorithms. ACM, New York, pp 856–857

10. Berman HM, Westbrook J, Feng Z, Gilliland G, Bhat TN, Weissig H, Shindyalov IN, Bourne PE (2000) The protein data bank. Nucleic Acids Res 28(1):235–242

11. Bomze I, Budinich M, Pardalos PM, Pelillo M (1999) The maximum clique problem. In: Du DZ, Pardalos PM (eds) Handbook of combinatorial optimization, vol. Supp. A. Kluwer, Dordrecht, pp 1–74

12. Bron C, Kerbosch J (1973) ACM algorithm 457: finding all cliques of an undirected graph. Comm ACM 16(9):575–577

13. Buss JF, Goldsmith J (1993) Nondeterminism within P. SIAM J Comput 22(3):560–572

14. Carraghan R, Pardalos PM (1990) An exact algorithm for the maximum clique problem. Oper Res Lett 9(6):375–382

15. Chandran LS, Grandoni F (2005) Refined memorisation for vertex cover. Inf Process Lett 93(3):125–131

16. Chang YH, Wang JS, Lee RCT (1994) Generating all maximal independent sets on trees in lexicographic order. Inf Sci 76(3–4):279–296

17. Chen J, Kanj IA, Jia W (2001) Vertex cover: further observations and further improvements. J Algorithms 41(2):280–301

18. Chen J, Kanj IA, Xia G (2010) Improved upper bounds for vertex cover. Theor Comput Sci 411(40–42):3736–3756

19. Chen J, Liu L, Jia W (2000) Improvement on vertex cover for low degree graphs. Networks 35(4):253–259

20. Downey RG, Fellows MR, Stege U (1999) Parameterized complexity: a framework for systematically confronting computational intractability. In: Roberts F, Kratochvil J, Nesetril J (eds) Contemporary trends in discrete mathematics, DIMACS: series in discrete mathematics and theoretical computer science, vol 49. American Mathematical Society, Providence, RI, pp 49–99

21. Eblen JD, Phillips CA, Rogers GL, Langston MA (2012) The maximum clique enumeration problem: Algorithms, applications, and implementations. BMC Bioinf 13(Suppl. 10):S5

22. Gerhards L, Lindenberg W (1979) Clique detection for nondirected graphs: two new algorithms. Computing 21(4):295–322

23. Gotoh O (1990) Consistency of optimal sequence alignments. Bull Math Biol 52(4):509–525

24. Gusfield D (1997) Algorithms on strings, trees, and sequences: computer science and computational biology. Cambridge University Press, Cambridge, England

25. Hickson RE, Simon C, Perrey SW (2000) The performance of several multiple protein-sequence alignment programs in relation to secondary-structure features for an rRNA sequence. Molecular Biol Evol 17(4):530–539

26. Jian T (1986) An $O(2^{0.304n})$ algorithm for solving maximum independent set problem. IEEE Trans Comput 35(9):847–851

27. Jiang H, Li CM, Manyá F (2017). In: Singh S, Markovitch S (eds) An exact algorithm for the maximum weight clique problem in large graphs. AAAI Press, pp 830–838

28. Johnson DS, Trick MA (eds) (1996) Cliques, coloring, and satisfiability: second dimacs implementation challenge, DIMACS: series in discrete mathematics and theoretical computer science, vol 26. American Mathematical Society, Providence RI

29. Johnston HJ (1976) Cliques of a graph—variations on the Bron-Kerbosh algorithm. Int J Comput Inf Sci 5(3):209–238

30. Jung H, Mehlhorn K (1988) Parallel algorithms for computing maximal independent sets in trees and for updating minimum spanning trees. Inf Process Lett 27(5):227–236

31. Kikusts P (1986) Another algorithm determining the independence number of a graph. Elektronische Informationsverarbeitung und Kybernetik 22(4):157–166

32. Knödel W (1968) Bestimmung aller maximalen, vollständigen Teilgraphen eines Graphen G nach Stoffers. Computing 3(3):239–240

33. Koch I (2001) Enumerating all connected maximal common subgraphs in two graphs. Theor Comput Sci 250(1–2):1–30

34. Konc J, Janežič D (2007) An improved branch and bound algorithm for the maximum clique problem. MATCH Commun Math Comput Chem 58(3):569–590

35. Kreher DL, Stinson DR (1999) Combinatorial algorithms: generation, enumeration, and search. CRC Press, Boca Raton FL

36. McClure MA, Vasi TK, Fitch WM (1994) Comparative analysis of multiple protein-sequence alignment methods. Molecular Biol Evol 11(4):571–592

37. Moon JW, Moser L (1965) On cliques in graphs. Israel J Math 3(1):23–28

38. Mulligan GD, Corneil DG (1972) Corrections to Bierstone's algorithm for generating cliques. J ACM 19(2):244–247

39. Niedermeier R, Rossmanith P (1999) Upper bounds for vertex cover further improved. In: Meinel C, Tison S (eds) Proceedings 16th symposium theoretical aspects of computer science, vol 1563. Lecture notes in computer science. Springer, Berlin Heidelberg, pp 561–570

40. Osteen RE (1974) Clique detection algorithms based on line addition and line removal. SIAM J Appl Math 26(1):126–135

41. Östergård PRJ (2002) A fast algorithm for the maximum clique problem. Discret Appl Math 120(1–3):197–207

42. Pardalos PM, Rodgers GP (1992) A branch and bound algorithm for the maximum clique problem. Comput Oper Res 19(5):363–375

43. Robson JM (1986) Algorithms for maximum independent sets. J Algorithms 7(3):425–440

44. Roytberg MA (1992) A search for common patterns in many sequences. Comput Appl Biosci 8(1):57–64

45. Sagan BE (1988) A note on independent sets in trees. SIAM J Discret Math 1(1):105–108

46. Sajith G, Saxena S (1994) Optimal parallel algorithms for coloring bounded degree graphs and finding maximal independent sets in rooted trees. Inf Process Lett 49(6):303–308

47. Sajith G, Saxena S (1995) Corrigendum: Optimal parallel algorithms for coloring bounded degree graphs and finding maximal independent sets in rooted trees. Inf Process Lett 54(5):305

48. Schuler GD, Altschul SF, Lipman DJ (1991) A framework for multiple sequence construction and analysis. Proteins: Struct Funct Genetics 9(1):180–190

49. Skiena SS (1998) The algorithm design manual, 1st edn. Springer, Berlin Heidelberg

50. Tarjan RE, Trojanowski AE (1977) Finding a maximum independent set. SIAM J Comput 6(3):537–546

51. Thompson JD, Higgins DG, Gibson TJ (1994) CLUSTAL W: Improving the sensitivity of progressive multiple sequence alignment through sequence weighting, positions-specific gap penalties and weight matrix choice. Nucleic Acids Res 22(22):4673–4680

52. Thompson JD, Linard B, Lecompte O, Poch O (2011) A comprehensive benchmark study of multiple sequence alignment methods: Current challenges and future perspectives. PLoS ONE 6(3):e18093

53. Thompson JD, Plewniak F, Poch O (1999) A comprehensive comparison of multiple sequence alignment programs. Nucleic Acids Res 27(13):2682–2690

54. Tomita E, Matsuzaki S, Nagao A, Ito H, Wakatsuki M (2017) A much faster algorithm for finding a maximum clique with computational experiments. J Inf Process 25(1):667–677
55. Tsukiyama S, Ide M, Ariyoshi H, Shirakawa I (1977) A new algorithm for generating all the maximal independent sets. SIAM J Comput 6(3):505–517
56. Vassilevska V (2009) Efficient algorithms for clique problems. Inf Process Lett 109(4):254–257
57. Vihinen M (1988) An algorithm for simultaneous comparison of several sequences. Comput Appl Biosci 4(1):89–92
58. Vingron M, Argos P (1991) Motif recognition and alignment of many sequences by comparison of dot-matrices. J Molecular Biol 218(1):33–43
59. Vingron M, Pevzner PA (1995) Multiple sequence comparison and consistency on multipartite graphs. Adv Appl Math 16(1):1–22
60. Wood DR (1997) An algorithm for finding a maximum clique in a graph. Oper Res Lett 21(5):211–217
61. Wu Q, Hao JK (2015) A review on algorithms for maximum clique problems. Euro J Oper Res 242(3):693–709

Graph Isomorphism

7

7.1 Graph Isomorphism

The question of equality or identity is fundamental for the objects of any universe of mathematical discourse. For example, a pair of fractions (which may look different) are the same if their difference is zero, two sets (which may be represented in quite different ways) are the same if they contain the same elements, etc. In the same vein, a pair of graphs may also look different but actually have the same structure.

Two graphs are isomorphic if there is a bijective correspondence between their vertex sets which preserves and reflects adjacencies—that is, such that two vertices of a graph are adjacent if and only if the corresponding vertices of the other graph are adjacent.

Definition 7.1 Two graphs $G_1 = (V_1, E_1)$ and $G_2 = (V_2, E_2)$ are **isomorphic**, denoted by $G_1 \cong G_2$, if there is a bijection $M \subseteq V_1 \times V_2$ such that, for every pair of vertices $v_i, v_j \in V_1$ and $w_i, w_j \in V_2$ with $(v_i, w_i) \in M$ and $(v_j, w_j) \in M$, $(v_i, v_j) \in E_1$ if and only if $(w_i, w_j) \in E_2$. In such a case, M is a graph isomorphism of G_1 to G_2.

Isomorphism expresses what, in less formal language, is meant when two graphs are said to be the same graph. Two isomorphic graphs may be depicted in such a way that they look very different—they are differently labeled, perhaps also differently drawn, and it is for this reason that they look different.

Example 7.1 The following two graphs are isomorphic, and $M = \{(v_1, w_1), (v_2, w_2), (v_3, w_3), (v_4, w_4), (v_5, w_5), (v_6, w_6)\}$ is a graph isomorphism of G_1 to G_2.

© The Author(s), under exclusive license to Springer Nature Switzerland AG 2021
G. Valiente, *Algorithms on Trees and Graphs*, Texts in Computer Science,
https://doi.org/10.1007/978-3-030-81885-2_7

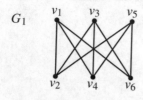

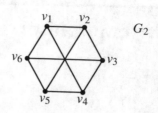

Nonisomorphism of graphs is not usually hard to prove, because several *invariants* or necessary conditions for isomorphism are not difficult to compute. These are properties that do not depend on the presentation or labeling of a graph. For instance, two graphs cannot be isomorphic if they differ in their order, size, or degree sequence.

Remark 7.1 Given two graphs $G_1 = (V_1, E_1)$ and $G_2 = (V_2, E_2)$ with $V_1 = \{u_1, \ldots, u_n\}$ and $V_2 = \{v_1, \ldots, v_n\}$, a necessary condition for $G_1 \cong G_2$ is that the multisets $\{\Gamma(u_i) \mid 1 \leqslant i \leqslant n\}$ and $\{\Gamma(v_i) \mid 1 \leqslant i \leqslant n\}$ be equal.

On the other hand, invariants are not sufficient conditions for isomorphism and nothing can be concluded about two graphs which share an invariant.

Example 7.2 The following two graphs are not isomorphic, although they have the same number of vertices, the same number of edges, and are both 4-regular.

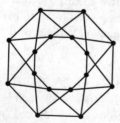

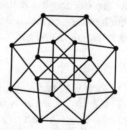

There are, in fact, only 10 paths of length 2 starting from any vertex (and only 5 vertices at distance 2 of any given vertex) of the graph to the left-hand side, but 12 paths of length 2 starting from any vertex (and 6 vertices at distance 2 of any given vertex) of the graph to the right-hand side:

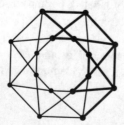

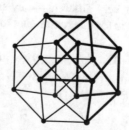

Isomorphism of graphs is usually much harder to prove than nonisomorphism of graphs, because all known *certificates* or necessary and sufficient conditions for graph isomorphism are as difficult to compute as graph isomorphism itself. The graph isomorphism problem is actually not only of practical interest but also of theoretical relevance. From a complexity-theoretical point of view, graph isomorphism is one of the few NP problems believed neither to be in P nor to be NP-complete.

7.1.1 An Algorithm for Graph Isomorphism

The general backtracking scheme presented in Sect. 2.2 can be instantiated to yield a simple algorithm for testing graph isomorphism based on the next result, which follows from Definition 7.1.

Notice first that two graphs are isomorphic if there is a bijective correspondence between their vertex sets such that two vertices of a graph are adjacent if and only if the corresponding vertices of the other graph are adjacent, while in most applications further information is attached to vertices and edges in the form of vertex and edge labels. Then, two labeled graphs are isomorphic if the underlying graphs are isomorphic and, furthermore, corresponding vertices and edges share the same label.

Recall that for a graph $G = (V, E)$, the label of a vertex $v \in V$ is stored in the attribute $v.label$ and the label of an edge $e \in E$ is stored in the attribute $e.label$. Unlabeled graphs are dealt with in the following algorithms as labeled graphs with all vertex and edge labels set to nil, that is, undefined.

Lemma 7.1 *Let $G_1 = (V_1, E_1)$ and $G_2 = (V_2, E_2)$ be two graphs, let $V \subseteq V_1$ and $W \subseteq V_2$, and let $M \subseteq V \times W$ be an isomorphism of the subgraph of G_1 induced by V to the subgraph of G_2 induced by W. For any vertices $v \in V_1 \setminus V$ and $w \in V_2 \setminus W$, $M \cup \{(v, w)\}$ is an isomorphism of the subgraph of G_1 induced by $V \cup \{v\}$ to the subgraph of G_2 induced by $W \cup \{w\}$ if and only if the following conditions*

- *$(x, v) \in E_1$ if and only if $(y, w) \in E_2$*
- *$(v, x) \in E_1$ if and only if $(w, y) \in E_2$*

hold for all vertices $x \in V$ and $y \in W$ with $(x, y) \in M$.

Proof Let $G_1 = (V_1, E_1)$ and $G_2 = (V_2, E_2)$ be two graphs, let $V \subseteq V_1$ and $W \subseteq V_2$, and let $M \subseteq V \times W$ be an isomorphism of the subgraph of G_1 induced by V to the subgraph of G_2 induced by W. See Fig. 7.1. Let also $v \in V_1 \setminus V$ and $w \in V_2 \setminus W$.

Suppose $(x, v) \in E_1$ if and only if $(y, w) \in E_2$ and $(v, x) \in E_1$ if and only if $(w, y) \in E_2$, for all $(x, y) \in M$. Then, by Definition 7.1, $M \cup \{(v, w)\}$ is an isomorphism of the subgraph of G_1 induced by $V \cup \{v\}$ to the subgraph of G_2 induced by $W \cup \{w\}$.

Fig. 7.1 Proof of Lemma 7.1

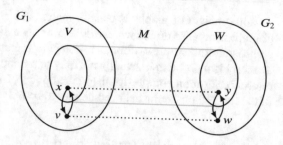

Conversely, let $M \cup \{(v, w)\}$ be an isomorphism of the subgraph of G_1 induced by $V \cup \{v\}$ to the subgraph of G_2 induced by $W \cup \{w\}$, and let $x \in V$ and $y \in W$ with $(x, y) \in M$. Since $(u, v), (x, y) \in M \cup \{(v, w)\}$, by Definition 7.1 it holds that $(x, v) \in E_1$ if and only if $(y, w) \in E_2$ and $(v, x) \in E_1$ if and only if $(w, y) \in E_2$. □

The following backtracking algorithm for graph isomorphism extends an (initially empty) vertex mapping M in all possible ways, in order to enumerate all graph isomorphisms of a graph G_1 to a graph G_2. Although the enumeration of all graph isomorphisms is sufficient but not necessary for testing whether or not two graphs are isomorphic, it will also be used in Sect. 7.2 for determining the automorphism group of a graph.

Each graph isomorphism mapping M is stored in a dictionary of vertices of G_1 to vertices of G_2.

> **function** *graph_isomorphism*(G_1, G_2)
>   let L be an empty list (of dictionaries of vertices to vertices)
>   let M be an empty dictionary (of vertices to vertices)
>   **if** G_1 and G_2 have the same number of vertices and edges **then**
>   let v be the first vertex of G_1
>   *extend_graph_isomorphism*(G_1, G_2, M, v, L)
>   **return** L

The actual extension of a mapping is done by the following recursive procedure, which extends a vertex mapping $M \subseteq V \times W$ of the subgraph of G_1 induced by V to the subgraph of G_2 induced by W by mapping vertex $v \in V_1 \setminus V$ to each vertex $w \in V_2 \setminus W$ in turn, that is, to each vertex which has not been mapped to yet, making a recursive call upon the successor of vertex v in G_1 whenever $M \cup \{(u, w)\}$ is a graph isomorphism of the subgraph of G_1 induced by $V \cup \{v\}$ to the subgraph of G_2 induced by $W \cup \{w\}$.

All such vertex mappings M which are defined for all vertices of G_1, that is, all isomorphisms of graph G_1 to graph G_2, are collected in a list L of dictionaries of vertices of G_1 to vertices of G_2.

procedure *extend_graph_isomorphism*(G_1, G_2, M, v, L)
    let M' be a copy of M
    $V_2 \leftarrow G_2.vertices()$
    for all vertices v' of G_1 **do**
        if $v' \in M'$ **then**
            delete $M[v']$ from V_2
    for all vertices w of G_2 **do**
        if $v.label \neq w.label$ **then**
            delete w from V_2
    for all vertices w in V_2 **do**
        if *preserves_adjacencies*(G_1, G_2, M', v, w) **then**
            $M'[v] \leftarrow w$
            if v is the last vertex of G_1 **then**
                append M' to L
            else
                let v' be the next vertex after v in G_1
                extend_graph_isomorphism(G_1, G_2, M', v', L)

The following function implements the obvious adjacency test: M may be extended by mapping vertex $v \in V_1$ to vertex $w \in V_2$ if and only if vertices v and w have the same indegree and the same outdegree, every (already mapped) predecessor x of vertex v was mapped to a predecessor of vertex w, and every (already mapped) successor x of vertex v was mapped to a successor of vertex w. See Fig. 7.2.

Fig. 7.2 Adjacency test for extending a graph isomorphism

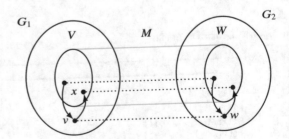

```
function preserves_adjacencies(G₁, G₂, M, v, w)
    if G₁.indeg(v) ≠ G₂.indeg(w) or G₁.outdeg(v) ≠ G₂.outdeg(w) then
        return false
    for all edges e coming into vertex v in G₁ do
        let x be the source vertex of edge e in G₁
        if x ∈ M and M[x] is not adjacent with w in G₂ then
            return false
    for all edges e going out of vertex v in G₁ do
        let x be the target vertex of edge e in G₁
        if x ∈ M and w is not adjacent with M[x] in G₂ then
            return false
    return true
```

Example 7.3 The undirected graphs of Fig. 7.3 are isomorphic. Let $[v_1, v_2, v_3, v_4, v_5]$ and $[w_1, w_2, w_3, w_4, w_5]$ be the order among the vertices of the graphs fixed by their representation. Then, the graph isomorphism M of G_1 to G_2 found by the backtracking algorithm for graph isomorphism is $M = \{(v_1, w_2), (v_2, w_3), (v_3, w_1), (v_4, w_5), (v_5, w_4)\}$. The sequence of calls to the recursive procedure *extend graph isomorphism* upon graphs G_1 and G_2, until a graph isomorphism M of G_1 to G_2 is found, shown in Fig. 7.4, corresponds to a preorder traversal of the search tree shown

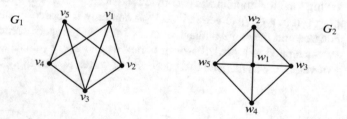

Fig. 7.3 Isomorphic graphs

M	v
{}	v_1
$\{(v_1, w_1)\}$	v_2
$\{(v_1, w_2)\}$	v_2
$\{(v_1, w_2), \ v_2, w_1)\}$	v_3
$\{(v_1, w_2), \ v_2, w_3)\}$	v_3
$\{(v_1, w_2), \ v_2, w_3), (v_3, w_1)\}$	v_4
$\{(v_1, w_2), \ v_2, w_3), (v_3, w_1), (v_4, w_4)\}$	v_5
$\{(v_1, w_2), \ v_2, w_3), (v_3, w_1), (v_4, w_5)\}$	v_5
$\{(v_1, w_2), \ v_2, w_3), (v_3, w_1), (v_4, w_5), (v_5, w_4)\}$	

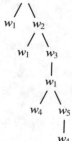

Fig. 7.4 Sequence of calls (left) and search tree (right) for graph isomorphism

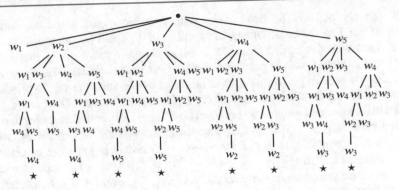

Fig. 7.5 Search tree for graph isomorphism

in Fig. 7.4. The levels of the search tree correspond to the vertices of G_1, and the nodes stand for vertices of G_2 to which the vertex of G_1 might be mapped to. The rightmost path from the root to a leaf corresponds to the graph isomorphism found, mapping vertices $[v_1, v_2, v_3, v_4, v_5]$ to vertices $[w_2, w_3, w_1, w_5, w_4]$, respectively. The whole search tree that corresponds to finding all isomorphic copies of G_1 in G_2 is shown in Fig. 7.5. Those leaves of the search tree representing an isomorphism of G_1 to G_2 are marked by a star.

Remark 7.2 Correctness of the backtracking algorithm for graph isomorphism follows from the fact that a graph isomorphism M of an (initially empty) subgraph of a graph G_1 to an (initially empty) subgraph of a graph G_2, is extended in all possible ways, one vertex pair at a time, until either M maps all the vertices of G_1 to all the vertices of G_2 (and $G_1 \cong G_2$) or no mapping M can be further extended (and $G_1 \not\cong G_2$).

Regarding computational complexity of the backtracking algorithm for graph isomorphism, the worst case arises when the graphs are complete and isomorphic. In this situation, the whole search tree will have to be traversed and, since vertex v_1 may be mapped to each of the n vertices of G_2, vertex v_2 may be mapped to $n - 1$ vertices (each of the n vertices of G_2 except the vertex of G_2 which vertex v_1 was mapped to), and so on, the search tree will have n nodes on the first level, $n(n - 1)$ nodes on the second level, and $n(n - 1) \cdots (n - (n - 1)) = n!$ nodes on the n-th level. Furthermore, every one of the $n!$ leaves of the search tree will correspond to a graph isomorphism.

Theorem 7.1 *The backtracking algorithm for graph isomorphism runs in worst-case $O((n + 1)!)$ time and $O(n^2)$ space, on connected graphs G_1 and G_2 with n vertices.*

Proof Let G_1 and G_2 be two graphs with n vertices and m edges, and let $[v_1, v_2, \ldots, v_n]$ be the order among the vertices of V_1 fixed by the representation of G_1. Since vertex v_1 can, in principle, be mapped to any vertex of G_2, there are at most $O(n)$ recursive calls to procedure *extend graph isomorphism*(G_1, G_2, v_1, L). Now, after having mapped vertex v_1 to some vertex of G_2, vertex v_2 can, in principle, be mapped to any remaining vertex of G_2 and there are at most $O(n(n-1))$ recursive calls to *extend graph isomorphism*(G_1, G_2, v_2, L). In general, after having mapped vertices $v_1, \ldots, v_{i-1}$ to (pairwise different) vertices of G_2, vertex v_i can be mapped, in principle, to any of the $n - (i-1)$ remaining vertices of G_2, and there are at most $O(n(n-1) \cdots (n-(i-1))) = O(n!/(n-i)!)$ recursive calls to procedure *extend graph isomorphism*(G_1, G_2, v_i, L), with $1 \leqslant i \leqslant n$. Therefore, the total number of recursive calls to procedure *extend graph isomorphism* is at most $\sum_{i=1}^{n} O(n!/(n-i)!) = O(\sum_{i=1}^{n} n!/(n-i)!) = O((n+1)!)$. The effort done on each recursive call to *extend graph isomorphism*(G_1, G_2, v, L) is not constant, though, but the total effort done in order to make $O(\deg(v))$ further recursive calls is at most $O(\deg(v))$. Therefore, the algorithm runs in $O((n+1)!)$ time.

Since there is at most one nested recursive call to the procedure *extend graph isomorphism* for each vertex of G_1, the number of nested recursive calls is at most $O(n)$ and given that for each recursive call, a local copy of mapping M is made, each nested recursive call uses $O(n)$ additional space and the algorithm uses $O(n^2)$ space. $\qquad \Box$

Several improvements to the backtracking algorithm for graph isomorphism will be introduced later, when discussing the more general problems of subgraph isomorphism in Sect. 7.3 and maximal common subgraph in Sect. 7.4.

7.2 Graph Automorphism

The automorphism group of a graph reveals information about the structure and symmetries of the graph.

Definition 7.2 An **automorphism** of a graph G is a graph isomorphism between G and itself.

An obvious automorphism is the identity mapping on the vertices of a graph. The inverse of an automorphism of a graph G is also an automorphism of G, and the composition of two automorphisms of G is an automorphism of G. As a matter of fact, the set of all automorphisms of a graph form a group under the operation of composition, called the *automorphism group* of the graph.

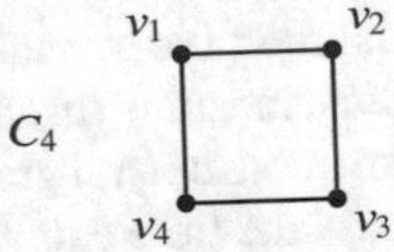

C_4

- (v_1, v_2, v_3, v_4)
- (v_2, v_3, v_4, v_1)
- (v_3, v_4, v_1, v_2)
- (v_4, v_1, v_2, v_3)

- (v_1, v_4, v_3, v_2)
- (v_2, v_1, v_4, v_3)
- (v_3, v_2, v_1, v_4)
- (v_4, v_3, v_2, v_1)

Fig. 7.6 Automorphism group of C_4

For instance, every permutation of the vertex set of the complete graph on n vertices K_n corresponds to an automorphism of K_n and then, the automorphism group of K_n is of order $n!$. The automorphism group of the cycle graph C_n on $n \geqslant 3$ vertices is a group of order $2n$ consisting of n rotations and n reflections, as is the automorphism group of the wheel graph W_{n+1} with $n \geqslant 4$ outer vertices.

Example 7.4 The automorphism group of the cycle graph on 4 vertices C_4 is of order 8. The permutations of the vertex set shown in Fig. 7.6 correspond to the graph automorphisms of C_4.

The backtracking algorithm for graph isomorphism can be used to enumerate all graph automorphisms of a graph G, that is, for determining the automorphism group of a graph G. Each graph automorphism mapping M is stored then in a dictionary of vertices of G to vertices of G.

function *graph_automorphism*(G)
    return *graph_isomorphism*(G, G)

7.3 Subgraph Isomorphism

An important generalization of graph isomorphism is known as subgraph isomorphism. The subgraph isomorphism problem is to determine whether a graph is isomorphic to a subgraph of another graph, and is a fundamental problem with a variety of applications in engineering sciences, organic chemistry, biology, and pattern matching.

Definition 7.3 A **subgraph isomorphism** of a graph $G_1 = (V_1, E_1)$ into a graph $G_2 = (V_2, E_2)$ is an injection $M \subseteq V_1 \times V_2$ such that, for every pair of vertices $v_i, v_j \in V_1$ and $w_i, w_j \in V_2$ with $(v_i, w_i) \in M$ and $(v_j, w_j) \in M$, $(w_i, w_j) \in E_2$ if $(v_i, v_j) \in E_1$. In such a case, M is a subgraph isomorphism of G_1 into G_2.

The problem of determining whether or not a graph is isomorphic to a subgraph of another graph belongs to the class of NP-complete problems, meaning that all known algorithms for solving the subgraph isomorphism problem upon general graphs (that is, graphs without any restriction on any graph parameter) take time exponential in the size of the graphs, and that it is highly unlikely that such an algorithm will be found which takes time polynomial in the size of the graphs.

Most practical applications of subgraph isomorphism require not only determining whether or not a given graph is isomorphic to a subgraph of another given graph, but also finding all subgraphs of a given graph which are isomorphic to another given graph. Already the number of subgraph isomorphisms of a graph into another graph can be exponential in the size of the graphs, though.

Example 7.5 There are $q!/(q-p)!$ different subgraph isomorphisms of the complete graph on p vertices $K_p = (V_1, E_1)$ into the complete graph on q vertices $K_q = (V_2, E_2)$, with $p \leqslant q$. (Since the graphs are complete, any injection $M \subseteq V_1 \times V_2$ will be a subgraph isomorphism of K_p into K_q, and there are $\binom{q}{p} = q!/(q-p)!/p!$ different injections of a set of p elements into a set of q elements, the complete graph on p vertices induced by each of which has $p!$ different automorphisms.) Table 7.1 gives some examples for small graphs.

Remark 7.3 Notice that in the definition of subgraph isomorphism, it is not required that $(v_i, v_j) \in E_1$ if $(w_i, w_j) \in E_2$. Under the additional condition of not only preserving the structure of graph G_1 but also reflecting in G_1 the structure of the subgraph of G_2 induced by M, the vertex mapping M is called an **induced subgraph isomorphism** of G_1 into G_2.

It follows from Definition 7.3 that if $(v, w) \in M$ in a subgraph isomorphism M of a graph G_1 into a graph G_2, then the degree of vertex v in G_1 cannot be greater than the degree of vertex w in G_2.

Lemma 7.2 *Let M be a subgraph isomorphism of a graph G_1 into a graph G_2. Then, $\deg(v) \leqslant \deg(w)$ for all $(v, w) \in M$.*

Proof Let $M \subseteq V_1 \times V_2$ be a subgraph isomorphism of a graph $G_1 = (V_1, E_1)$ into a graph $G_2 = (V_2, E_2)$, and let $(v, w) \in M$. Let also $\Gamma(v) = \{x \in V_1 \mid (x, v) \in E_1\} \cup \{x \in V_1 \mid (v, x) \in E_1\}$, and let $\Gamma(w) = \{y \in V_2 \mid (y, w) \in E_2\} \cup \{y \in V_2 \mid (w, y) \in E_2\}$. Then, $\deg(v) = |\Gamma(v)|$ and $\deg(w) = |\Gamma(w)|$. By Definition 7.3, for all $(x, y) \in M$ with $x \in \Gamma(v)$ it must be $y \in \Gamma(w)$. Then, M being an injection, it must be $|\Gamma(v)| \leqslant |\Gamma(w)|$. Therefore, $\deg(v) \leqslant \deg(w)$. $\square$

Table 7.1 Number of different subgraph isomorphisms of the complete graph on p vertices K_p into the complete graph on q vertices K_q, with $1 \leqslant p \leqslant q \leqslant 12$

	K_1	K_2	K_3	K_4	K_5	K_6	K_7	K_8	K_9	K_{10}	K_{11}	K_{12}
K_1	1	2	3	4	5	6	7	8	9	10	11	12
K_2		2	6	12	20	30	42	56	72	90	110	132
K_3			6	24	60	120	210	336	504	720	990	1,320
K_4				24	120	360	840	1,680	3,024	5,040	7,920	11,880
K_5					120	720	2,520	6,720	15,120	30,240	55,440	95,040
K_6						720	5,040	20,160	60,480	151,200	332,640	665,280
K_7							5,040	40,320	181,440	604,800	1,663,200	3,991,680
K_8								40,320	362,880	1,814,400	6,652,800	19,958,400
K_9									362,880	3,628,800	19,958,400	79,833,600
K_{10}										3,628,800	39,916,800	239,500,800
K_{11}											39,916,800	479,001,600
K_{12}												479,001,600

7.3.1 An Algorithm for Subgraph Isomorphism

The general backtracking scheme presented in Sect. 2.2 can also be instantiated to yield a simple algorithm for subgraph isomorphism based on the next result, which follows from Definition 7.3.

Notice first that there is a subgraph isomorphism between two labeled graphs if there is a subgraph isomorphism between the underlying graphs and, furthermore, corresponding vertices and edges share the same label.

Lemma 7.3 *Let $G_1 = (V_1, E_1)$ and $G_2 = (V_2, E_2)$ be two graphs, let $V \subseteq V_1$ and $W \subseteq V_2$, and let $M \subseteq V \times W$ be an isomorphism of the subgraph of G_1 induced by V to a subgraph of G_2 with vertex set W. For any vertices $v \in V_1 \setminus V$ and $w \in V_2 \setminus W$, $M \cup \{(v, w)\}$ is an isomorphism of the subgraph of G_1 induced by $V \cup \{v\}$ to a subgraph of G_2 with vertex set $W \cup \{w\}$ if and only if the following conditions*

- *if $(x, v) \in E_1$, then $(y, w) \in E_2$*
- *if $(v, x) \in E_1$, then $(w, y) \in E_2$*

hold for all vertices $x \in V_1$ and $y \in V_2$ with $(x, y) \in M$.

Proof Let $G_1 = (V_1, E_1)$ and $G_2 = (V_2, E_2)$ be two graphs, let $V \subseteq V_1$ and $W \subseteq V_2$, and let $M \subseteq V \times W$ be an isomorphism of the subgraph of G_1 induced by V to a subgraph (W, E) of G_2. See Fig. 7.7. Let also $v \in V_1 \setminus V$ and $w \in V_2 \setminus W$.

Suppose $(y, w) \in E_2$ if $(x, v) \in E_1$ and $(w, y) \in E_2$ if $(v, x) \in E_1$, for all $(x, y) \in M$. Then, by Definition 7.1, $M \cup \{(v, w)\}$ is an isomorphism of the subgraph of G_1 induced by $V \cup \{v\}$ to the subgraph $(W \cup \{w\}, E \cup \{(y, w) \in E_2 \mid y \in W, (x, y) \in M, (x, v) \in E_1\}) \cup \{(w, y) \in E_2 \mid y \in W, (x, y) \in M, (v, x) \in E_1\})$ of G_2.

Fig. 7.7 Proof of Lemma 7.3

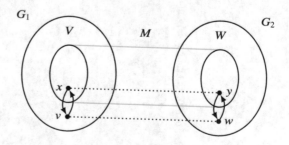

Conversely, let $M \cup \{(v, w)\}$ be an isomorphism of the subgraph of G_1 induced by $V \cup \{v\}$ to a subgraph $(W \cup \{w\}, E)$ of G_2, and let $x \in V$ and $y \in W$ with $(x, y) \in M$. Since $(v, w), (x, y) \in M \cup \{(v, w)\}$, by Definition 7.1 it holds that $(x, v) \in E_1$ if and only if $(y, w) \in E \subseteq E_2$ and $(v, x) \in E_1$ if and only if $(w, y) \in E \subseteq E_2$. $\square$

The following backtracking algorithm for subgraph isomorphism extends an (initially empty) vertex mapping M in all possible ways, in order to enumerate all subgraph isomorphisms of a graph G_1 into a graph G_2. Each subgraph isomorphism mapping M is stored in a dictionary of vertices of G_1 to vertices of G_2.

> **function** *subgraph_isomorphism*(G_1, G_2)
>   let C be an empty dictionary (of vertices to sets of vertices)
>   **for all** vertices v of G_1 **do**
>   let $C[v]$ be an empty set (of vertices)
>   **for all** vertices w of G_2 **do**
>   **if** $G_1.indeg(v) \leqslant G_2.indeg(w)$ and
> $G_1.outdeg(v) \leqslant G_2.outdeg(w)$ and
> $v.label = w.label$ **then**
> $C[v] \leftarrow C[v] \cup \{w\}$
>   let v be the first vertex of G_1
>   let M be an empty dictionary (of vertices to vertices)
>   let L be an empty list (of dictionaries of vertices to vertices)
>   *extend_subgraph_isomorphism*(G_1, G_2, C, v, M, L)
>   **return** L

The vertices of graph G_2 which a vertex of graph G_1 might be mapped to are represented by a set of candidate vertices $C[v]$, for each vertex v of G_1. Lemma 7.2 implies that only those vertices w of G_2 with $\deg(v) \leqslant \deg(w)$ are candidates which vertex v of G_1 might be mapped to.

The actual extension of a subgraph isomorphism mapping is done by the following recursive procedure, which extends a vertex mapping M of a subgraph of G_1 to a subgraph of G_2 by mapping vertex $v \in V_1$ to each candidate vertex $w \in V_2$ in turn which has not been mapped to yet, making a recursive call upon the successor of vertex v in G_1 whenever the extension of M mapping v to w preserves adjacencies. All such mappings M which are defined for all vertices of G_1, that is, all isomorphisms of graph G_1 to a subgraph of G_2, are collected in a list L of mappings.

procedure *extend_subgraph_isomorphism*(G_1, G_2, C, v, M, L)
  for all vertices w of G_2 **do**
    if $w \in C[v]$ **then**
      $M[v] \leftarrow w$
      let N be an empty dictionary (of vertices to sets of vertices)
      for all vertices x of G_1 **do**
        $N[x] \leftarrow C[x]$
        if $x \neq v$ **then**
          $N[x] \leftarrow N[x] \setminus \{w\}$
      for all vertices y of G_2 **do**
        if $y \neq w$ **then**
          $N[v] \leftarrow N[v] \setminus \{y\}$
      for all vertices w in V_2 **do**
        if *refine_subgraph_isomorphism*(G_1, G_2, C, v, w) **then**
          if v is the last vertex of G_1 **then**
            append a copy of M to L
          else
            let v' be the next vertex after v in G_1
            extend_subgraph_isomorphism(G_1, G_2, N, v', M, L)

The obvious adjacency test, that a subgraph isomorphism mapping M may be extended by mapping vertex $v \in V_1$ to vertex $w \in V_2$ if and only if indeg(v) $\leqslant$ indeg(w), outdeg(v) $\leqslant$ outdeg(w), every (already mapped) predecessor x of vertex v was mapped to a predecessor of vertex w, and every (already mapped) successor x of vertex v was mapped to a successor of vertex w, can be improved in order to earlier detect whether such an extension will lead to a later impossibility to further extend it in any way, because it will cause some unmapped vertex x to have no candidate vertex y which it could be mapped to, based the following result.

Lemma 7.4 *Let $G_1 = (V_1, E_1)$ and $G_2 = (V_2, E_2)$ be two graphs, let $V \subseteq V_1$ and $W \subseteq V_2$, and let $M \subseteq V \times W$ be a graph isomorphism of the subgraph of G_1 induced by V to the subgraph of G_2 induced by W. A necessary condition for M to be contained in some subgraph isomorphism of G_1 into G_2 is that for all vertices $v \in V_1 \setminus V$, there is some vertex $w \in V_2 \setminus W$ such that $M \cup \{(v, w)\}$ is a graph isomorphism of the subgraph of G_1 induced by $V \cup \{v\}$ to the subgraph of G_2 induced by $W \cup \{w\}$.*

Proof Let $M \subseteq V \times W$ be a graph isomorphism of the subgraph of $G_1 = (V_1, E_1)$ induced by $V \subseteq V_1$ to the subgraph of $G_2 = (V_2, E_2)$ induced by $W \subseteq V_2$, and let $v \in V_1 \setminus V$ such that, for all vertices $w \in V_2 \setminus W$, $M \cup \{(v, w)\}$ is not an isomorphism of the subgraph of G_1 induced by $V \cup \{v\}$ to the subgraph of G_2 induced by $W \cup \{w\}$, that is, either $G_1[v] \neq G_2[w]$, or there is some $(x, y) \in M$ such that either $(x, v) \in E_1$ but $(y, w) \notin E_2$, or $(v, x) \in E_1$ but $(w, y) \notin E_2$, or $(x, v) \in E_1$ and $(y, w) \in E_2$ but $G_1[x, v] \neq G_2[y, w]$, or $(v, x) \in E_1$ and $(w, y) \in E_2$ but $G_1[v, x] \neq G_2[w, y]$. Then, by Lemma 7.3, $M \cup \{(v, w)\}$ is not an isomorphism of the subgraph of G_1 induced by $V \cup \{v\}$ to the subgraph of G_2 induced by $W \cup \{w\}$. Since this holds for all vertices $w \in V_2 \setminus W$, the vertex mapping M cannot be extended by mapping vertex $v \in V_1 \setminus V$ to any vertex $w \in V_2 \setminus W$ and therefore, M cannot be extended to a subgraph isomorphism of G_1 into G_2. $\square$

The extended adjacency test given by the previous lemma turns out to be computationally too expensive to be performed upon every potential extension of a subgraph isomorphism mapping, meaning that it does not pay off in practice to systematically perform such an extended adjacency test. A weaker form of the lemma, consisting of testing only those potential extensions of a subgraph isomorphism mapping upon the neighborhood of the vertex last added to the mapping, turns out to perform much better in practice (and in theory).

Lemma 7.5 *Let $G_1 = (V_1, E_1)$ and $G_2 = (V_2, E_2)$ be two graphs, let $V \subseteq V_1$ and $W \subseteq V_2$, and let $M \subseteq V \times W$ be a graph isomorphism of the subgraph of G_1 induced by V to the subgraph of G_2 induced by W. A necessary condition for the extension of a mapping $M \cup \{(x, y)\} \subseteq V \times W$ to be contained in some subgraph isomorphism of G_1 into G_2 is that for all vertices $v \in V_1 \setminus V$ which are adjacent with vertex $x \in V$, there is some vertex $w \in V_2 \setminus W$ which is adjacent with vertex $y \in W$ and such that $M \cup \{(x, y), (v, w)\}$ is a graph isomorphism of the subgraph of G_1 induced by $V \cup \{x, v\}$ to the subgraph of G_2 induced by $W \cup \{y, w\}$.*

The following function allows to detect if the extension of M by mapping vertex v to some vertex $w \in V_2 \setminus W$ cannot lead to any subgraph isomorphism of G_1 into G_2, that is, if such an extension would leave no candidate vertices $y \in V_2 \setminus (W \cup \{w\})$ which some other vertex $x \in V_1 \setminus (V \cup \{v\})$ could be mapped to, by updating the set of candidate vertices $C[x]$ which each vertex $x \in V_1 \setminus (V \cup \{v\})$ might be mapped to, based on the adjacencies of vertex v in G_1 and of vertex w in G_2, and returns true if and only if all these sets of candidate vertices remain nonempty.

```
function refine_subgraph_isomorphism(G₁, G₂, C, v, w)
  if G₁.indeg(v) > G₂.indeg(w) or G₁.outdeg(v) > G₂.outdeg(w) then
    return false
  for all edges e coming into vertex v in G₁ do
    let x be the source vertex of edge e in G₁
    for all vertices y of G₂ do
      if y is not adjacent with w in G₂ then
        C[x] ← C[x] \ {y}
      else
        let e' be the edge from vertex y to vertex w in G₂
        if e.label ≠ e'.label then
          C[x] ← C[x] \ {y}
  for all edges e going out of vertex v in G₁ do
    let x be the target vertex of edge e in G₁
    for all vertices y of G₂ do
      if w is not adjacent with y in G₂ then
        C[x] ← C[x] \ {y}
      else
        let e' be the edge from vertex w to vertex y in G₂
        if e.label ≠ e'.label then
          C[x] ← C[x] \ {y}
  for all vertices x of G₁ do
    if C[x] is empty then
      return false
  return true
```

7.4 Maximal Common Subgraph Isomorphism

Another important generalization of graph isomorphism which also generalizes subgraph isomorphism is known as maximal common subgraph isomorphism. The maximal common subgraph isomorphism problem consists in finding a common subgraph between two graphs which is not a proper subgraph of any other common subgraph between the two graphs, and is also a fundamental problem with a variety of applications in engineering sciences, organic chemistry, biology, and pattern matching.

Definition 7.4 A **common induced subgraph isomorphism** of a graph G_1 to a graph G_2 is a structure (S_1, S_2, M), where S_1 is an induced subgraph of G_1, S_2 is an induced subgraph of G_2, and M is a graph isomorphism of S_1 to S_2. A common induced subgraph isomorphism is **maximal** if there is no common induced subgraph isomorphism (S_1', S_2', M') of G_1 to G_2 such that S_1 is a proper (induced) subgraph of G_1 and S_2 is a proper (induced) subgraph of G_2.

Replacing induced subgraph by subgraph in Definition 7.4 leads to the notion of maximal common subgraph isomorphism. Further, a maximum common (induced) subgraph isomorphism of two graphs is a largest maximal common (induced) subgraph isomorphism of the two graphs.

Example 7.6 A maximum common induced subgraph isomorphism of the following two graphs is $((V_s, E_s), (V_t, E_t), M)$ with

$$V_s = \{v_1, v_2, v_3, v_4\},$$
$$E_s = \{(v_1, v_2), (v_1, v_3), (v_1, v_4), (v_2, v_3), (v_2, v_4), (v_3, v_4),$$
$$(v_2, v_1), (v_3, v_1), (v_4, v_1), (v_3, v_2), (v_4, v_2), (v_4, v_3)\},$$
$$V_t = \{w_2, w_3, w_5, w_6\},$$
$$E_t = \{(w_2, w_3), (w_2, w_5), (w_2, w_6), (w_3, w_5), (w_3, w_6), (w_5, w_6),$$
$$(w_3, w_2), (w_5, w_2), (w_6, w_2), (w_5, w_3), (w_6, w_3), (w_6, w_5)\},$$
$$M = \{(v_1, w_5), (v_2, w_6), (v_3, w_3), (v_4, w_2)\},$$

while a maximum common subgraph isomorphism of the two graphs is $((V_s \cup \{v_5\}, E_s \cup \{(v_4, v_5), (v_5, v_4)\}), (V_t \cup \{w_1\}, E_t \cup \{(w_1, w_2), (w_2, w_1)\}), M \cup \{(v_5, w_1)\})$.

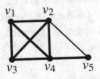 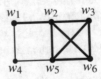

Computation of all maximal common induced subgraph isomorphisms of two graphs can be reduced to finding all maximal cliques in the graph product of the two graphs. The vertices in the graph product are ordered pairs (v, w) with $v \in V_1$ and $w \in V_2$, and there is an edge from vertex (v_i, w_i) to vertex (v_j, w_j) in the graph product if and only if $v_i \neq v_j$, $w_i \neq w_j$, and either there is an edge from vertex v_i to vertex v_j in graph G_1 and an edge from vertex w_i to vertex w_j in graph G_2, or there is no edge from vertex v_i to vertex v_j in graph G_1 and no edge from vertex w_i to vertex w_j in graph G_2.

Definition 7.5 Let $G_1 = (V_1, E_1)$ and $G_2 = (V_2, E_2)$ be two graphs. The **graph product** of G_1 and G_2, denoted by $G_1 \times G_2$, is the graph $G = (V, E)$ with vertex set $V = V_1 \times V_2$ and edge set $E = \{((v_i, w_i), (v_j, w_j)) \in V \times V \mid v_i \neq v_j, w_i \neq w_j, (v_i, v_j) \in E_1, (w_i, w_j) \in E_2\} \cup \{((v_i, w_i), (v_j, w_j)) \in V \times V \mid v_i \neq v_j, w_i \neq w_j, (v_i, v_j) \notin E_1, (w_i, w_j) \notin E_2\}$.

Example 7.7 Let $G_1 = (V_1, E_1)$ and $G_2 = (V_2, E_2)$ be two graphs with $V_1 = \{v_1, v_2\}$, $E_1 = \{(v_1, v_2), (v_2, v_2)\}$, $V_2 = \{w_1, w_2, w_3\}$, $E_2 = \{(w_1, w_2), (w_3, w_2)\}$.

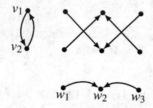

The graph product $G_1 \times G_2$ has vertex set $\{(v_1, w_1), (v_1, w_2), (v_1, w_3), (v_2, w_1), (v_2, w_2), (v_2, w_3)\}$ and edge set $\{((v_1, w_1), (v_2, w_2)), ((v_1, w_3), (v_2, w_2)), ((v_2, w_1), (v_1, w_2)), ((v_2, w_3), (v_1, w_2))\}$.

Complete subgraphs of the graph product of two graphs can be projected onto induced subgraphs of the respective graphs.

Definition 7.6 Let $G_1 = (V_1, E_1)$ and $G_2 = (V_2, E_2)$ be two graphs, and let $C = (V, E)$ be a complete subgraph of $G_1 \times G_2$. The **projection** of C onto G_1 is the subgraph of G_1 induced by the vertices $v \in V_1$ such that $(v, w) \in V$ for some vertex $w \in V_2$. The **projection** of C onto G_2 is the subgraph of G_2 induced by the vertices $w \in V_2$ such that $(v, w) \in V$ for some vertex $v \in V_1$.

Remark 7.4 Notice that self-loops are ignored in Definition 7.5, for otherwise there would be two counter-parallel edges in the graph product between every loopless vertex of G_1 and every loopless vertex of G_2. In order to keep the graph product small, graph are thus assumed to have no self-loops.

Now, maximal common induced subgraph isomorphisms of a graph G_1 to another graph G_2 correspond to maximal cliques in the graph product $G_1 \times G_2$.

Theorem 7.2 *Let G_1 and G_2 be two graphs, and let (S_1, S_2, M) be a maximal common induced subgraph isomorphism of G_1 to G_2. Then, there is a maximal clique C of $G_1 \times G_2$ whose projection onto G_1 is S_1 and whose projection onto G_2 is S_2. Conversely, let $C = (X, E)$ be a maximal clique of $G_1 \times G_2$, let S_1 be the projection of C onto G_1, and let S_2 be the projection of C onto G_2. Then, there is a maximal common induced subgraph isomorphism (S_1, S_2, M) of G_1 to G_2.*

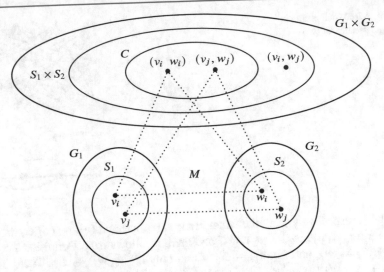

Fig. 7.8 Proof of Theorem 7.2

Proof Let $G_1 = (V_1, E_1)$ and $G_2 = (V_2, E_2)$ be two graphs, and let (S_1, S_2, M) be a common induced subgraph isomorphism of G_1 to G_2. Let also $C = (X, E)$ be the subgraph of $S_1 \times S_2$ induced by the vertex set $X = \{(v, w) \in V_1 \times V_2 \mid (v, w) \in M\}$. See Fig. 7.8. Since M is a bijection, $v_i \neq v_j$ and $w_i \neq w_j$ for all distinct vertices (v_i, w_i) and (v_j, w_j) of C. Furthermore, there is an edge from v_i to v_j in S_1 if and only if there is an edge from w_i to w_j in S_2, for all (v_i, w_i) and (v_j, w_j) in M, because M is an isomorphism of S_1 to S_2. Then, by Definition 7.5, there is an edge from (v_i, w_i) to (v_j, w_j) in $S_1 \times S_2$, for all distinct vertices (v_i, w_i) and (v_j, w_j) of $S_1 \times S_2$ which are also in M and, being C an induced subgraph of $S_1 \times S_2$, there is an edge from (v_i, w_i) to (v_j, w_j) in C, for all distinct vertices (v_i, w_i) and (v_j, w_j) of C. Therefore, C is a clique. Furthermore, since S_1 is an induced subgraph of G_1 and S_2 is an induced subgraph of G_2, it follows from Definition 7.6 that the projection of C onto G_1 is S_1 and the projection of C onto G_2 is S_2.

Suppose now that the clique $C = (X, E)$ is not maximal, and let $S_1 = (U_1, F_1)$ and $S_2 = (U_2, F_2)$. Then, there is some vertex $(v, w) \in V_1 \times V_2 \setminus X$ such that the subgraph of $G_1 \times G_2$ induced by $X \cup \{(v, w)\}$ is a clique. For all $(x, y) \in X$, since there are edges from (x, y) to (v, w) and from (v, w) to (x, y) in the new clique, it follows from Definition 7.5 that $x \neq v, y \neq w, (x, v) \in E_1$ if and only if $(y, w) \in E_2$, and $(v, x) \in E_1$ if and only if $(w, y) \in E_2$. Let T_1 be the subgraph of G_1 induced by $U_1 \cup \{v\}$, and let T_2 be the subgraph of G_2 induced by $U_2 \cup \{w\}$. Then, $M \cup \{(v, w)\}$ is a graph isomorphism of T_1 to T_2 and $(T_1, T_2, M \cup \{(v, w)\})$ is a common induced subgraph isomorphism of G_1 to G_2, contradicting the hypothesis that (S_1, S_2, M) is a maximal common induced subgraph isomorphism of G_1 to G_2. Therefore, the clique C is maximal.

Conversely, let $G_1 = (V_1, E_1)$ and $G_2 = (V_2, E_2)$ be two graphs, let $C = (X, E)$ be a maximal clique of $G_1 \times G_2$, and let S_1 and S_2 be the projections of C onto

Fig. 7.9 A maximal clique in the graph product of two graphs

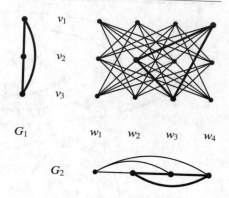

G_1 and G_2, respectively. For all distinct vertices (v_i, w_i) and (v_j, w_j) of C, there is an edge from (v_i, w_i) to (v_j, w_j) in C and then, it follows from Definition 7.5 that $v_i \neq v_j$, $w_i \neq w_j$, and $(v_i, v_j) \in E_1$ if and only if $(w_i, w_j) \in E_2$. Then, $M = \{(v, w) \in V_1 \times V_2 \mid (v, w) \in X\}$ is a graph isomorphism of S_1 to S_2 and therefore, since projections of cliques are induced subgraphs by Definition 7.6, (S_1, S_2, M) is a common induced subgraph isomorphism of G_1 to G_2.

Suppose now that the common induced subgraph isomorphism (S_1, S_2, M) of G_1 to G_2 is not maximal, and let $S_1 = (U_1, F_1)$ and $S_2 = (U_2, F_2)$. Then, there are vertices $v \in V_1 \setminus U_1$ and $w \in V_2 \setminus U_2$ such that $(T_1, T_2, M \cup \{(v, w)\})$ is a common induced subgraph isomorphism of G_1 to G_2, where T_1 is the subgraph of G_1 induced by $U_1 \cup \{v\}$ and T_2 is the subgraph of G_2 induced by $U_2 \cup \{w\}$. By Definition 7.1 and Definition 7.4, $(x, v) \in E_1$ if and only if $(y, w) \in E_2$ and $(v, x) \in E_1$ if and only if $(w, y) \in E_2$, for all vertices $x \in U_1$ and $y \in U_2$ and then, by Definition 7.5 and since $x \neq v$ and $y \neq w$, there are edges from (x, y) to (v, w) and from (v, w) to (x, y) in the subgraph of $G_1 \times G_2$ induced by $X \cup \{(v, w)\}$. Then, the subgraph of $G_1 \times G_2$ induced by $X \cup \{(v, w)\}$ is a clique, contradicting the hypothesis that $C = (X, E)$ is a maximal clique. Therefore, (S_1, S_2, M) is a maximal common induced subgraph isomorphism of G_1 to G_2. $\qquad \square$

Example 7.8 Let $G_1 = (V_1, E_1)$ and $G_2 = (V_2, E_2)$ be the two graphs shown in Fig. 7.9, with $V_1 = \{v_1, v_2, v_3\}$ and $V_2 = \{w_1, w_2, w_3, w_4\}$ and where pairs of counter-parallel edges be depicted as undirected edges. The projections of the maximal clique of $G_1 \times G_2$ induced by $\{(v_1, w_4), (v_2, w_2), (v_3, w_3)\}$ are isomorphic, and correspond to a maximal common induced subgraph isomorphism of G_1 to G_2.

Notice that Theorem 7.2 does not hold anymore if maximal common subgraph isomorphisms are not induced subgraph isomorphisms. See Exercise 7.6.

7.4.1 An Algorithm for Maximal Common Subgraph Isomorphism

Computation of all maximal common induced subgraph isomorphisms of two graphs proceeds by first computing the graph product of the two graphs, and then finding all maximal cliques in the graph product.

Notice first that the vertices in the graph product of two labeled graphs $G_1 = (V_1, E_1)$ and $G_2 = (V_2, E_2)$ are ordered pairs (v, w) in the underlying graph product $G_1 \times G_2$ with $v \in V_1$ and $w \in V_2$ sharing the same vertex label, and there is an edge from vertex (v_i, w_i) to vertex (v_j, w_j) in the labeled graph product if and only there is such an edge in the graph product of the underlying graphs and if there are edges from vertex v_i to vertex v_j in graph G_1 and from vertex w_i to vertex w_j in graph G_2 sharing the same edge label. That is, the graph product of two labeled graphs $G_1 = (V_1, E_1)$ and $G_2 = (V_2, E_2)$ is the graph $G = (V, E)$ with vertex set $V = \{(v, w) \in V_1 \times V_2 \mid G_1[v] = G_2[w]\}$ and edge set $E = \{((v_i, w_i), (v_j, w_j)) \in V \times V \mid v_i \neq v_j, w_i \neq w_j, (v_i, v_j) \in E_1, (w_i, w_j) \in E_2, G_1[v_i, v_j] = G_2[w_i, w_j]\} \cup \{((v_i, w_i), (v_j, w_j)) \in V \times V \mid v_i \neq v_j, w_i \neq w_j, (v_i, v_j) \notin E_1, (w_i, w_j) \notin E_2\}$.

```
function graph_product(G_1, G_2)
    let G be an empty graph
    for all vertices v of G_1 do
        for all vertices w of G_2 do
            if v.label = w.label then
                x ← G.new_vertex()
                x.label ← (v, w)
    for all vertices v of G do
        (v_1, v_2) ← v.label
        for all vertices w of G do
            (w_1, w_2) ← w.label
            if v_1 ≠ w_1 and v_2 ≠ w_2 then
                if v_1 and w_1 are adjacent in G_1 then
                    let e_1 be the edge from vertex v_1 to vertex w_1 in G_1
                    if v_2 and w_2 are adjacent in G_2 then
                        let e_2 be the edge from vertex v_2 to vertex w_2 in G_2
                        if e_1.label = e_2.label then
                            G.new_edge(v, w)
                else if v_2 and w_2 are not adjacent in G_2 then
                    G.new_edge(v, w)
    return G
```

The following function implements the reduction of maximal common induced subgraph isomorphism to maximal clique.

function *maximal_common_subgraph_isomorphism*(G_1, G_2)
  $G \leftarrow$ *graph_product*(G_1, G_2)
  let L be an empty list (of sets of vertices)
  for all sets C in *all_maximal_cliques*(G) **do**
    let M be an empty dictionary (of vertices to vertices)
    for all vertices v in C **do**
      $(v_1, v_2) \leftarrow v.label$
      $M[v_1] \leftarrow v_2$
    append M to L
  return L

7.5 Applications

Isomorphism problems on graphs find application whenever structures represented by graphs need to be identified or compared. In most application areas, graph isomorphism, subgraph isomorphism, and maximal common (induced) subgraph isomorphism can be seen as some form of pattern matching or information retrieval.

Roughly stated, given a *pattern* graph and a *text* graph, the pattern matching problem consists of finding all subgraphs of the text that are isomorphic to the pattern. When vertices and edges of the pattern may be labeled with variables, the pattern matching problem consists of finding all subgraphs of the text that are isomorphic to some extension of the pattern, obtained by replacing these variables by appropriate subgraphs of the text. While the former corresponds to the graph or subgraph isomorphism problem, the latter is nothing but a different formulation of the maximal common (induced) subgraph isomorphism problem. All of these pattern matching problems are a form of information retrieval.

The application of the different isomorphism problems on graphs to the retrieval of information from databases of molecular graphs is discussed next, as it is representative of these pattern matching problems on graphs. See the bibliographic notes below for further applications of graph isomorphism and related problems.

A molecular graph is a connected, labeled, and undirected graph that represents the constitution of a molecular structure. The vertices represent atoms, and the edges represent chemical bonds. Vertices are labeled with the atom type (the atomic symbol of the chemical element), and edges are labeled with the bond order or bond type (single, double, triple, or aromatic).

In schematic diagrams of molecular graphs, a single bond between two atoms is represented by an undirected edge between the atoms, a double bond is represented by a pair of parallel edges, and a triple bond is represented by three parallel edges between the atoms. An aromatic bond is represented by a circle inscribed in the corresponding cyclic chain.

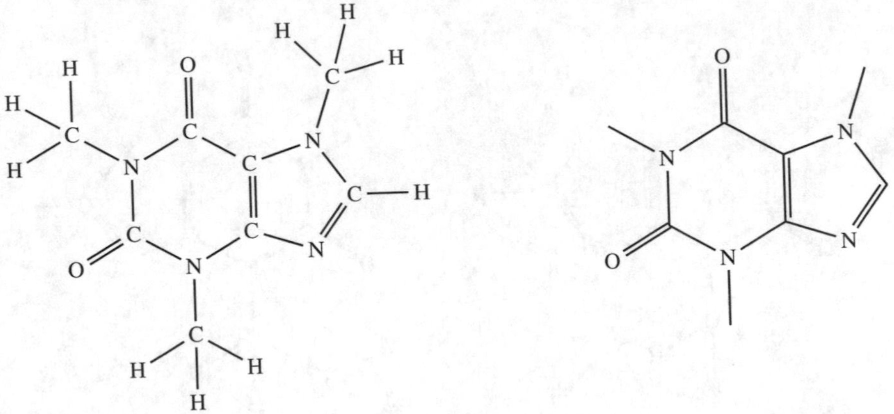

Fig. 7.10 Molecular graph (left) and simplified representation (right) of the caffeine chemical structure

Example 7.9 The chemical structure of the caffeine or 1,3,7-trimethylxanthine, which has the chemical formula $C_8H_{10}N_4O_2$, is represented by the molecular graph shown in Fig. 7.10 (left).

Remark 7.5 A simplified notation for drawing chemical structures is used by chemists in which carbon atoms are represented as a joint between two or more edges, and hydrogen atoms together with their incident edges are left off. Furthermore, methyl (CH_3) groups are also left off while keeping their incident edges. For instance, the chemical structure of the caffeine from Example 7.9 has the simplified representation shown in Fig. 7.10 (right).

Isomorphism of molecular graphs finds application in the *structure search* of a database of compounds for the presence or absence of a specified compound: when there is a need to retrieve data associated with some existing compound, or when a new compound has been discovered or synthesized.

As a matter of fact, chemical structure search cannot be made solely at the level of chemical formula, because different compounds (called isomers) may share the same chemical formula but exhibit a different structure, as shown by the following example.

Example 7.10 The molecular graphs of the 1,3-dimethylxanthine (theophylline) and the 3,7-dimethylxanthine (theobromine), shown in Fig. 7.11, are not isomorphic, although both compounds share the chemical formula $C_7H_8N_4O_2$.

Subgraph isomorphism of molecular graphs finds application in the *substructure search* of a database of compounds for all compounds containing some specified compound as a substructure. Furthermore, maximal common subgraph isomorphism

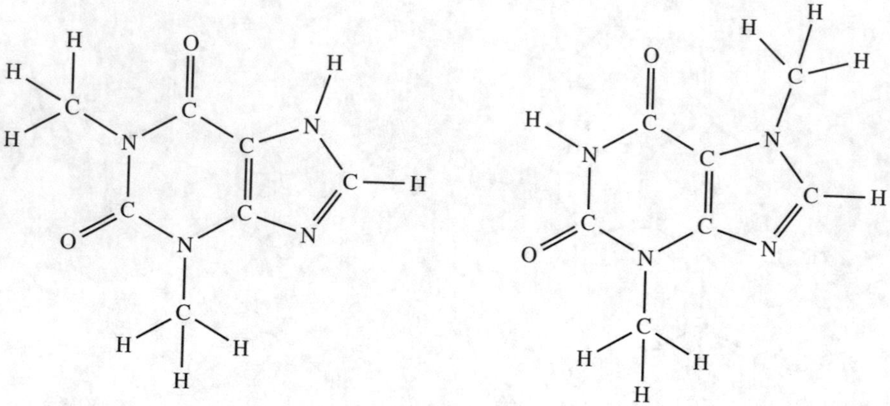

Fig. 7.11 Molecular graph of the theophylline (left) and molecular graph of the theobromine (right)

of molecular graphs finds application in the *similarity search* of a database of compounds for those compounds that are most similar to some specified compound, according to some quantitative definition of structural similarity. The similarity measure used is most often based upon the substructures in common between the specified compound and each compound in the database.

Summary

Several isomorphism problems on graphs were addressed in this chapter which form a hierarchy of pattern matching problems, where maximal common induced subgraph isomorphism generalizes (induced) subgraph isomorphism, which in turn generalizes graph isomorphism. Simple algorithms are given in detail for enumerating all solutions to these problems. The algorithms for graph and subgraph isomorphism are based on the backtracking technique presented in Sect. 2.2, while the maximal common induced subgraph isomorphism problem is reduced to the maximal clique problem and solved using the algorithms discussed in Chap. 6 for finding maximal cliques. References to more sophisticated algorithms are given in the bibliographic notes below. Chemical structure search is also discussed as a prototypical application of the different isomorphism problems on (molecular) graphs.

Bibliographic Notes

A comprehensive survey of the so-called *graph isomorphism disease* can be found in [29,68]. Complexity aspects of the graph isomorphism problem are treated in much detail in [17,43,73]. See also [2,4,28,32].

Several backtracking algorithms have been proposed for the graph isomorphism problem. The first ever backtracking algorithm for graph isomorphism was presented in [67], followed by [9,19]. One of the best-known backtracking algorithms for graph and subgraph isomorphism, presented in [79], introduced a constraint satisfaction technique that is now known as *really full look ahead*. The backtracking algorithm for subgraph isomorphism is a modern presentation of that algorithm. The algorithms presented in [56] for graph and subgraph isomorphism and in [57] for maximal common subgraph isomorphism used the constraint satisfaction technique that is now known as *forward checking*. Again, the backtracking algorithm for maximal common subgraph isomorphism is a modern presentation of that algorithm. Good surveys of constraint satisfaction techniques can be found in [18,54,78].

A practical backtracking algorithm for computing the automorphism group of a graph is described in [59,60]. An alternative approach to the graph isomorphism problem is to iterate a partitioning of the vertex sets (or the edge sets) of the two graphs into classes, based on necessary conditions for graph isomorphism, and to refine the partitioned classes until no further partitioning is possible. The first partitioning algorithm for graph isomorphism was presented in [80], followed by [16,52,62,72]. For a modern description, see [49, Chap. 7].

The reduction of maximum common subgraph isomorphism to clique was proposed in [1,6,30,51] and further exploited in [47]. Further, the equivalence of graph isomorphisms and maximal cliques in the graph product was proved in [48]. The graph product of two graphs [7] is also known as conjunction [38], direct product [46, pp. 167–170], cartesian product [65, pp. 33–36], and Kronecker product [84]. Several distance measures between objects represented by graphs were proposed which are based on the maximum common subgraph [13,15,26]. See also [41,81].

Applications of graph isomorphism and related problems to chemistry were first addressed in [77], and despite applications to chemical computation such as modeling of chemical reactions [58] the most important practical application lies in information retrieval from chemical databases, since every database of chemical compounds must offer some kind of chemical structure and substructure search system, let alone bibliographic searches in specialized databases or over the Internet. Several such chemical information retrieval systems based on graph isomorphism, subgraph isomorphism, and maximal common subgraph isomorphism algorithms were developed for the CAS (Americal Chemical Society) Registry database [42], which holds over 18 million compounds and 13 million biological sequences, and for the Beilstein database [39,40], which holds over 8 million compounds and 5 million chemical reactions. See also [3,5,75,86,87]. More recent algorithms for maximal common subgraph isomorphism of molecular graphs are described in [20,21,24,25,27,76,83,89].

Applications of graph isomorphism and related problems to computational biology include the comparison of two-dimensional and three-dimensional protein structures, as well as the discovery of common structural motifs [31,61,71,82]. See also [10].

Another important application area of subgraph isomorphism is object recognition and computer vision. In structural and syntactical pattern recognition, objects are represented by symbolic data structures, such as strings, trees, or graphs, and relational descriptions of complex objects are also used in computer vision. See [12,37] for an overview. Similar object recognition problems arise in computer-aided design [11,53,64,66]. Isomorphism and related problems on graphs can also be formulated as *consistent labeling* problems [35,36] and, as a matter of fact, the *relaxation* and *look-ahead* operators [33] used to improve efficiency of relational matching [34,44,45,74,88] constitute one of the most important constraint satisfaction techniques.

Last, but not least, graph grammars and graph transformation [22,23,69] constitute an important area of application for subgraph isomorphism algorithms [14,50,70, 90], because a graph rewriting or graph transformation step succeeds by first finding a subgraph isomorphism of the left-hand side graph of some graph transformation rule into the graph to be rewritten or transformed, and then performing the rewriting or transformation specified by the rule.

Problems

7.1 Name at least three invariants for graph isomorphism. Are they also invariants for subgraph isomorphism and for maximal common subgraph isomorphism? Justify your answer.

7.2 Enumerate all the automorphisms of the wheel graph W_5.

7.3 Give upper and lower bounds on the number of subgraph isomorphisms of a connected graph with n_1 vertices into a connected graph with n_2 vertices, where $n_1 \leqslant n_2$.

7.4 Give upper and lower bounds on the number of maximal common induced subgraph isomorphisms between two connected graphs with n_1 and n_2 vertices.

7.5 Enumerate all maximal common induced subgraph isomorphisms between the complete bipartite graph $K_{3,4}$ and the wheel graph W_5.

7.6 Show that computation of all maximal common subgraphs of two graphs can be reduced to finding all maximal cliques in the graph product of the line graphs of the two graphs. See Exercise 7.6.

Exercises

Exercise 7.1 Reordering the adjacency lists of the graphs by non-decreasing vertex degree tends to produce deeper search trees, and backtracking tends to perform better on deeper trees than on shallower trees, because *pruning* a search tree at a level close to the root cuts off a larger subtree of a deep search tree than of a shallow search tree. Study the effect of vertex ordering on the performance of the backtracking algorithms for graph and subgraph isomorphism.

Exercise 7.2 Implement a variant of the backtracking algorithm for graph isomorphism to decide whether or not two graphs are isomorphic, without enumerating all graph isomorphism mappings between the two graphs.

Exercise 7.3 Replace the refinement procedure in the backtracking algorithm for subgraph isomorphism, which implements the simplified necessary condition for extending a subgraph isomorphism mapping given in Lemma 7.5, by the necessary condition given in Lemma 7.4 and implement the modified algorithm for subgraph isomorphism.

Exercise 7.4 Perform experiments on random graphs to compare the efficiency of the backtracking algorithm for subgraph isomorphism using the two necessary conditions of Exercise 7.3.

Exercise 7.5 Implement a variant of the backtracking algorithm for subgraph isomorphism to find one isomorphism of a graph as a subgraph of another graph, instead of enumerating all such subgraph isomorphism mappings.

Exercise 7.6 Computation of all maximal common subgraphs of two graphs can also be reduced to finding all maximal cliques in the graph product, but in the graph product of the line graphs of the two graphs, as discussed in [47]. As a matter of fact, with the exception of K_3 and $K_{1,3}$, any two connected graphs with isomorphic line graphs are isomorphic [63,85]. Implement a variant of the algorithm for maximal common induced subgraph isomorphism to find a maximum common subgraph isomorphism of two graphs. Notice that the line graph product can grow very large, even for small graphs.

Exercise 7.7 Implement a variant of the algorithm for maximal common subgraph isomorphism to find a maximum common subgraph isomorphism of two graphs, instead of enumerating all maximal common subgraph isomorphisms and then choosing a largest one.

Exercise 7.8 The most widely used representation of molecular graphs is the connection table, which contains a list of all of the atoms within a chemical structure together with bond information. Hydrogen atoms are often excluded, since their presence or absence can be deduced from the bond orders and atomic types. For instance, the standard representation of a molecular graph adopted by the NIST Chemistry WebBook [55], which is available at http://webbook.nist.gov/chemistry/, is a text file consisting of three lines followed by a *count line* containing the number n of nonhydrogen atoms and the number m of bonds, separated by a space and followed by n lines containing the *atom block* and m lines containing the *bond block*. A full description of the CTAB format can be found at http://www.mdli.com/downloads/literature/ctfile.pdf. Implement procedures to read the connection table of a molecular graph and to write a graph as a connection table.

Exercise 7.9 A more comprehensive connection table format has been adopted by the Protein Data Bank [8], which is available at http://www.rcsb.org/pdb/, in order to represent crystallographic as well as structural information. A full description of the PDB connection table format can be found at the same Internet address. Implement procedures to read the PDB file of a molecular graph and to write a graph as a PDB file.

Exercise 7.10 Perform isomorphism experiments on molecular graphs extracted from the NIST Chemistry WebBook and the Protein Data Bank. See Exercises 7.8 and 7.9.

References

1. Akinniyi FA, Wong AKC, Stacey DA (1986) A new algorithm for graph monomorphism based on the projections of the product graph. IEEE Trans Syst Man Cybern 16(5):740–751
2. Arvind V, Köbler J, Rattan G, Verbitsky O (2017) Graph isomorphism, color refinement, and compactness. Comput Complex 26(3):627–685
3. Ash JE, Warr WA, Willett P (eds) (1991) Chem Struct Syst. Ellis Horwood, Chichester, England
4. Babai L (2016) Graph isomorphism in quasipolynomial time. In: Wichs D, Mansour Y (eds) Proceedings 48th annual ACM symposium theory of computing. ACM, New York, pp 684–697
5. Barnard JM (1993) Substructure searching methods: old and new. J Chem Inf Comput Sci 33(4):532–538
6. Barrow HG, Burstall RM (1976) Subgraph isomorphism, matching relational structures and maximal cliques. Inf Process Lett 4(4):83–84
7. Berge C (1985) Graphs and hypergraphs, 2nd edn. North-Holland, Amsterdam
8. Berman HM, Westbrook J, Feng Z, Gilliland G, Bhat TN, Weissig H, Shindyalov IN, Bourne PE (2000) The protein data bank. Nucleic Acids Res 28(1):235–242
9. Berztiss AT (1973) A backtrack procedure for isomorphism of directed graphs. J ACM 20(3):365–377
10. Bonnici V, Giugno R, Pulvirenti A, Shasha D, Ferro A (2013) A subgraph isomorphism algorithm and its application to biochemical data. BMC Bioinf 14(Suppl. 7):S13
11. Brown AD, Thomas PR (1988) Goal-oriented subgraph isomorphism technique for IC device recognition. IEE Proc 135(6):141–150
12. Bunke H (1993) Structural and syntactic pattern recognition. In: Chen CH, Pau LF, Wang PSP (eds) Handbook of pattern recognition and computer vision, Chap. 1.5. World Scientific, Singapore, pp 163–209
13. Bunke H (1997) On a relation between graph edit distance and maximum common subgraph. Pattern Recognit Lett 18(8):689–694
14. Bunke H, Glauser T, Tran TH (1991) An efficient implementation of graph grammars based on the RETE matching algorithm. In: Ehrig H, Kreowski HJ, Rozenberg G (eds) Graph-grammars and their application to computer science, vol 532. Lecture notes in computer science. Springer, Berlin Heidelberg, pp 174–189
15. Bunke H, Shearer K (1998) A graph distance metric based on the maximal common subgraph. Pattern Recognit Lett 19(3–4):255–259
16. Corneil DG, Gotlieb CC (1970) An efficient algorithm for graph isomorphism. J ACM 17(1):51–64

17. Corneil DG, Kirkpatrick DG (1980) A theoretical analysis of various heuristics for the graph isomorphism problem. SIAM J Comput 9(2):281–297

18. Dechter R (1992) Constraint networks. In: Shapiro SC (ed) Encyclopedia of artificial intelligence, vol 1. Wiley, New York NY, pp 276–285

19. Deo NJ, Davis JM, Lord RE (1977) A new algorithm for digraph isomorphism. BIT 17(2):16–30

20. Duesbury E, Holliday J, Willett P (2015) Maximum common substructure-based data fusion in similarity searching. J Chem Inf Model 55(2):222–230

21. Durand PJ, Pasari R, Baker JW, Tsai CC (1999) An efficient algorithm for the similarity analysis of molecules. Internet J Chem 2(17)

22. Ehrig, H., Engels, G., Kreowski, H.J., Rozenberg, G. (eds.): Handbook of graph grammars and computing by graph transformation, vol 2: Applications, languages, and tools. World Scientific, Singapore (1999)

23. Ehrig H, Kreowski HJ, Montanari U, Rozenberg G (1999) (eds.): Handbook of graph grammars and computing by graph transformation, volume 3: Concurrency, parallelism. and distribution. World Scientific, Singapore

24. Fan BT, Panaye A, Doucet JP (1999) Comment on "Isomorphism, automorphism partitioning, and canonical labeling can be solved in polynomial-time for molecular graphs.ï¿½ J Chem Inf Comput Sci 39(4):630–631

25. Faulon JL (1998) Isomorphism, automorphism partitioning, and canonical labeling can be solved in polynomial-time for molecular graphs. J Chem Inf Comput Sci 38(3):432–444

26. Fernández ML, Valiente G (2001) A graph distance measure combining maximum common subgraph and minimum common supergraph. Pattern Recognit Lett 22(6–7):753–758

27. Gardiner EJ, Artymiuk PJ, Willett P (1997) Clique-detection algorithms for matching three-dimensional molecular structures. J Molecular Graph Model 15(4):245–253

28. Garey MR, Johnson DS (1979) Computers and intractability: a guide to NP-completeness. W. H. Freeman, San Francisco CA

29. Gati G (1979) Further annotated bibliography on the isomorphism disease. J Graph Theory 3:95–109

30. Ghahraman DE, Wong AKC, Au T (1980) Graph monomorphism algorithms. IEEE Trans Syst Man Cybern 10(4):189–197

31. Grindley HM, Artymiuk PJ, Rice DW, Willett P (1993) Identification of tertiary structure resemblance in proteins using a maximal common subgraph isomorphism algorithm. J Molecular Biol 229(3):707–721

32. Grohe M, Schweitzer P (2020) The graph isomorphism problem. Commun ACM 63(11):128–134

33. Haralick RM, Elliott GL (1980) Increasing tree search efficiency for constraint satisfaction problems. Artif Intell 14(3):263–313

34. Haralick RM, Kartus JS (1978) Arrangements, homomorphisms, and discrete relaxation. IEEE Trans Systems Man Cybern 8(8):600–612

35. Haralick RM, Shapiro LG (1979) The consistent labeling problem: Part I. IEEE Trans Pattern Anal Mach Intell 1(2):173–184

36. Haralick RM, Shapiro LG (1980) The consistent labeling problem: Part II. IEEE Trans Pattern Anal Mach Intell 2(3):193–203

37. Haralick RM, Shapiro LG (1993) Computer robot vision, vol 2. Addison-Wesley, Reading MA

38. Harary F, Wilcox GW (1967) Boolean operations on graphs. Math Scand 20(1):41–51

39. Heller SR (ed) (1990) The Beilstein online database: implementation, content, and retrieval, ACS symposium series, vol 436. American Chemical Society, Washington DC

40. Heller SR (ed) (1998) The Beilstein system: strategies for effective searching. American Chemical Society, Washington DC

41. Hidović D, Pelillo M (2004) Metrics for attributed graphs based on the maximal similarity common subgraph. Int J Pattern Recognit Artif Intell 18(3):299–313

42. Kasparek SV (1990) Computer graphics and chemical structures: database management systems. Wiley, New York

43. Köbler J, Schöning U, Turán J (1993) The graph isomorphism problem: its structural complexity. Progress in theoretical computer science. Birkhäuser, Boston NY

44. Kitchen L (1980) Relaxation applied to matching quantitative relational structures. IEEE Trans Syst Man Cybern 10(2):96–101

45. Kitchen L, Rosenfeld A (1979) Discrete relaxation for matching relational structures. IEEE Trans Systems Man Cybern 9(12):869–874

46. Knuth DE (1993) The stanford graphbase: a platform for combinatorial computing. ACM Press, New York NY

47. Koch I (2001) Enumerating all connected maximal common subgraphs in two graphs. Theor Comput Sci 250(1–2):1–30

48. Kozen D (1978) A clique problem equivalent to graph isomorphism. ACM SIGACT News 10(1):50–52

49. Kreher DL, Stinson DR (1999) Combinatorial algorithms: generation, enumeration, and search. CRC Press, Boca Raton FL

50. Larrosa J, Valiente G (2002) Constraint satisfaction algorithms for graph pattern matching. Math Struct Comput Sci 12(4):403–422

51. Levi G (1972) A note on the derivation of maximal common subgraphs of two directed or undirected graphs. Calcolo 9(4):1–12

52. Levi G (1974) Graph isomorphism: A heuristic edge-partitioning oriented program. Computing 12(4):291–313

53. Ling Z, Yun DYY (1996) An efficient subcircuit extraction algorithm by resource management. In: Qian-Ling Z, Ting-A0 T, Huihua Y (eds) Proceedings 2nd international conference on application specific integrated circuits. IEEE, Piscataway NJ, pp 9–14

54. Mackworth AK (1992) Constraint satisfaction. In: Shapiro SC (ed) Encyclopedia of artificial intelligence, vol 1. Wiley, New York, pp 285–293

55. Mallard WG, Linstrom PJ (eds) (2000) NIST Chemistry WebBook. No. 69 in NIST Standard Reference Database. National Institute of Standards and Technology, Gaithersburg MD

56. McGregor JJ (1979) Relational consistency algorithms and their application in finding subgraph and graph isomorphisms. Inf Sci 19(3):229–250

57. McGregor JJ (1982) Backtrack search algorithms and the maximal common subgraph problem. Softw Practice Exp 12(1):23–34

58. McGregor JJ, Willett P (1981) Use of a maximal common subgraph algorithm in the automatic identification of the ostensible bond changes occurring in chemical reactions. J Chem Inf Comput Sci 21(3):137–140

59. McKay BD (1981) Practical graph isomorphism. Congressus Numerantium 30(1):45–87

60. McKay BD, Piperno A (2014) Practical graph isomorphism, II. J Symbol Comput 60(1):94–112

61. Mitchell EM, Artymiuk PJ, Rice DW, Willett P (1989) Use of techniques derived from graph theory to compare secondary structure motifs in proteins. J Molecular Biol 212(1):151–166

62. Mittal HB (1988) A fast backtrack algorithm for graph isomorphism. Inf Process Lett 29(2):105–110

63. Nicholson V, Tsai CC, Johnson M, Naim M (1987) A subgraph isomorphism theorem for molecular graphs. In: King RB, King DHRRB, Rouvray DH (eds) Graph theory and topology in chemistry, no. 51 in studies in physical and theoretical chemistry. Elsevier, New York, pp 226–230

64. Ohlrich M, Ebeling C, Ginting E, Sather L (1993) SubGemini: Identifying subcircuits using a fast subgraph isomorphism algorithm. In: Dunlop AE (ed) Proceedings 30th annual ACM/IEEE design automation conference. IEEE, Piscataway NJ, pp 31–37

65. Ore O (1962) Theory of graphs, colloquium publications, vol 38. American Mathematical Society, Providence RI

66. Perkowski MA, Chrzanowska-Jeske M, Pierzchala E, Coppola A (1994) An exact algorithm for the fitting problem in the application specific state machine device. J Circuits Syst Comput 4(2):173–190

67. Ray LC, Kirsch RA (1957) Finding chemical records by digital computers. Science 126(3278):814–819

68. Read RC, Corneil DG (1977) The graph isomorphism disease. J Graph Theory 1(4):339–363

69. Rozenberg G (ed) (1997) Handbook of graph grammars and computing by graph transformation, volume 1: Foundations. World Scientific, Singapore

70. Rudolf M (2000) Utilizing constraint satisfaction techniques for efficient graph pattern matching. In: Ehrig H, Engels G, Kreowski HJ, Rozenberg G (eds) Theory and application of graph transformations, vol 1764. Lecture notes in computer science. Springer, Berlin Heidelberg, pp 215–227

71. Samudrala R, Moult J (1998) A graph-theoretic algorithm for comparative modeling of protein structure. J Molecular Biol 279(1):287–302

72. Schmidt DC, Druffel LE (1976) A fast backtracking algorithm to test directed graphs for isomorphism using distance matrices. J ACM 23(3):433–445

73. Schöning U (1988) Graph isomorphism is in the low hierarchy. J Comput Syst Sci 37(3):312–323

74. Shapiro LG, Haralick RM (1987) Relational matching. Appl Opt 26(10):1845–1851

75. Stobaugh RE (1985) Chemical substructure searching. J Chem Inf Comput Sci 25(1):271–275

76. Takahashi Y, Satoh Y, Suzuki H, Sasaki SI (1987) Recognition of largest common structural fragment among a variety of chemical structures. Anal Sci 3(1):23–28

77. Tarjan RE (1977) Graph algorithms in chemical computation. In: Christoffersen RE (ed) Algorithms for chemical computations, ACS symposium series, vol 46, chap. 1. Americal Chemical Society, Washington DC, pp 1–19

78. Tsang E (1993) Foundations of constraint satisfaction. Academic Press, London, England

79. Ullmann JR (1976) An algorithm for subgraph isomorphism. J ACM 23(1):31–42

80. Unger SH (1964) GIT—a heuristic program for testing pairs of directed line graphs for isomorphism. Commun ACM 7(1):26–34

81. Valiente G (2007) Efficient algorithms on trees and graphs with unique node labels. In: Kandel A, Bunke H, Last M (eds) Applied graph theory in computer vision and pattern recognition, Studies in computational intelligence, vol 52. Springer, pp 137–149

82. Viksna J, Gilbert D (2001) Pattern matching and pattern discovery algorithms for protein topologies. In: Gascuel O, Moret BME (eds) Algorithms in bioinformatics, vol 2149. Lecture notes in computer Science. Springer, Berlin Heidelberg, pp 98–111

83. Wang T, Zhou J (1997) EMCSS: a new method for maximal common substructure search. J Chem Inf Comput Sci 37(5):828–834

84. Weichsel PM (1963) The Kronecker product of graphs. Proc Am Math Soc 13(1):47–52

85. Whitney H (1932) Congruent graphs and the connectivity of graphs. Am J Math 54(1):150–168

86. Willett P (1991) Three-dimensional chemical structure handling. Wiley, New York

87. Willett P, Barnard JM, Downs GM (1998) Chemical similarity searching. J Chem Inf Comput Sci 38(6):983–996

88. Wilson RC, Evans AN, Hancock ER (1995) Relational matching by discrete relaxation. Image Vis Comput 13(5):411–421

89. Xu J (1996) GMA: A generic match algorithm for structural homomorphism, isomorphism, and maximal common substructure match and its applications. J Chem Inf Comput Sci 36(1):25–34

90. Zündorf A (1996) Graph pattern matching in PROGRES. In: Cuny J, Ehrig H, Engels G, Rozenberg G (eds) Graph grammars and their application to computer science, vol 1073. Lecture notes in computer Science. Springer, Berlin Heidelberg, pp 454–468

Implementation of the Algorithms in Python

A.1 Introduction

This appendix assumes a basic knowledge of programming using the Python programming language, including concepts such as mutable and immutable objects, and local, global, and nonlocal declarations. See, for example, [3–6, 10] for an introduction to programming in Python, [9] for an introduction to the Python programming language, [1, 8] for more advanced aspects of the Python programming language, and [7] for the implementation of data structures and algorithms in Python.

The following declaration is needed for the function annotations used in the Python code throughout this appendix.

Type hints

```
from typing import Iterator, Dict, List, Set, Tuple
```

A.1.1 Basic Data Structures

Representation of stacks

```
class Stack:

    def __init__(self) -> None:
        self.data = list()

    def empty(self) -> bool:
        return self.size() == 0
```

© The Editor(s) (if applicable) and The Author(s), under exclusive license
to Springer Nature Switzerland AG 2021
G. Valiente, *Algorithms on Trees and Graphs*, Texts in Computer Science,
https://doi.org/10.1007/978-3-030-81885-2

```python
    def size(self) -> int:
        return len(self.data)

    def top(self):
        if self.empty():
            return None
        else:
            return self.data[-1]

    def push(self, x) -> None:
        self.data.append(x)

    def pop(self):
        return self.data.pop()
```

Representation of queues

```python
from collections import deque

class Queue:

    def __init__(self) -> None:
        self.data = deque()

    def empty(self) -> bool:
        return self.size() == 0

    def size(self) -> int:
        return len(self.data)

    def front(self):
        if self.empty():
            return None
        else:
            return self.data[0]

    def enqueue(self, x) -> None:
        self.data.append(x)

    def dequeue(self):
        return self.data.popleft()
```

Representation of priority queues

```python
import heapq

class PriorityQueue:

    def __init__(self) -> None:
        self.data = list()

    def empty(self) -> bool:
        return self.size() == 0

    def size(self) -> int:
        return len(self.data)

    def front(self):
        if self.empty():
            return None
        else:
            return self.data[0]

    def enqueue(self, x) -> None:
        heapq.heappush(self.data, x)

    def dequeue(self):
        return heapq.heappop(self.data)
```

A.1.2 Representation of Trees and Graphs

The following class implements the adjacency map representation of a graph, using Python lists, dictionaries, and iterators over the lists of vertices, edges, incoming edges, and outgoing edges, borrowing ideas from [12]. Let $G = (V, E)$ be a graph with n vertices and m edges. Operations $G.vertices()$, $G.incoming()$, $G.outgoing()$, $G.adjacent(v, w)$, $G.source(e)$, $G.target(e)$, and $G.opposite(v, e)$ take $O(1)$ time. Operations $G.first\_vertex()$, $G.first\_edge()$, $G.first\_in\_edge(v)$, $G.first\_adj\_edge(v)$, $G.new\_vertex()$, $G.new\_edge(v, w)$, and $G.del\_edge(e)$ also take $O(1)$ time. Operations $G.indeg(v)$, $G.last\_in\_edge(v)$, $G.in\_pred(e)$, and $G.in\_succ(e)$ take $O(indeg(v))$ time, where $e = (v, w)$. Operations $G.outdeg(v)$, $G.last\_adj\_edge(v)$, $G.adj\_pred(e)$, and $G.adj\_succ(e)$ take $O(outdeg(v))$ time, where $e = (v, w)$. Operation $G.del\_vertex(v)$, takes $O(deg(v))$ time. Operations $G.number\_of\_vertices()$, $G.last\_vertex()$, $G.pred\_vertex(v)$, and $G.succ\_vertex(v)$ take $O(n)$ time. Finally, operations $G.edges()$, $G.number\_of\_edges()$, $G.last\_edge()$, $G.pred\_edge(e)$, and $G.succ\_edge(e)$ take $O(m)$ time. These are all expected time bounds.

Representation of graphs

```python
class Graph:

    class Vertex:

        def __init__(self, label) -> None:
            self._label = label

        def label(self):
            return self._label

        def __str__(self) -> str:
            return str(self._label)

        def __hash__(self) -> int:
            return hash(id(self))

        def __lt__(self, other) -> bool:
            return hash(self) < hash(other)

    class Edge:

        def __init__(
                self,
                source: 'Vertex',
                target: 'Vertex',
                label) -> None:
            self._source = source
            self._target = target
            self._label = label

        def source(self) -> 'Vertex':
            return self._source

        def target(self) -> 'Vertex':
            return self._target

        def label(self):
            return self._label

        def __str__(self) -> str:
            return str(self._label)

        def __hash__(self) -> int:
            return hash(
                (id(self._source), id(self._target)))

        def opposite(self, v) -> 'Vertex':
            if v is self._source:
```

```
                    return self._target
            else:
                return self._source

    def __init__(self) -> None:
        self._data = dict()

    def __str__(self) -> str:
        edges = sum([len(self._data[v]._outgoing)
                    for v in self._data])
        return '{0} {1} {2}'.format(
            'graph', len(self._data), edges)

    def empty(self) -> bool:
        return len(self._data) == 0

    def vertices(self) -> Iterator:
        return iter(self._data.keys())

    def edges(self) -> Iterator:
        edges = [self._data[v]._outgoing[w]
                for v in self._data
                for w in self._data[v]._outgoing]
        return iter(edges)

    def incoming(self, v: Vertex) -> Iterator:
        return iter(self._data[v]._incoming.values())

    def outgoing(self, v: Vertex) -> Iterator:
        return iter(self._data[v]._outgoing.values())

    def source(self, e: Edge) -> 'Vertex':
        return e.source()

    def target(self, e: Edge) -> 'Vertex':
        return e.target()

    def label(self, v: Vertex):
        return v.label()

    def label(self, e: Edge):
        return e.label()

    def new_vertex(self, label=None) -> 'Vertex':
        v = self.Vertex(label)
        self._data[v] = type(str(), (), {})
        self._data[v]._incoming = dict()
        self._data[v]._outgoing = dict()
        return v
```

```python
def new_edge(
        self,
        source: Vertex,
        target: Vertex,
        label=None) -> 'Edge':
    e = self.Edge(source, target, label)
    self._data[source]._outgoing[target] = e
    self._data[target]._incoming[source] = e
    return e

def del_vertex(self, v: Vertex) -> None:
    for e in [self._data[v]._incoming[u]
              for u in self._data[v]._incoming]:
        self.del_edge(e)
    for e in [self._data[v]._outgoing[w]
              for w in self._data[v]._outgoing]:
        self.del_edge(e)
    del self._data[v]

def del_edge(self, e: Edge) -> None:
    v = e.source()
    w = e.target()
    del self._data[v]._outgoing[w]
    del self._data[w]._incoming[v]

def number_of_vertices(self) -> int:
    return len([v for v in self.vertices()])

def number_of_edges(self) -> int:
    return len([e for e in self.edges()])

def indeg(self, v: Vertex) -> int:
    return len([e for e in self.incoming(v)])

def outdeg(self, v: Vertex) -> int:
    return len([e for e in self.outgoing(v)])

def adjacent(self, v: Vertex, w: Vertex) -> bool:
    return w in self._data[v]._outgoing

def opposite(self, v: Vertex, e: Edge) -> 'Vertex':
    return e.opposite(v)

def first_vertex(self) -> 'Vertex':
    return next(self.vertices())

def last_vertex(self) -> 'Vertex':
    for v in self.vertices():
        pass
    return v
```

```python
    def pred_vertex(self, v: Vertex) -> 'Vertex':
        it = self.vertices()
        x, y = None, next(it)
        while y is not v:
            x, y = y, next(it)
        return x

    def succ_vertex(self, v: Vertex) -> 'Vertex':
        it = self.vertices()
        try:
            while next(it) is not v:
                pass
            return next(it)
        except StopIteration:
            return None

    def first_edge(self) -> 'Edge':
        return next(self.edges())

    def last_edge(self) -> 'Edge':
        for e in self.edges():
            pass
        return e

    def pred_edge(self, e: Edge) -> 'Edge':
        it = self.edges()
        x, y = None, next(it)
        while y is not e:
            x, y = y, next(it)
        return x

    def succ_edge(self, e: Edge) -> 'Edge':
        it = self.edges()
        try:
            while next(it) is not e:
                pass
            return next(it)
        except StopIteration:
            return None

    def first_in_edge(self, v: Vertex) -> 'Edge':
        return next(self.incoming(v))

    def last_in_edge(self, v: Vertex) -> 'Edge':
        for e in self.incoming(v):
            pass
        return e

    def in_pred(self, e: Edge) -> 'Edge':
```

```python
            it = self.incoming(self.target(e))
            x, y = None, next(it)
            while y is not e:
                x, y = y, next(it)
            return x

    def in_succ(self, e: Edge) -> 'Edge':
        it = self.incoming(self.target(e))
        try:
            while next(it) is not e:
                pass
            return next(it)
        except StopIteration:
            return None

    def first_adj_edge(self, v: Vertex) -> 'Edge':
        return next(self.outgoing(v))

    def last_adj_edge(self, v: Vertex) -> 'Edge':
        for e in self.outgoing(v):
            pass
        return e

    def adj_pred(self, e: Edge) -> 'Edge':
        it = self.outgoing(self.source(e))
        x, y = None, next(it)
        while y is not e:
            x, y = y, next(it)
        return x

    def adj_succ(self, e: Edge) -> 'Edge':
        it = self.outgoing(self.source(e))
        try:
            while next(it) is not e:
                pass
            return next(it)
        except StopIteration:
            return None
```

Algorithm for the generation of graphs from lists of edges

```python
def graph_from_edge_list(
        edges: list,
        undirected=False) -> Graph:
    vertices = set(v for (v, w) in edges).union(
        set(w for (v, w) in edges))
    G = Graph()
    vertices = dict()
    for (v, w) in edges:
        if v not in vertices:
            vertices[v] = G.new_vertex(v)
        if w not in vertices:
            vertices[w] = G.new_vertex(w)
        G.new_edge(vertices[v], vertices[w])
        if undirected:
            G.new_edge(vertices[w], vertices[v])
    return G
```

Algorithm for the generation of complete graphs with n vertices labeled $1, \ldots, n$

```python
def complete_graph(n: int) -> Graph:
    G = Graph()
    vertices = dict()
    for i in range(1, n + 1):
        vertices[i] = G.new_vertex(i)
    for i in range(1, n + 1):
        for j in range(i + 1, n + 1):
            G.new_edge(vertices[i], vertices[j])
            G.new_edge(vertices[j], vertices[i])
    return G
```

The following class implements the graph-based representation of a tree. Let $T = (V, E)$ be a tree with n nodes. Operations $T.is\_root(v)$ and $T.parent(v)$ take expected $O(1)$ time. Operations $T.number\_of\_children(v)$, $T.children(v)$, $T.is\_leaf(v)$, $T.first\_child(v)$, $T.last\_child(v)$, $T.previous\_sibling(v)$, $T.next\_sibling(v)$, $T.is\_first\_child(v)$, $T.is\_last\_child(v)$, and take expected $O(children[v])$ time. Operations $T.number\_of\_nodes()$ and $T.root()$ take expected $O(n)$ time.

Representation of trees

```python
class Tree (Graph):

    def __init__(self) -> None:
        super(Tree, self).__init__()
        self._root = None

    def __str__(self) -> str:
        return '{0} {1} {2}'.format(
            'tree',
            self.number_of_nodes(),
            self.number_of_edges())

    def number_of_nodes(self) -> int:
        return self.number_of_vertices()

    def root(self) -> 'Vertex':
        if self._root is None:
            for v in self.vertices():
                if self.indeg(v) == 0:
                    self._root = v
                    break
        return self._root

    def is_root(self, v) -> bool:
        return self.indeg(v) == 0

    def number_of_children(self, v) -> int:
        return self.outdeg(v)

    def parent(self, v) -> 'Vertex':
        return next(self.incoming(v)).source()

    def children(self, v) -> List['Vertex']:
        return [self.target(e) for e in self.outgoing(v)]

    def is_leaf(self, v) -> bool:
        return self.outdeg(v) == 0

    def first_child(self, v) -> 'Vertex':
        if self.is_leaf(v):
            return None
        else:
            return self.first_adj_edge(v).target()

    def last_child(self, v) -> 'Vertex':
        if self.is_leaf(v):
            return None
        else:
```

```
                return self.last_adj_edge(v).target()

    def previous_sibling(self, v) -> 'Vertex':
        if self.is_root(v):
            return None
        e = next(self.incoming(v))
        if self.adj_pred(e) is None:
            return None
        return self.adj_pred(e).target()

    def next_sibling(self, v) -> 'Vertex':
        if self.is_root(v):
            return None
        e = next(self.incoming(v))
        if self.adj_succ(e) is None:
            return None
        return self.adj_succ(e).target()

    def is_first_child(self, v) -> bool:
        if self.is_root(v):
            return True
        return self.previous_sibling(v) is None

    def is_last_child(self, v) -> bool:
        if self.is_root(v):
            return True
        return self.next_sibling(v) is None

    def new_node(self, label=None) -> 'Vertex':
        return self.new_vertex(label)
```

Algorithm for the generation of trees from lists of edges

```
def tree_from_edge_list(edges: list) -> Tree:
    T = Tree()
    nodes = dict()
    for (v, w) in edges:
        if v not in nodes:
            nodes[v] = T.new_node(v)
        if w not in nodes:
            nodes[w] = T.new_node(w)
        T.new_edge(nodes[v], nodes[w])
    T.root()
    return T
```

Algorithm for the generation of complete binary trees with n nodes labeled $1, \ldots, n$

```python
def complete_binary_tree(n: int) -> Tree:
    T = Tree()
    nodes = dict()
    for i in range(1, n + 1):
        nodes[i] = T.new_node(i)
    for i in range(2, n + 1):
        T.new_edge(nodes[i // 2], nodes[i])
    T.root()
    return T
```

A.2 Algorithmic Techniques

A.2.1 The Tree Edit Distance Problem

Simple algorithm for the edit distance between ordered trees

```python
def tree_edit(T1: Tree, T2: Tree) -> List[Graph.Edge]:

    def tree_edit_graph() -> Graph:
        G = Graph()
        n1 = T1.number_of_nodes()
        n2 = T2.number_of_nodes()
        preorder_tree_traversal(T1)
        preorder_tree_traversal(T2)
        preorder_tree_depth(T1)
        preorder_tree_depth(T2)
        d1 = [0 for _ in range(n1 + 1)]
        d2 = [0 for _ in range(n2 + 1)]
        for v in T1.vertices():
            d1[v.order] = v.depth
        for w in T2.vertices():
            d2[w.order] = w.depth
        A = [[0 for _ in range(n2 + 1)]
             for _ in range(n1 + 1)]
        for i in range(n1 + 1):
            for j in range(n2 + 1):
                A[i][j] = G.new_vertex((i, j))
        for i in range(n1):
            G.new_edge(A[i][n2], A[i + 1][n2], 'del')
        for j in range(n2):
            G.new_edge(A[n1][j], A[n1][j + 1], 'ins')
```

```
                for i in range(n1):
                    for j in range(n2):
                        if d1[i + 1] >= d2[j + 1]:
                            G.new_edge(A[i][j], A[i + 1][j], 'del')
                        if d1[i + 1] == d2[j + 1]:
                            G.new_edge(
                                A[i][j], A[i + 1][j + 1], 'sub')
                        if d1[i + 1] <= d2[j + 1]:
                            G.new_edge(A[i][j], A[i][j + 1], 'ins')
            return G

    G = tree_edit_graph()
    s = G.first_vertex()
    t = G.last_vertex()
    (W, S) = breadth_first_spanning_subtree(G, s)
    return acyclic_shortest_path(G, s, t, W, S)
```

A.2.2 Backtracking

Backtracking algorithm for the edit distance between ordered trees

```
def backtracking_tree_edit() ->
        List[Dict[Tree.Vertex, Tree.Vertex]]:

    def set_up_candidate_nodes() ->
            Dict[Tree.Vertex, List[Tree.Vertex]]:
        preorder_tree_depth(T1)
        preorder_tree_depth(T2)
        T2.dummy = T2.new_node()
        C = dict()
        for v in T1.vertices():
            C[v] = [T2.dummy]
            for w in T2.vertices():
                if w != T2.dummy and v.depth == w.depth:
                    C[v].append(w)
        return C

    def extend_tree_edit(
            M: Dict[Tree.Vertex, Tree.Vertex],
            L: List[Dict[Tree.Vertex, Tree.Vertex]],
            C: Dict[Tree.Vertex, List[Tree.Vertex]],
            v: Tree.Vertex) ->
            List[Dict[Tree.Vertex, Tree.Vertex]]:
        for w in C[v]:
            M[v] = w
```

```
                    if v == preorder[1][T1.number_of_nodes()]:
                        L.append(M.copy())
                    else:
                        N = {key: value[:]
                             for key, value in C.items()}
                        refine_candidate_nodes(N, v, w)
                        L = extend_tree_edit(
                            M, L, N, preorder[1][v.order + 1])
            return L

        def refine_candidate_nodes(
                C: Dict[Tree.Vertex, List[Tree.Vertex]],
                v: Tree.Vertex,
                w: Tree.Vertex) -> None:
            if w != T2.dummy:
                for x in T1.vertices():
                    C[x][:] = [y for y in C[x] if y != w]
            for x in T1.children(v):
                C[x][:] = [
                    y for y in C[x] if not (
                        y != T2.dummy and T2.parent(y) != w)]
            if not T1.is_root(v) and T2.label(w) is not None:
                for x in T1.children(T1.parent(v)):
                    if v.order < x.order:
                        C[x][:] = [
                            y for y in C[x] if not (
                                y != T2.dummy and
                                w.order > y.order)]

    preorder_tree_traversal(T1)
    preorder_tree_traversal(T2)
    preorder = (None, dict(), dict())
    for v in T1.vertices():
        preorder[1][v.order] = v
    for w in T2.vertices():
        preorder[2][w.order] = w
    M = dict()
    L = list()
    C = set_up_candidate_nodes()
    v = preorder[1][1]
    return extend_tree_edit(M, L, C, v)
```

A.2.3 Branch-and-Bound

Branch-and-bound algorithm for the edit distance between ordered trees

```python
def branch_and_bound_tree_edit(
        T1: Tree, T2: Tree) -> Dict[Tree.Vertex, Tree.Vertex]:

    def set_up_candidate_nodes() ->
            Dict[Tree.Vertex, List[Tree.Vertex]]:
        preorder_tree_depth(T1)
        preorder_tree_depth(T2)
        T2.dummy = T2.new_node()
        C = dict()
        for v in T1.vertices():
            C[v] = [T2.dummy]
            for w in T2.vertices():
                if w != T2.dummy and v.depth == w.depth:
                    C[v].append(w)
        return C

    def extend_branch_and_bound_tree_edit(
            M: Dict[Tree.Vertex, Tree.Vertex],
            A: Dict[Tree.Vertex, Tree.Vertex],
            C: Dict[Tree.Vertex, List[Tree.Vertex]],
            v: Tree.Vertex) ->
            Dict[Tree.Vertex, Tree.Vertex]:
        for w in C[v]:
            M[v] = w
            if w == T2.dummy:
                T1.cost += 1
            if T1.cost < T1.low_cost:
                if v == preorder[1][T1.number_of_nodes()]:
                    A = M.copy()
                    T1.low_cost = T1.cost
                else:
                    N = {key: value[:]
                            for key, value in C.items()}
                    refine_candidate_nodes(T1, T2, N, v, w)
                    A = extend_branch_and_bound_tree_edit(
                        M, A, N, preorder[1][v.order + 1])
            if w == T2.dummy:
                T1.cost -= 1
        return A

    def refine_candidate_nodes(
            C: Dict[Tree.Vertex, List[Tree.Vertex]],
            v: Tree.Vertex,
            w: Tree.Vertex) -> None:
        if w != T2.dummy:
```

```
            for x in T1.vertices():
                C[x][:] = [y for y in C[x] if not (y == w)]
        for x in T1.children(v):
            C[x][:] = [
                y for y in C[x] if not (
                    y != T2.dummy and T2.parent(y) != w)]
        if not T1.is_root(v) and T2.label(w) is not None:
            for x in T1.children(T1.parent(v)):
                if v.order < x.order:
                    C[x][:] = [
                        y for y in C[x] if not (
                            y != T2.dummy and
                            w.order > y.order)]

    preorder_tree_traversal(T1)
    preorder_tree_traversal(T2)
    preorder = (None, dict(), dict())
    for v in T1.vertices():
        preorder[1][v.order] = v
    for w in T2.vertices():
        preorder[2][w.order] = w
    M = dict()
    A = dict()
    C = set_up_candidate_nodes()
    v = preorder[1][1]
    T1.cost = 0
    T1.low_cost = T1.number_of_nodes()
    return extend_branch_and_bound_tree_edit(M, A, C, v)
```

A.2.4 Divide-and-Conquer

Divide-and-conquer algorithm for the edit distance between ordered trees

```
def divide_and_conquer_tree_edit(
        T1: Tree, T2: Tree) ->
        Tuple[int, Dict[Tree.Vertex, Tree.Vertex]]:

    def tree_edit(
            v1: Tree.Vertex,
            v2: Tree.Vertex,
            w1: Tree.Vertex,
            w2: Tree.Vertex,
            M: Dict[Tree.Vertex, Tree.Vertex]) ->
            Tuple[int, Dict[Tree.Vertex, Tree.Vertex]]:
```

```python
        if v1 == v2:
            if w1 == w2:
                M[v1] = w1
                dist = 1
            else:
                k2 = predecessor_of_last_child(
                    T2, w1, w2, preorder[2])
                (dist, M) = tree_edit(v1, v2, w1, k2, M)
        else:
            if w1 == w2:
                k1 = predecessor_of_last_child(
                    T1, v1, v2, preorder[1])
                (dist, M) = tree_edit(v1, k1, w1, w2, M)
            else:
                k1 = predecessor_of_last_child(
                    T1, v1, v2, preorder[1])
                M1 = M.copy()
                (dist_del, M1) = tree_edit(v1, k1, w1, w2, M1)
                k2 = predecessor_of_last_child(
                    T2, w1, w2, preorder[2])
                M2 = M.copy()
                (dist_ins, M2) = tree_edit(v1, v2, w1, k2, M2)
                M3 = M.copy()
                (dist_pre, M3) = tree_edit(v1, k1, w1, k2, M3)
                kk1 = preorder[1][k1.order + 1]
                kk2 = preorder[2][k2.order + 1]
                (dist_pos, M3) = tree_edit(
                    kk1, v2, kk2, w2, M3)
                dist = max(
                    dist_del,
                    dist_ins,
                    dist_pre +
                    dist_pos)
                if dist == dist_del:
                    M = M1
                elif dist == dist_ins:
                    M = M2
                else:
                    M = M3
    return (dist, M)


def predecessor_of_last_child(
        T: Tree,
        v1: Tree.Vertex,
        v2: Tree.Vertex,
        preorder: Dict[int, Graph.Vertex]) ->
        Tree.Vertex:
    for k in reversed(T.children(v1)):
        if k.order <= v2.order:
            break
```

```
                return preorder[k.order - 1]

    preorder_tree_traversal(T1)
    preorder_tree_traversal(T2)
    preorder = (None, dict(), dict())
    for v in T1.vertices():
        preorder[1][v.order] = v
    for w in T2.vertices():
        preorder[2][w.order] = w
    M = dict()
    v1 = preorder[1][1]
    v2 = preorder[1][T1.number_of_nodes()]
    w1 = preorder[2][1]
    w2 = preorder[2][T2.number_of_nodes()]
    return tree_edit(v1, v2, w1, w2, M)
```

A.2.5 Dynamic Programming

Top-down dynamic programming algorithm for the edit distance between ordered trees

```
def top_down_dynamic_programming_tree_edit(
        T1: Tree,
        T2: Tree) ->
        Tuple[int, Dict[Tree.Vertex, Tree.Vertex]]:

    def tree_edit(
            v1: Tree.Vertex,
            v2: Tree.Vertex,
            w1: Tree.Vertex,
            w2: Tree.Vertex,
            M: Dict[Tree.Vertex, Tree.Vertex],
            S: Dict[Tuple[Tree.Vertex],
                Tuple[int, Dict[Tree.Vertex, Tree.Vertex]]]) ->
            Tuple[int, Dict[Tree.Vertex, Tree.Vertex]]:
        if (v1, v2, w1, w2) in S:
            (dist, M) = S[(v1, v2, w1, w2)]
        else:
            if v1 == v2:
                if w1 == w2:
                    M[v1] = w1
                    dist = 1
                else:
                    k2 = predecessor_of_last_child(
                        T2, w1, w2)
```

```
                    (dist, M) = tree_edit(
                        v1, v2, w1, k2, M, S)
        else:
            if w1 == w2:
                k1 = predecessor_of_last_child(
                    T1, v1, v2)
                (dist, M) = tree_edit(
                    v1, k1, w1, w2, M, S)
            else:
                k1 = predecessor_of_last_child(
                    T1, v1, v2)
                M1 = M.copy()
                (dist_del, M1) = tree_edit(
                    v1, k1, w1, w2, M1, S)
                k2 = predecessor_of_last_child(
                    T2, w1, w2)
                M2 = M.copy()
                (dist_ins, M2) = tree_edit(
                    v1, v2, w1, k2, M2, S)
                M3 = M.copy()
                (dist_pre, M3) = tree_edit(
                    v1, k1, w1, k2, M3, S)
                kk1 = preorder[1][k1.order + 1]
                kk2 = preorder[2][k2.order + 1]
                (dist_pos, M3) = tree_edit(
                    kk1, v2, kk2, w2, M3, S)
                dist = max(
                    dist_del,
                    dist_ins,
                    dist_pre +
                    dist_pos)
                if dist == dist_del:
                    M = M1
                elif dist == dist_ins:
                    M = M2
                else:
                    M = M3
        S[(v1, v2, w1, w2)] = (dist, M)
    return (dist, M)


def predecessor_of_last_child(
        T: Tree,
        v1: Tree.Vertex,
        v2: Tree.Vertex,
        preorder: Dict[int, Graph.Vertex]) ->
        Tree.Vertex:
    for k in reversed(T.children(v1)):
        if k.order <= v2.order:
            break
```

```
        return preorder[k.order - 1]

    preorder_tree_traversal(T1)
    preorder_tree_traversal(T2)
    preorder = (None, dict(), dict())
    for v in T1.vertices():
        preorder[1][v.order] = v
    for w in T2.vertices():
        preorder[2][w.order] = w
    M = dict()
    S = dict()
    v1 = preorder[1][1]
    v2 = preorder[1][T1.number_of_nodes()]
    w1 = preorder[2][1]
    w2 = preorder[2][T2.number_of_nodes()]
    return tree_edit(v1, v2, w1, w2, M, S)
```

Bottom-up dynamic programming algorithm for the edit distance between ordered trees

```
def bottom_up_dynamic_programming_tree_edit(
        T1: Tree,
        T2: Tree) ->
        Tuple[int, Dict[Tree.Vertex, Tree.Vertex]]:

    def tree_edit(
            v: Tree.Vertex,
            w: Tree.Vertex) ->
            Tuple[int, Dict[Tree.Vertex, Tree.Vertex]]:
        n1 = T1.number_of_children(v)
        n2 = T2.number_of_children(w)
        D = [[0 for _ in range(n2 + 1)]
             for _ in range(n1 + 1)]
        E = [[[] for _ in range(n2 + 1)]
             for _ in range(n1 + 1)]
        for i in range(1, n1 + 1):
            D[i][0] = D[i - 1][0]
            E[i][0] = E[i - 1][0]
            E[i][0].append((k_th_child(T1, v, i), None))
        for j in range(1, n2 + 1):
            D[0][j] = D[0][j - 1]
            E[0][j] = E[0][j - 1]
            E[0][j].append((None, k_th_child(T2, w, j)))
        for i in range(1, n1 + 1):
            vv = k_th_child(T1, v, i)
            for j in range(1, n2 + 1):
                ww = k_th_child(T2, w, j)
```

```
                    d_del = D[i - 1][j]
                    d_ins = D[i][j - 1]
                    (d_sub, L) = tree_edit(vv, ww)
                    if d_del >= D[i - 1][j - 1] + d_sub:
                        D[i][j] = d_del
                        E[i][j] = E[i - 1][j] + [(vv, None)]
                    else:
                        if d_ins >= D[i - 1][j - 1] + d_sub:
                            D[i][j] = d_ins
                            E[i][j] = E[i][j - 1] + [(None, ww)]
                        else:
                            D[i][j] = D[i - 1][j - 1] + d_sub
                            E[i][j] = E[i - 1][j - 1] + L
        return (D[n1][n2] + 1, E[n1][n2] + [(v, w)])


    def k_th_child(
            T: Tree,
            v: Tree.Vertex,
            k: int) -> Tree.Vertex:
        i = 0
        for w in T.children(v):
            i += 1
            if i == k:
                break
        return w

    (d, L) = tree_edit(T1.root(), T2.root())
    M = dict()
    for (v, w) in L:
        if v is not None:
            M[v] = w
    return (d, M)
```

A.3   Tree Traversal

A.3.1   Preorder Traversal of a Tree

Recursive algorithm for the preorder traversal of a tree

```
def preorder_tree_traversal(T: Tree) -> None:

    def preorder_tree_traversal(v: Tree.Vertex) -> None:
        T.num += 1
        v.order = T.num
```

```
        for w in T.children(v):
            preorder_tree_traversal(w)

    T.num = 0
    preorder_tree_traversal(T.root())
```

Iterative algorithm for the preorder traversal of a tree

```
def preorder_tree_traversal(T: Tree) -> None:
    S = Stack()
    S.push(T.root())
    num = 0
    while not S.empty():
        v = S.pop()
        num += 1
        v.order = num
        for w in reversed(T.children(v)):
            S.push(w)
```

Iterative algorithm for the preorder traversal of a tree giving a list of nodes

```
def preorder_tree_list_traversal(
        T: Tree) -> List[Graph.Vertex]:
    L = list()
    S = Stack()
    S.push(T.root())
    while not S.empty():
        v = S.pop()
        L.append(v)
        for w in reversed(T.children(v)):
            S.push(w)
    return L
```

A.3.2   Postorder Traversal of a Tree

Recursive algorithm for the postorder traversal of a tree

```python
def postorder_tree_traversal(T: Tree) -> None:

    def postorder_tree_traversal(v: Tree.Vertex) -> None:
        for w in T.children(v):
            postorder_tree_traversal(w)
        T.num += 1
        v.order = T.num

    T.num = 0
    postorder_tree_traversal(T.root())
```

Iterative algorithm for the postorder traversal of a tree

```python
def postorder_tree_traversal(T: Tree) -> None:
    S = Stack()
    S.push(T.root())
    num = 0
    while not S.empty():
        v = S.pop()
        num += 1
        v.order = T.number_of_nodes() - num + 1
        for w in T.children(v):
            S.push(w)
```

Recursive algorithm for the postorder traversal of a tree giving a list of nodes

```python
def postorder_tree_list_traversal(
        T: Tree) -> List[Graph.Vertex]:
    L = list()
    S = Stack()
    S.push(T.root())
    while not S.empty():
        v = S.pop()
        L.append(v)
        for w in T.children(v):
            S.push(w)
    return reversed(L)
```

A.3.3 Top-Down Traversal of a Tree

Iterative algorithm for the top-down traversal of a tree

```python
def top_down_tree_traversal(T: Tree) -> None:
    Q = Queue()
    Q.enqueue(T.root())
    num = 0
    while not Q.empty():
        v = Q.dequeue()
        num += 1
        v.order = num
        for w in T.children(v):
            Q.enqueue(w)
```

Iterative algorithm for the top-down traversal of a tree giving a list of nodes

```python
def top_down_tree_list_traversal(
        T: Tree) -> List[Graph.Vertex]:
    L = list()
    Q = Queue()
    Q.enqueue(T.root())
    while not Q.empty():
        v = Q.dequeue()
        L.append(v)
        for w in T.children(v):
            Q.enqueue(w)
    return L
```

A.3.4 Bottom-Up Traversal of a Tree

Iterative algorithm for the bottom-up traversal of a tree

```python
def bottom_up_tree_traversal(T: Tree) -> None:
    Q = Queue()
    for v in T.vertices():
        v.children = 0
    for v in T.vertices():
        for w in T.children(v):
            v.children += 1
```

```
R = Queue()
R.enqueue(T.root())
while not R.empty():
    v = R.dequeue()
    for w in T.children(v):
        if T.is_leaf(w):
            Q.enqueue(w)
        else:
            R.enqueue(w)
num = 0
while not Q.empty():
    v = Q.dequeue()
    num += 1
    v.order = num
    if not T.is_root(v):
        T.parent(v).children -= 1
        if T.parent(v).children == 0:
            Q.enqueue(T.parent(v))
```

Iterative algorithm for the bottom-up traversal of a tree giving a list of nodes

```
def bottom_up_tree_list_traversal(
        T: Tree) -> List[Graph.Vertex]:
    L = list()
    Q = Queue()
    for v in T.vertices():
        v.children = 0
    for v in T.vertices():
        for w in T.children(v):
            v.children += 1
    R = Queue()
    R.enqueue(T.root())
    while not R.empty():
        v = R.dequeue()
        for w in T.children(v):
            if T.is_leaf(w):
                Q.enqueue(w)
            else:
                R.enqueue(w)
    while not Q.empty():
        v = Q.dequeue()
        L.append(v)
        if not T.is_root(v):
            T.parent(v).children -= 1
            if T.parent(v).children == 0:
                Q.enqueue(T.parent(v))
    return L
```

A.3.5 Applications

Algorithm for computing the depth of the nodes of a tree

```python
def preorder_tree_depth(T: Tree) -> None:
    deepest = T.first_vertex()
    S = Stack()
    S.push(T.root())
    while not S.empty():
        v = S.pop()
        if T.is_root(v):
            v.depth = 0
        else:
            v.depth = T.parent(v).depth + 1
            if v.depth > deepest.depth:
                deepest = v
        for w in reversed(T.children(v)):
            S.push(w)
    T.depth = deepest.depth
```

Algorithm for computing the height of the nodes of a tree

```python
def bottom_up_tree_height(T: Tree) -> None:
    for v in T.vertices():
        v.children = 0
    for v in T.vertices():
        for w in T.children(v):
            v.children += 1
    Q = Queue()
    for v in T.vertices():
        if v.children == 0:
            Q.enqueue(v)
    while not Q.empty():
        v = Q.dequeue()
        v.height = 0
        if not T.is_leaf(v):
            for w in T.children(v):
                v.height = max(v.height, w.height)
            v.height += 1
        if not T.is_root(v):
            w = T.parent(v)
            w.children -= 1
            if w.children == 0:
                Q.enqueue(w)
    T.height = T.root().height
```

Algorithm for computing the layered layout of a tree

```python
def layered_tree_layout(T: Tree) -> None:
    S = Stack()
    S.push(T.root())
    while not S.empty():
        v = S.pop()
        if T.is_root(v):
            v.depth = 0
        else:
            v.depth = T.parent(v).depth + 1
        w = T.last_child(v)
        while w is not None:
            S.push(w)
            w = T.previous_sibling(w)

    for v in T.vertices():
        v.breadth = 0
        v.children = 0
    for v in T.vertices():
        for w in T.children(v):
            v.children += 1
    Q = Queue()
    for v in T.vertices():
        if v.children == 0:
            Q.enqueue(v)
    while not Q.empty():
        v = Q.dequeue()
        if T.is_leaf(v):
            v.breadth = 1
        if not T.is_root(v):
            T.parent(v).breadth += v.breadth
            T.parent(v).children -= 1
            if T.parent(v).children == 0:
                Q.enqueue(T.parent(v))

    Q = Queue()
    Q.enqueue(T.root())
    while not Q.empty():
        v = Q.dequeue()
        if T.is_root(v):
            v.x = 0
        else:
            if T.is_first_child(v):
                v.x = T.parent(v).x
            else:
                w = T.previous_sibling(v)
                v.x = w.x + w.breadth
        v.y = -v.depth
        for w in T.children(v):
```

```
                Q.enqueue(w)

        for v in T.vertices():
            v.children = 0
        for v in T.vertices():
            for w in T.children(v):
                v.children += 1
        Q = Queue()
        for v in T.vertices():
            if v.children == 0:
                Q.enqueue(v)
        while not Q.empty():
            v = Q.dequeue()
            if not T.is_leaf(v):
                v.x = (T.first_child(v).x +
                        T.last_child(v).x) / 2
            if not T.is_root(v):
                T.parent(v).children -= 1
                if T.parent(v).children == 0:
                    Q.enqueue(T.parent(v))
```

A.4 Tree Isomorphism

A.4.1 Tree Isomorphism

Simple algorithm for testing isomorphism of ordered trees

```
def simple_ordered_tree_isomorphism(
        T1: Tree, T2: Tree) -> bool:
    if T1.number_of_nodes() != T2.number_of_nodes():
        return False
    top_down_tree_traversal(T1)
    top_down_tree_traversal(T2)
    n = T1.number_of_nodes()
    disorder = [0 for _ in range(n + 1)]
    for w in T2.vertices():
        disorder[w.order] = w
    for v in T1.vertices():
        v.M = disorder[v.order]
    for v in T1.vertices():
        w = v.M
```

```
            if T1.label(v) != T2.label(w):
                return False
            if not T1.is_leaf(v) and (
                    T2.is_leaf(w) or T1.first_child(v).M !=
                    T2.first_child(w)):
                return False
            if not T1.is_last_child(v) and (T2.is_last_child(
                    w) or T1.next_sibling(v).M !=
                    T2.next_sibling(w)):
                return False
        return True
```

Algorithm for testing isomorphism of ordered trees

```
def ordered_tree_isomorphism(T1: Tree, T2: Tree) -> bool:

    def map_ordered_tree(
            r1: Tree.Vertex,
            r2: Tree.Vertex) -> bool:
        if r1.label() != r2.label():
            return False
        r1.M = r2
        if T1.number_of_children(
                r1) != T2.number_of_children(r2):
            return False
        if not T1.is_leaf(r1):
            v1 = T1.first_child(r1)
            v2 = T2.first_child(r2)
            if not map_ordered_tree(v1, v2):
                return False
            while not T1.is_last_child(v1):
                v1 = T1.next_sibling(v1)
                v2 = T2.next_sibling(v2)
                if not map_ordered_tree(v1, v2):
                    return False
        return True

    if T1.number_of_nodes() != T2.number_of_nodes():
        return False
    return map_ordered_tree(T1.root(), T2.root())
```

Algorithm for testing isomorphism of unordered trees

```python
def unordered_tree_isomorphism(T1: Tree, T2: Tree) -> bool:

    def assign_isomorphism_codes(T: Tree) -> None:
        for v in postorder_tree_list_traversal(T):
            v.code = list()
            if T.is_leaf(v):
                v.code.append(1)
            else:
                L = list()
                size = 1
                for w in T.children(v):
                    size += w.code[0]
                    L.append(w.code)
                L.sort()
                v.code.append(size)
                for x in L:
                    for y in x:
                        v.code.append(y)

    def build_tree_isomorphism_mapping() -> None:
        for w in T2.vertices():
            w.mapped_to = False
        T1.root().M = T2.root()
        T2.root().mapped_to = True
        for v in preorder_tree_list_traversal(T1):
            if not T1.is_root(v):
                for w in T2.children(T1.parent(v).M):
                    if v.code == w.code and not w.mapped_to:
                        if T2.is_root(w) or T1.parent(
                                v).M == T2.parent(w):
                            v.M = w
                            w.mapped_to = True
                            break

    if T1.number_of_nodes() != T2.number_of_nodes():
        return False
    assign_isomorphism_codes(T1)
    assign_isomorphism_codes(T2)
    if T1.root().code == T2.root().code:
        build_tree_isomorphism_mapping()
        return True
    else:
        return False
```

A.4.2 Subtree Isomorphism

Algorithm for finding a top-down subtree isomorphism of ordered trees

```python
def top_down_ordered_subtree_isomorphism(
        T1: Tree,
        T2: Tree) -> bool:

    def map_ordered_subtree(
            r1: Tree.Vertex,
            r2: Tree.Vertex) -> bool:
        if r1.label() != r2.label():
            return False
        r1.M = r2
        if T1.number_of_children(
                r1) > T2.number_of_children(r2):
            return False
        if not T1.is_leaf(r1):
            v1 = T1.first_child(r1)
            v2 = T2.first_child(r2)
            if not map_ordered_subtree(v1, v2):
                return False
            while not T1.is_last_child(v1):
                v1 = T1.next_sibling(v1)
                v2 = T2.next_sibling(v2)
                if not map_ordered_subtree(v1, v2):
                    return False
        return True

    if T1.number_of_nodes() > T2.number_of_nodes():
        return False
    return map_ordered_subtree(T1.root(), T2.root())
```

Algorithm for finding a top-down subtree isomorphism of unordered trees

```python
def top_down_unordered_subtree_isomorphism(
        T1: Tree, T2: Tree) -> bool:

    def compute_height_and_size_of_all_nodes(
            T: Tree) -> None:
        for v in postorder_tree_list_traversal(T):
            v.height = 0
            v.size = 1
            if not T.is_leaf(v):
                for w in T.children(v):
```

```
                    v.height = max(v.height, w.height)
                    v.size += w.size
                v.height += 1

    def top_down_unordered_subtree_isomorphism(
            r1: Tree.Vertex,
            r2: Tree.Vertex) -> bool:
        if T1.label(r1) != T2.label(r2):
            return False
        if T1.is_leaf(r1):
            return True
        if T1.number_of_nodes() > T2.number_of_nodes(
        ) or r1.height > r2.height or r1.size > r2.size:
            return False
        G = Graph()
        for v1 in T1.children(r1):
            v = G.new_vertex(T1.label(v1))
            v.GT = v1
            v1.T1G = v
        for v2 in T2.children(r2):
            w = G.new_vertex(T2.label(v1))
            w.GT = v2
            v2.T2G = w
        for v1 in T1.children(r1):
            for v2 in T2.children(r2):
                if top_down_unordered_subtree_isomorphism(
                        v1, v2):
                    G.new_edge(v1.T1G, v2.T2G)
        L = maximum_cardinality_bipartite_matching(G)
        if len(L) == T1.number_of_children(r1):
            for e in L:
                G.source(e).GT.B.add(G.target(e).GT)
            return True
        else:
            return False

    def reconstruct_unordered_subtree_isomorphism() -> None:
        T1.root().M = T2.root()
        for v in preorder_tree_list_traversal(T1):
            if not T1.is_root(v):
                for w in v.B:
                    if T1.parent(v).M == T2.parent(w):
                        v.M = w
                        break

compute_height_and_size_of_all_nodes(T1)
compute_height_and_size_of_all_nodes(T2)
for v1 in T1.vertices():
    v1.B = set()
iso = top_down_unordered_subtree_isomorphism(
```

```
                    T1.root(), T2.root())
        if iso:
            reconstruct_unordered_subtree_isomorphism()
        return iso
```

Algorithm for enumerating bottom-up subtree isomorphisms of ordered trees

```python
def bottom_up_ordered_subtree_isomorphism(
        T1: Tree,
        T2: Tree) ->
        List[List[Tuple[Tree.Vertex, Tree.Vertex]]]:

    def compute_height_and_size_of_all_nodes(
            T: Tree) -> None:
        L = postorder_tree_list_traversal(T)
        for v in L:
            v.height = 0
            v.size = 1
            if not T.is_leaf(v):
                for w in T.children(v):
                    v.height = max(v.height, w.height)
                    v.size += w.size
                v.height += 1

    def map_ordered_subtree(
            r1: Tree.Vertex,
            r2: Tree.Vertex) -> bool:
        if T1.label(r1) != T2.label(r2):
            return False
        r1.M = r2
        if T1.number_of_children(
                r1) > T2.number_of_children(r2):
            return False
        if not T1.is_leaf(r1):
            v1 = T1.first_child(r1)
            v2 = T2.first_child(r2)
            if not map_ordered_subtree(v1, v2):
                return False
            while not T1.is_last_child(v1):
                v1 = T1.next_sibling(v1)
                v2 = T2.next_sibling(v2)
                if not map_ordered_subtree(v1, v2):
                    return False
        return True

    compute_height_and_size_of_all_nodes(T1)
    compute_height_and_size_of_all_nodes(T2)
```

```
    L = list()
    for v2 in T2.vertices():
        if T1.root().height == v2.height and T1.root(
                ).size == v2.size:
            if map_ordered_subtree(T1.root(), v2):
                L.append([(v1, v1.M)
                            for v1 in T1.vertices()])
    return L
```

Algorithm for enumerating bottom-up subtree isomorphisms of unordered trees

```
def bottom_up_unordered_subtree_isomorphism(
        T1: Tree,
        T2: Tree) ->
        List[List[Tuple[Tree.Vertex, Tree.Vertex]]]:

    def partition_tree_in_isomorphism_equivalence_classes(
            T: Tree) -> None:
        nonlocal num
        for v in postorder_tree_list_traversal(T):
            if T.is_leaf(v):
                v.code = 1
            else:
                L = sorted([w.code for w in T.children(v)])
                if tuple(L) in D:
                    v.code = D[tuple(L)]
                else:
                    num += 1
                    D[tuple(L)] = num
                    v.code = num

    def map_unordered_subtree(
            r1: Tree.Vertex,
            r2: Tree.Vertex) -> None:
        L2 = [v2 for v2 in T2.children(r2)]
        for v1 in T1.children(r1):
            for v2 in L2:
                if v1.code == v2.code:
                    v1.M = v2
                    L2.remove(v2)
                    map_unordered_subtree(v1, v2)
                    break

    num = 1
    D = dict()
    partition_tree_in_isomorphism_equivalence_classes(T1)
```

```
        partition_tree_in_isomorphism_equivalence_classes(T2)
        L = list()
        for v2 in postorder_tree_list_traversal(T2):
            if T1.root().code == v2.code:
                T1.root().M = v2
                map_unordered_subtree(T1.root(), v2)
                L.append([(v1, v1.M) for v1 in T1.vertices()])
    return L
```

A.4.3 Maximum Common Subtree Isomorphism

**Algorithm for finding a top-down maximum common subtree isomorphism of
ordered trees**

```
def top_down_ordered_max_common_subtree_isomorphism(
        T1: Tree,
        T2: Tree) -> None:

    def map_ordered_common_subtree(
            r1: Tree.Vertex,
            r2: Tree.Vertex) -> None:
        if T1.label(r1) != T2.label(r2):
            return False
        else:
            r1.M = r2
            if not T1.is_leaf(r1) and not T2.is_leaf(r2):
                v1 = T1.first_child(r1)
                v2 = T2.first_child(r2)
                while map_ordered_common_subtree(v1, v2):
                    if T1.is_last_child(v1):
                        break
                    v1 = T1.next_sibling(v1)
                    if T2.is_last_child(v2):
                        break
                    v2 = T2.next_sibling(v2)
            return True

    for v in T1.vertices():
        v.M = None
    map_ordered_common_subtree(T1.root(), T2.root())
```

Algorithm for finding a top-down maximum common subtree isomorphism of unordered trees

```python
def top_down_unordered_max_common_subtree_isomorphism(
        T1: Tree,
        T2: Tree) -> None:

    def max_common_subtree_isomorphism(
            r1: Tree.Vertex,
            r2: Tree.Vertex) -> None:
        if T1.label(r1) != T2.label(r2):
            return 0
        if T1.is_leaf(r1) or T2.is_leaf(r2):
            return 1
        G = Graph()
        for v1 in T1.children(r1):
            v = G.new_vertex(T1.label(v1))
            v.GT = v1
            v1.T1G = v
        for v2 in T2.children(r2):
            w = G.new_vertex(T2.label(v2))
            w.GT = v2
            v2.T2G = w
        for v1 in T1.children(r1):
            for v2 in T2.children(r2):
                res = max_common_subtree_isomorphism(v1, v2)
                if res != 0:
                    e = G.new_edge(v1.T1G, v2.T2G)
                    e.weight = res
        L = maximum_weight_bipartite_matching(G)
        res = 1
        for e in L:
            G.source(e).GT.B.add(G.target(e).GT)
            res += e.weight
        return res

    r1 = T1.root()
    r2 = T2.root()
    for v1 in T1.vertices():
        v1.T1G = None
    for v2 in T2.vertices():
        v2.T2G = None
    if T1.label(r1) == T2.label(r2):
        for v1 in T1.vertices():
            v1.B = set()
        r1.B.add(r2)
        max_common_subtree_isomorphism(r1, r2)
        for v1 in T1.vertices():
            v1.M = None
        r1.M = r2
```

```
                for v in preorder_tree_list_traversal(T1):
                    if not T1.is_root(v):
                        for w in v.B:
                            if T1.parent(v).M == T2.parent(w):
                                v.M = w
                                break
```

Algorithm for finding a bottom-up maximum common subtree isomorphism of ordered trees

```
def bottom_up_ordered_max_common_subtree_isomorphism(
        T1: Tree,
        T2: Tree) -> None:

    def partition_tree_in_isomorphism_equivalence_classes(
            T: Tree) -> None:
        nonlocal num
        for v in postorder_tree_list_traversal(T):
            if T.is_leaf(v):
                v.code = 1
            else:
                LL = [w.code for w in T.children(v)]
                if tuple(LL) in D:
                    v.code = D[tuple(LL)]
                else:
                    num += 1
                    D[tuple(LL)] = num
                    v.code = num

    def find_largest_common_subtree(
            ) -> Tuple[Tree.Vertex, Tree.Vertex]:
        Q1 = PriorityQueue()
        Q2 = PriorityQueue()
        for v1 in postorder_tree_list_traversal(T1):
            v1.size = 1
            for w1 in T1.children(v1):
                v1.size += w1.size
            Q1.enqueue(((-v1.size, v1.code), v1))
        for v2 in postorder_tree_list_traversal(T2):
            v2.size = 1
            for w2 in T2.children(v2):
                v2.size += w2.size
            Q2.enqueue(((-v2.size, v2.code), v2))
        while Q1 and Q2:
            (prio_v, v) = Q1.front()
            (prio_w, w) = Q2.front()
            if v.code == w.code:
```

```
                            break
                    if prio_v < prio_w:
                        Q1.dequeue()
                    else:
                        Q2.dequeue()
            v.M = w
            return (v, w)

    def map_ordered_subtree(
            r1: Tree.Vertex,
            r2: Tree.Vertex) -> None:
        L2 = [v2 for v2 in T2.children(r2)]
        for v1 in T1.children(r1):
            for v2 in L2:
                if v1.code == v2.code:
                    v1.M = v2
                    L2.remove(v2)
                    map_ordered_subtree(v1, v2)
                    break

    num = 1
    D = dict()
    partition_tree_in_isomorphism_equivalence_classes(T1)
    partition_tree_in_isomorphism_equivalence_classes(T2)
    for v1 in T1.vertices():
        v1.M = None
    (v, w) = find_largest_common_subtree()
    map_ordered_subtree(v, w)
```

Algorithm for finding a bottom-up maximum common subtree isomorphism of unordered trees

```
def bottom_up_unordered_max_common_subtree_isomorphism(
        T1: Tree, T2: Tree) -> None:

    def partition_tree_in_isomorphism_equivalence_classes(
            T: Tree) -> None:
        nonlocal num
        for v in postorder_tree_list_traversal(T):
            if T.is_leaf(v):
                v.code = 1
            else:
                L = sorted([w.code for w in T.children(v)])
                if tuple(L) in D:
                    v.code = D[tuple(L)]
                else:
                    num += 1
```

```
                            D[tuple(L)] = num
                            v.code = num

        def find_largest_common_subtree(
                ) -> Tuple[Tree.Vertex, Tree.Vertex]:
            Q1 = PriorityQueue()
            Q2 = PriorityQueue()
            for v1 in postorder_tree_list_traversal(T1):
                v1.size = 1
                for w1 in T1.children(v1):
                    v1.size += w1.size
                Q1.enqueue(((-v1.size, v1.code), v1))
            for v2 in postorder_tree_list_traversal(T2):
                v2.size = 1
                for w2 in T2.children(v2):
                    v2.size += w2.size
                Q2.enqueue(((-v2.size, v2.code), v2))
            while not Q1.empty() and not Q2.empty():
                (prio_v, v) = Q1.front()
                (prio_w, w) = Q2.front()
                if v.code == w.code:
                    break
                if prio_v < prio_w:
                    Q1.dequeue()
                else:
                    Q2.dequeue()
            v.M = w
            return (v, w)

        def preorder_subtree_list_traversal(
                T: Tree, r: Tree.Vertex):
            L = list()
            S = Stack()
            S.push(r)
            while not S.empty():
                v = S.pop()
                L.append(v)
                for w in reversed(T.children(v)):
                    S.push(w)
            return L

        num = 1
        D = dict()
        partition_tree_in_isomorphism_equivalence_classes(T1)
        partition_tree_in_isomorphism_equivalence_classes(T2)
        for v1 in T1.vertices():
            v1.M = None
        for w2 in T2.vertices():
            w2.mapped_to = False
        (v, w) = find_largest_common_subtree()
```

```
                v.M = w
                w.mapped_to = True
                for v1 in preorder_subtree_list_traversal(T1, v):
                    if v1 != v:
                        for w1 in T2.children(T1.parent(v1).M):
                            if v1.code == w1.code and not w1.mapped_to:
                                v1.M = w1
                                w1.mapped_to = True
                                break
```

A.5 Graph Traversal

A.5.1 Depth-First Traversal of a Graph

Algorithm for the depth-first traversal of a graph

```
def depth_first_traversal(G: Graph) -> None:
    for v in G.vertices():
        v.order = -1
    num = 0
    S = Stack()
    for v in G.vertices():
        if v.order == -1:
            S.push(v)
            while not S.empty():
                v = S.pop()
                if v.order == -1:
                    num += 1
                    v.order = num
                    for e in reversed(list(G.outgoing(v))):
                        w = G.target(e)
                        if w.order == -1:
                            S.push(w)
```

Algorithm for finding a depth-first spanning tree of a graph

```
def depth_first_spanning_tree(
        G: Graph, v: Graph.Vertex) ->
        Tuple[Set[Graph.Vertex], Set[Graph.Edge]]:
    W = set()
```

```
S = set()
for w in G.vertices():
    w.order = -1
W.add(v)
num = 1
v.order = num
Z = Stack()
for e in reversed(list(G.outgoing(v))):
    w = G.target(e)
    Z.push(e)
while not Z.empty():
    e = Z.pop()
    v = G.target(e)
    if v.order == -1:
        W.add(v)
        S.add(e)
        num += 1
        v.order = num
        for e in reversed(list(G.outgoing(v))):
            w = G.target(e)
            if w.order == -1:
                Z.push(e)
return (W, S)
```

Algorithm for the leftmost depth-first traversal of a bidirected graph

```
def leftmost_depth_first_traversal(
        G: Graph, e: Graph.Edge) -> List[Graph.Edge]:

    def leftmost_depth_first_traversal(
            e: Graph.Edge) -> List[Graph.Edge]:
        L = list()
        v = G.target(e)
        for erev in G.incoming(G.source(e)):
            if G.source(erev) == v:
                break
        L.append(e)
        if v.visited:
            if erev.visited:
                eprime = erev
                while True:
                    eprime = G.adj_succ(eprime)
                    if eprime is None:
                        eprime = G.first_adj_edge(
                            G.source(erev))   # cyclic
                    if eprime == erev or not eprime.visited:
                        break
```

```
                    if eprime.visited:
                        return L
                else:
                    eprime = erev
            else:
                eprime = G.adj_succ(erev)
                if eprime is None:
                    eprime = G.first_adj_edge(
                        G.source(erev))  # cyclic
            e.visited = True
            v.visited = True
            L += leftmost_depth_first_traversal(eprime)
            return L

    for v in G.vertices():
        v.visited = False
    for ee in G.edges():
        ee.visited = False
    G.source(e).visited = True
    return leftmost_depth_first_traversal(e)
```

A.5.2   Breadth-First Traversal of a Graph

Algorithm for the breadth-first traversal of a graph

```
def breadth_first_traversal(G: Graph) -> None:
    for v in G.vertices():
        v.order = -1
    num = 0
    Q = Queue()
    for v in G.vertices():
        if v.order == -1:
            Q.enqueue(v)
            num += 1
            v.order = num
            while not Q.empty():
                v = Q.dequeue()
                for e in G.outgoing(v):
                    w = G.target(e)
                    if w.order == -1:
                        Q.enqueue(w)
                        num += 1
                        w.order = num
```

Algorithm for finding a breadth-first spanning tree of a graph

```python
def breadth_first_spanning_tree(
        G: Graph, v: Graph.Vertex) ->
        Tuple[Set[Graph.Vertex], Set[Graph.Edge]]:
    W = set()
    S = set()
    for w in G.vertices():
        w.order = -1
    W.add(v)
    num = 1
    v.order = num
    Q = Queue()
    for e in G.outgoing(v):
        if G.target(e) != v:
            Q.enqueue(e)
    while not Q.empty():
        e = Q.dequeue()
        w = e.target()
        if w not in W:
            num += 1
            w.order = num
            W.add(w)
            S.add(e)
            for e in G.outgoing(w):
                if G.target(e) not in W:
                    Q.enqueue(e)
    return (W, S)
```

A.5.3 Applications

Algorithm for finding a shortest path in a breadth-first spanning tree of a graph

```python
def acyclic_shortest_path(
        G: Graph,
        s: Graph.Vertex,
        t: Graph.Vertex,
        W: Set[Graph.Vertex],
        S: Set[Graph.Edge]) -> List[Graph.Edge]:
    d = dict()
    for e in S:
        d[G.target(e)] = e
    P = list()
    v = t
```

```
    while v != s:
        e = d[v]
        P.append(e)
        v = G.source(e)
    P.reverse()
    return P
```

Algorithm for finding a maximum cardinality matching in a bipartite graph

```
def maximum_cardinality_bipartite_matching(
        G: Graph) -> Set[Graph.Edge]:
    D = dict()  # of pairs of vertices to edges
    X = set([G.source(e) for e in G.edges()])
    Y = set([G.target(e) for e in G.edges()])
    M = set()  # of edges
    for x in X:
        for e in G.outgoing(x):
            D[(G.source(e), G.target(e))] = e
    while True:
        P = augmenting_path(G, X, Y, M, D)
        M.symmetric_difference_update(P)
        if len(P) == 0:
            break
    return M
```

Algorithm for finding an augmenting path in a bipartite graph with respect to a matching

```
def augmenting_path(G: Graph,
                    X: Set[Graph.Vertex],
                    Y: Set[Graph.Vertex],
                    M: Set[Graph.Edge],
                    D: Dict[Tuple[Graph.Vertex],
                            Graph.Edge]) -> Set[Graph.Edge]:

    def matched(x: Graph.Vertex) -> bool:
        for e in M:
            if G.source(e) == x or G.target(e) == x:
                return True
        return False

    def depth_first_traversal(
            x: Graph.Vertex) -> Graph.Vertex:
        x.visited = True
```

```
                    y = None
                    for e in G.outgoing(x):
                        w = G.target(e)
                        if not w.visited and e not in M:
                            w.visited = True
                            w.pred = x
                            if not matched(w):
                                return w
                            else:
                                for ee in G.incoming(w):
                                    z = G.source(ee)
                                    if not z.visited and ee in M:
                                        z.visited = True
                                        z.pred = w
                                        y = depth_first_traversal(z)
                                        if y is not None:
                                            return y
                    return y

        P = set()
        for v in G.vertices():
            v.visited = False
            v.pred = None
        for x in X:
            if not matched(x):
                y = depth_first_traversal(x)
                if y is not None:
                    z = y
                    while z != x:
                        if (z, z.pred) in D:
                            P.add(D[(z, z.pred)])
                        else:
                            P.add(D[(z.pred, z)])
                        z = z.pred
                    break
        return P
```

Algorithm for finding a maximum weight matching in a bipartite graph

```
def maximum_weight_bipartite_matching(
        G: Graph) -> Set[Graph.Edge]:

    def matched(x: Graph.Vertex) -> bool:
        for e in M:
            if G.source(e) == x or G.target(e) == x:
                return True
        return False
```

```
D = dict()   # of pairs of vertices to edges
X = set([G.source(e) for e in G.edges()])
Y = set([G.target(e) for e in G.edges()])
M = set()   # of edges
for xx in X:
    for e in G.outgoing(xx):
        D[(G.source(e), G.target(e))] = e
for xx in X:
    xx.f = max([e.weight for e in G.outgoing(xx)])
for yy in Y:
    yy.f = 0
while len(M) != min(len(X), len(Y)):
    for x in sorted(X):
        if not matched(x):
            break
    S = set([x])
    T = set()
    for yy in Y.difference(T):
        yy.slack = x.f + yy.f - D[x, yy].weight
    while True:
        slack = min(
            [yy.slack for yy in Y.difference(T)])
        if slack != 0:
            epsilon = slack
            for xx in S:
                xx.f -= epsilon
            for yy in T:
                yy.f += epsilon
            for yy in Y.difference(T):
                yy.slack -= epsilon
                slack = min(slack, yy.slack)
        if slack == 0:
            for y in Y.difference(T):
                if y.slack == slack:
                    break
            if not matched(y):
                for e in G.edges():
                    e.deleted = G.source(
                        e).f + G.target(e).f != e.weight
                P = augmenting_path(G, M, D, x, y)
                M.symmetric_difference_update(P)
                break
            else:
                for z in X:
                    if D[z, y] in M:
                        break
                S.add(z)
                T.add(y)
                for yy in Y.difference(T):
                    yy.slack = min(yy.slack,
```

```
                                      z.f + yy.f - D[z, yy].weight)

        return M
```

Algorithm for finding an augmenting path between two vertices in a bipartite graph with respect to a matching

```
def augmenting_path(G: Graph,
                    M: Set[Graph.Edge],
                    D: Dict[Tuple[Graph.Vertex], Graph.Edge],
                    x: Graph.Vertex,
                    y: Graph.Vertex) -> Set[Graph.Edge]:


    def depth_first_traversal(
            x: Graph.Vertex,
            y: Graph.Vertex) -> None:
        x.visited = True
        for e in G.outgoing(x):
            if not e.deleted:
                w = G.target(e)
                if not w.visited and e not in M:
                    w.visited = True
                    w.pred = x
                    if w == y:
                        return
                    else:
                        for ee in G.incoming(w):
                            if not ee.deleted:
                                z = G.source(ee)
                                if not z.visited and ee in M:
                                    z.visited = True
                                    z.pred = w
                                    depth_first_traversal(z, y)


    for v in G.vertices():
        v.visited = False
        v.pred = None
    depth_first_traversal(x, y)
    P = set()
    z = y
    while z != x:
        if (z, z.pred) in D:
            P.add(D[(z, z.pred)])
        else:
            P.add(D[(z.pred, z)])
        z = z.pred
    return P
```

Algorithm for testing isomorphism of ordered graphs

```python
def ordered_graph_isomorphism(G1: Graph, G2: Graph):

    def obtain_order_in_which_vertices_were_visited(
            G: Graph, L: List[Graph.Edge]):
        for v in G.vertices():
            v.order = -1
        e = L[0]
        num = 1
        G.source(e).order = num
        for e in L:
            if G.target(e).order == -1:
                num += 1
                G.target(e).order = num

    def build_ordered_graph_isomorphism_mapping():
        n = G1.number_of_vertices()
        disorder1 = [0 for _ in range(n + 1)]
        for v1 in G1.vertices():
            disorder1[v1.order] = v1
        disorder2 = [0 for _ in range(n + 1)]
        for v2 in G2.vertices():
            disorder2[v2.order] = v2
        for i in range(1, n + 1):
            disorder1[i].M = disorder2[i]

    def test_mapping_for_ordered_graph_isomorphism():
        for v in G1.vertices():
            L1 = [G1.target(e) for e in G1.outgoing(v)]
            L2 = [G2.target(e) for e in G2.outgoing(v.M)]
            if (len(L1) != len(L2)):
                return False
            for offset in range(len(L2)):
                if L1[0].M == L2[offset]:
                    break
            for i in range(len(L1)):
                if L1[i].M != L2[(i + offset) % len(L2)]:
                    return False
        return True

    if G1.number_of_vertices() != G2.number_of_vertices():
        return False
    if G1.number_of_edges() != G2.number_of_edges():
        return False
    if G1.number_of_edges() == 0:
        v = G1.first_vertex()
        w = G2.first_vertex()
        while v is not None:
```

```
                    v.M = w
                    v = G1.succ_vertex(v)
                    w = G2.succ_vertex(w)
                return True
    e1 = G1.first_edge()
    L1 = leftmost_depth_first_traversal(G1, e1)
    obtain_order_in_which_vertices_were_visited(G1, L1)
    for e2 in G2.edges():
        L2 = leftmost_depth_first_traversal(G2, e2)
        obtain_order_in_which_vertices_were_visited(G2, L2)
        build_ordered_graph_isomorphism_mapping()
        if test_mapping_for_ordered_graph_isomorphism():
            return True
    return False
```

A.6   Clique, Independent Set, and Vertex Cover

A.6.1   Cliques, Maximal Cliques, and Maximum Cliques

Backtracking algorithm for enumerating cliques of a graph

```
def all_cliques(G: Graph) -> List[Set[Graph.Vertex]]:

    def next_clique(
            C: Set[Graph.Vertex],
            P: Set[Graph.Vertex]) ->
            List[Set[Graph.Vertex]]:
        L = list()
        if len(C) > 2:
            L.append(C.copy())  # to collect the clique
        for v in P.copy():
            P.remove(v)
            PP = set()
            for e in G.outgoing(v):
                w = G.target(e)
                if w in P:
                    PP.add(w)
            C.add(v)
            L += next_clique(C, PP)
            C.remove(v)
        return L

    C = set()
    P = set(G.vertices())
    return next_clique(C, P)
```

Simple backtracking algorithm for enumerating maximal cliques of a graph

```python
def simple_all_maximal_cliques(
        G: Graph) -> List[Set[Graph.Vertex]]:

    def simple_next_maximal_clique(
            C: Set[Graph.Vertex],
            P: Set[Graph.Vertex],
            S: Set[Graph.Vertex]) ->
            List[Set[Graph.Vertex]]:
        L = list()
        if not P and not S:
            L.append(C.copy())  # to collect the clique
        else:
            for v in P.copy():
                P.remove(v)
                PP = set()
                SS = set()
                for e in G.outgoing(v):
                    w = G.target(e)
                    if w in P:
                        PP.add(w)
                    if w in S:
                        SS.add(w)
                C.add(v)
                L += simple_next_maximal_clique(C, PP, SS)
                C.remove(v)
                S.add(v)
        return L

    C = set()
    P = set(G.vertices())
    S = set()
    return simple_next_maximal_clique(C, P, S)
```

Backtracking algorithm for enumerating maximal cliques of a graph

```python
def all_maximal_cliques(
        G: Graph) -> List[Set[Graph.Vertex]]:

    def next_maximal_clique(
            C: Set[Graph.Vertex],
            P: Set[Graph.Vertex],
            S: Set[Graph.Vertex]) ->
            List[Set[Graph.Vertex]]:
        L = list()
        if not P:
```

```
                if not S:
                    L.append(C.copy())   # to collect the clique
            else:
                u = next(iter(P))   # an arbitrary element from P
                for v in P.copy():
                    if not G.adjacent(u, v) or
                            not G.adjacent(v, u):
                        P.remove(v)
                        PP = set()
                        SS = set()
                        for e in G.outgoing(v):
                            w = G.target(e)
                            if w in P:
                                PP.add(w)
                            if w in S:
                                SS.add(w)
                        C.add(v)
                        L += next_maximal_clique(C, PP, SS)
                        C.remove(v)
                        S.add(v)
        return L

    C = set()
    P = set(G.vertices())
    S = set()
    return next_maximal_clique(C, P, S)
```

Branch-and-bound algorithm for finding a maximum clique of a graph

```
def maximum_clique(G: Graph) -> Set[Graph.Vertex]:

    def next_maximum_clique(
            C: Set[Graph.Vertex],
            P: Set[Graph.Vertex],
            S: Set[Graph.Vertex],
            max_deg: int,
            M: set) -> Set[Graph.Vertex]:
        if len(M) < len(C):
            M = C.copy()
        if P:
            u = next(iter(P))   # an arbitrary element from P
            for v in P.copy():
                if not G.adjacent(u, v) or
                        not G.adjacent(v, u):
                    P.remove(v)
                    PP = set()
                    SS = set()
```

```
                        for e in G.outgoing(v):
                            w = G.target(e)
                            if w in P:
                                PP.add(w)
                            if w in S:
                                SS.add(w)
                        C.add(v)
                        if len(M) < len(C) + \
                                len(PP) and len(M) < max_deg + 1:
                            M = next_maximum_clique(
                                C, PP, SS, max_deg, M)
                        C.remove(v)
                        S.add(v)
                return M.copy()  # to collect the maximum clique

    C = set()
    P = set()
    S = set()
    max_deg = 0
    for v in G.vertices():
        P.add(v)
        max_deg = max(max_deg, G.outdeg(v))
    M = set()
    return next_maximum_clique(C, P, S, max_deg, M)
```

A.6.2　Maximal and Maximum Independent Sets

Algorithm for finding a maximum independent set of a tree

```
def maximum_independent_set(T: Tree) -> Set[Graph.Vertex]:
    I = set()
    for v in T.vertices():
        v.independent = True
    for v in postorder_tree_list_traversal(T):
        for w in T.children(v):
            if w.independent:
                v.independent = False
                break
        if v.independent:
            I.add(v)
    return I
```

Algorithm for computing the complement of a graph

```python
def graph_complement(G: Graph) -> Graph:
    H = Graph()
    for v in G.vertices():
        w = H.new_vertex(v.label())
        v.M = w
        w.M = v
    for v in G.vertices():
        for w in G.vertices():
            if v is not w and not G.adjacent(v, w):
                H.new_edge(v.M, w.M)
    return H
```

Algorithm for finding a maximum independent set of a graph

```python
def maximum_independent_set(G: Graph) -> Set[Graph.Vertex]:
    H = graph_complement(G)
    C = maximum_clique(H)
    return set([v.M for v in C])
```

A.6.3 Minimal and Minimum Vertex Covers

Algorithm for finding a minimum vertex cover of a tree

```python
def minimum_vertex_cover(T: Tree) -> Set[Graph.Vertex]:
    I = maximum_independent_set(T)
    return set(T.vertices()).difference(I)
```

Algorithm for finding a minimum vertex cover of a graph

```python
def minimum_vertex_cover(G: Graph) -> Set[Graph.Vertex]:
    I = maximum_independent_set(G)
    return set(G.vertices()).difference(I)
```

A.7 Graph Isomorphism

A.7.1 Graph Isomorphism

Algorithm for enumerating graph isomorphisms

```python
def graph_isomorphism(
        G1: Graph, G2: Graph) ->
        List[Dict[Graph.Vertex, Graph.Vertex]]:

    def extend_graph_isomorphism(
            M: Dict[Graph.Vertex, Graph.Vertex],
            v: Graph.Vertex,
            L: List[Dict[Graph.Vertex, Graph.Vertex]]) ->
            None:
        V2 = list(G2.vertices())
        for vv in G1.vertices():
            if vv in M:
                V2.remove(M[vv])
        for w in V2:
            if preserves_adjacencies(M, v, w):
                M[v] = w
                if v == G1.last_vertex():
                    L.append(M)
                else:
                    extend_graph_isomorphism(
                        M.copy(), G1.succ_vertex(v), L)

    def preserves_adjacencies(
            M: Dict[Graph.Vertex, Graph.Vertex],
            v: Graph.Vertex,
            w: Graph.Vertex) -> bool:
        if G1.indeg(v) != G2.indeg(
                w) or G1.outdeg(v) != G2.outdeg(w):
            return False
        for e in G1.incoming(v):
            x = e.opposite(v)
            if x in M and not G2.adjacent(M[x], w):
                return False
        for e in G1.outgoing(v):
            x = e.opposite(v)
            if x in M and not G2.adjacent(w, M[x]):
                return False
        return True

    L = list()
    M = dict()
    if G1.number_of_vertices() == G2.number_of_vertices(
    ) and G1.number_of_edges() == G2.number_of_edges():
```

```
            v = G1.first_vertex()
            extend_graph_isomorphism(M, v, L)
        return L
```

A.7.2 Graph Automorphism

Algorithm for enumerating graph automorphisms

```
def graph_automorphism(
        G: Graph) -> List[Dict[Graph.Vertex, Graph.Vertex]]:
    return graph_isomorphism(G, G)
```

A.7.3 Subgraph Isomorphism

Algorithm for enumerating subgraph isomorphisms

```
def subgraph_isomorphism(
        G1: Graph,
        G2: Graph) ->
        List[Dict[Graph.Vertex, Graph.Vertex]]:

    def extend_subgraph_isomorphism(
            C: Dict[Graph.Vertex, Set[Graph.Vertex]],
            v: Graph.Vertex,
            M: Dict[Graph.Vertex, Graph.Vertex],
            L: List[Dict[Graph.Vertex, Graph.Vertex]]) ->
            None:
        for w in G2.vertices():
            if w in C[v]:
                M[v] = w
                N = dict()
                for x in G1.vertices():
                    N[x] = C[x].copy()
                    if x != v:
                        N[x].discard(w)
                for y in G2.vertices():
                    if y != w:
                        N[v].discard(y)
                if refine_subgraph_isomorphism(N, v, w):
                    if v == G1.last_vertex():
                        L.append(M.copy())
                    else:
                        extend_subgraph_isomorphism(
                            N, G1.succ_vertex(v), M, L)
```

```python
def refine_subgraph_isomorphism(
        C: Dict[Graph.Vertex, Set[Graph.Vertex]],
        v: Graph.Vertex,
        w: Graph.Vertex) -> bool:
    if G1.indeg(v) > G2.indeg(
            w) or G1.outdeg(v) > G2.outdeg(w):
        return False
    for e1 in G1.incoming(v):
        x = G1.source(e1)
        for y in G2.vertices():
            if G2.adjacent(y, w):
                for e2 in G2.incoming(w):
                    if G2.source(e2) == y:
                        break
                if G1.label(e1) != G2.label(e2):
                    C[x].discard(y)
            else:
                C[x].discard(y)
    for e1 in G1.outgoing(v):
        x = G1.target(e1)
        for y in G2.vertices():
            if G2.adjacent(w, y):
                for e2 in G2.outgoing(w):
                    if G2.target(e2) == y:
                        break
                if G1.label(e1) != G2.label(e2):
                    C[x].discard(y)
            else:
                C[x].discard(y)
    for x in G1.vertices():
        if len(C[x]) == 0:
            return False
    return True

C = dict()
for v in G1.vertices():
    C[v] = set()
    for w in G2.vertices():
        if G1.indeg(v) <= G2.indeg(w) and G1.outdeg(
                v) <= G2.outdeg(w) and G1.label(
                v) == G2.label(w):
            C[v].add(w)
L = list()
M = dict()
v = G1.first_vertex()
extend_subgraph_isomorphism(C, v, M, L)
return L
```

A.7.4 Maximal Common Subgraph Isomorphism

Algorithm for enumerating maximal common subgraph isomorphisms

```python
def maximal_common_subgraph_isomorphism(
        G1: Graph,
        G2: Graph) ->
        List[Dict[Graph.Vertex, Graph.Vertex]]:

    def graph_product() -> Graph:
        G = Graph()
        for v in G1.vertices():
            for w in G2.vertices():
                if G1.label(v) == G2.label(w):
                    x = G.new_vertex((v, w))
        for v in G.vertices():
            (v1, v2) = G.label(v)
            for w in G.vertices():
                (w1, w2) = G.label(w)
                if v1 != w1 and v2 != w2:
                    if G1.adjacent(v1, w1):
                        for e1 in G1.outgoing(v1):
                            if G1.target(e1) == w1:
                                break
                        if G2.adjacent(v2, w2):
                            for e2 in G2.outgoing(v2):
                                if G2.target(e2) == w2:
                                    break
                            if G1.label(e1) == G2.label(e2):
                                G.new_edge(v, w)
                    elif not G2.adjacent(v2, w2):
                        G.new_edge(v, w)
        return G

    G = graph_product()
    L = list()
    for C in all_maximal_cliques(G):
        M = dict()
        for v in C:
            (v1, v2) = G.label(v)
            M[v1] = v2
        L.append(M)
    return L
```

References

1. Beazley D, Jones BK (2013) Python cookbook, 3rd edn. O'Reilly, Sebastopol, CA
2. Goodrich MT, Tamassia R, Goldwasser MH (2013) Data structures and algorithms in python. Wiley, Hoboken, NJ
3. Hunt J (2019) Advanced guide to python 3 programming. Undergraduate topics in computer science. Springer Nature, Cham, Switzerland
4. Hunt J (2019) A beginners guide to python 3 programming. Undergraduate topics in computer science. Springer Nature, Cham, Switzerland
5. Langtangen HP (2008) Python scripting for computational science, Texts in computational science and engineering, vol 3, 3rd edn. Springer Nature, Berlin Heidelberg
6. Lee KD (2014) Python programming fundamentals, 2nd edn. Undergraduate topics in computer science. Springer Nature, London, England
7. Lee KD, Hubbard S (2015) Data structures and algorithms with python. Undergraduate topics in computer science. Springer Nature, Cham, Switzerland
8. Lutz M (2010) Programming python, 4th edn. O'Reilly, Sebastopol, CA
9. Lutz M (2013) Learning python, 5th edn. O'Reilly, Sebastopol, CA
10. Uçoluk G, Kalkan S (2012) Introduction to programming concepts with case studies in python. Springer Nature, Wien, Switzerland

B.1 Introduction

1.1 There is an edge in K_n between each vertex and each other vertex and thus, the size of K_n is $n!$. On the other hand, there is an edge between each of the p edges of the first subset and each of the q vertices of the second subset and thus, the size of $K_{p,q}$ is pq.

1.2 Since a graph is bipartite if and only if it does not contain a cycle of odd length [10], C_n is bipartite for n even and K_n is bipartite only for $n = 1, 2$.

1.3 The depth-first and breadth-first spanning trees, with each of the vertices of the given graph in turn as the initial vertex, are shown in Fig. B.1, with the only exception of the breadth-first spanning forest with v_2 as the initial vertex, which is not a spanning tree of the given graph. The number of spanning trees of the given graph, which can be obtained by computing the determinant of the Laplacian matrix of the graph, which is the difference between a diagonal matrix of vertex indegrees and the adjacency matrix of the graph [15], is 45. The number of spanning trees of the underlying undirected graph, which can also be obtained by computing the determinant of the Laplacian matrix of the graph, where the diagonal matrix consists of vertex degrees [9, 10], is 288. See also [1], and see [2] for a review of spanning tree enumeration algorithms.

1.4 Let $G = (V, E)$ be a graph with n vertices, let us assume the vertices are numbered $1, 2, \ldots, n$ in some arbitrary manner and their number is stored in the attribute *index*, and let A be the adjacency matrix representation of G.

- $G.del\_edge(v, w)$ can be implemented by setting $A[v.index, w.index]$ to false.
- $G.edges()$ is the list of pairs of vertices (v, w) of G such that $A[v.index, w.index]$ is true.

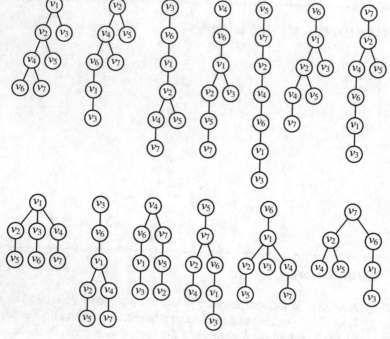

Fig. B.1 Depth-first spanning trees (top) and breadth-first spanning trees (bottom) in the solution to Problem 1.3.

- $G.incoming(v)$ is the set of vertices u of G such that $A[u.index, v.index]$ is true.
- $G.outgoing(v)$ is the set of vertices w of G such that $A[v.index, w.index]$ is true.
- $G.source(v, w)$ is vertex v of G.
- $G.target(v, w)$ is vertex w of G.

1.5 Let $T = (V, E)$ be a tree with n nodes, and let (F, N) be the first child, next-sibling representation of T. Let also P be the array-of-parents representation of T, and let (L, S) be the last-child, previous-sibling representation of T, that is, L and S are arrays of n nodes, indexed by the nodes of T, such that $L[v] = last[v]$ if v is not a leaf node, $L[v] = nil$ otherwise, $S[v] = previous[v]$ is v is not a first-child n-ode, and $S[v] = nil$ otherwise. Operations $T.number\_of\_nodes()$, $T.is\_root(v)$, and $T.parent(v)$ take $O(1)$ time using the array-of-parents representation of T. Operations $T.is\_leaf(v)$, $T.first\_child(v)$, $T.next\_sibling(v)$, and $T.is\_last\_child(v)$ take $O(1)$ time using the first child, next-sibling representation of T. The remaining operations can be implemented to take $O(1)$ time as follows:

- $T.last\_child(v)$ is given by $L[v]$.
- $T.previous\_sibling(v)$ is given by $S[v]$.
- $T.is\_first\_child(v)$ is true if $S[v] = nil$, and false otherwise.

1.6 Assuming the representation of the tree is a connected graph, the absence of cycles can be determined by checking that the number of nodes is one plus the number of edges. A simple procedure to determine that the representation of the tree is a connected graph consists of following the path of parent nodes for all nodes of the tree, checking that they all end up at the root. A more efficient procedure consists of traversing the graph in either depth-first or breadth-first order starting from the root, checking that all the nodes are reachable from the root, as follows:

> **function** *is_tree*(G)
>   **if** the number of vertices is not one plus the number of edges of G **then**
>   **return** false
>   **for all** vertices v of G **do**
> $v.visited \leftarrow$ false
>   let S be an empty stack (of vertices)
>   let v be the first vertex of G
>   **while** $indeg[r] \neq 0$ **do**
>   let v be the source vertex of the first edge coming into vertex v in G
>   root node
>   push v onto S
>   **while** S is not empty **do**
>   pop from S the top vertex v
>   **if** not $v.visited$ **then**
> $v.visited \leftarrow$ true
>   **for all** vertices w adjacent with vertex v in G **do**
>   **if** not $w.visited$ **then**
>   push w onto S
>   **for all** vertices v of T **do**
>   **if** not $v.visited$ **then**
>   **return** false
>   **return** true

B.2   Algorithmic Techniques

2.1 A least-cost transformation between the trees of Fig. 2.15 is given by the bijection $M = \{(v_1, w_1), (v_2, w_3), (v_3, w_4), (v_4, w_5), (v_5, w_6), (v_6, w_7)\}$. There are 40 nontrivial valid transformations between the trees, as shown in Fig. B.2.

2.2 In the general tree edit distance problem, deletions and insertions are allowed on leaves and nonleaves as well, and a sequence of elementary edit operations corresponds to a valid transformation of an ordered tree T_1 into an ordered tree T_2 if it preserves ancestor and sibling order, that is, if an ancestor of a node of T_1, which is substituted or replaced with a node of T_2, is substituted or replaced with an ancestor

v_1	v_2	v_3	v_4	v_5	v_6	v_7	v_8	v_9	v_{10}	v_{11}	v_{12}	v_{13}
w_1	λ	λ	λ	λ	λ	λ	λ	λ	λ	λ	λ	λ
w_1	λ	λ	λ	λ	λ	λ	w_2	λ	λ	λ	λ	λ
w_1	λ	λ	λ	λ	λ	λ	w_3	λ	λ	λ	λ	λ
w_1	λ	λ	λ	λ	λ	λ	w_3	λ	w_4	λ	λ	λ
w_1	λ	λ	λ	λ	λ	λ	w_3	λ	w_4	w_5	λ	λ
w_1	λ	λ	λ	λ	λ	λ	w_3	λ	w_4	w_6	λ	λ
w_1	λ	λ	λ	λ	λ	λ	w_3	λ	w_7	λ	λ	λ
w_1	λ	λ	λ	λ	λ	λ	w_3	w_4	λ	λ	λ	λ
w_1	λ	λ	λ	λ	λ	λ	w_3	w_4	w_7	λ	λ	λ
w_1	λ	λ	λ	λ	λ	λ	w_3	w_7	λ	λ	λ	λ
w_1	w_2	λ	λ	λ	λ	λ	λ	λ	λ	λ	λ	λ
w_1	w_2	λ	λ	λ	λ	λ	w_3	λ	λ	λ	λ	λ
w_1	w_2	λ	λ	λ	λ	λ	w_3	λ	w_4	λ	λ	λ
w_1	w_2	λ	λ	λ	λ	λ	w_3	λ	w_4	w_5	λ	λ
w_1	w_2	λ	λ	λ	λ	λ	w_3	λ	w_4	w_6	λ	λ
w_1	w_2	λ	λ	λ	λ	λ	w_3	λ	w_7	λ	λ	λ
w_1	w_2	λ	λ	λ	λ	λ	w_3	w_4	λ	λ	λ	λ
w_1	w_2	λ	λ	λ	λ	λ	w_3	w_4	w_7	λ	λ	λ
w_1	w_2	λ	λ	λ	λ	λ	w_3	w_7	λ	λ	λ	λ
w_1	w_3	λ	λ	λ	λ	λ	λ	λ	λ	λ	λ	λ

v_1	v_2	v_3	v_4	v_5	v_6	v_7	v_8	v_9	v_{10}	v_{11}	v_{12}	v_{13}
w_1	w_3	λ	λ	λ	w_4	λ	λ	λ	λ	λ	λ	λ
w_1	w_3	λ	λ	λ	w_4	λ	w_5	λ	λ	λ	λ	λ
w_1	w_3	λ	λ	λ	w_4	λ	w_6	λ	λ	λ	λ	λ
w_1	w_3	λ	λ	λ	w_4	w_5	λ	λ	λ	λ	λ	λ
w_1	w_3	λ	λ	λ	w_4	w_5	w_6	λ	λ	λ	λ	λ
w_1	w_3	λ	λ	λ	w_4	w_6	λ	λ	λ	λ	λ	λ
w_1	w_3	λ	λ	λ	w_7	λ	λ	λ	λ	λ	λ	λ
w_1	w_3	w_4	λ	λ	λ	λ	λ	λ	λ	λ	λ	λ
w_1	w_3	w_4	λ	λ	w_7	λ	λ	λ	λ	λ	λ	λ
w_1	w_3	w_4	λ	w_5	λ	λ	λ	λ	λ	λ	λ	λ
w_1	w_3	w_4	λ	w_5	w_7	λ	λ	λ	λ	λ	λ	λ
w_1	w_3	w_4	λ	w_6	λ	λ	λ	λ	λ	λ	λ	λ
w_1	w_3	w_4	λ	w_6	w_7	λ	λ	λ	λ	λ	λ	λ
w_1	w_3	w_4	w_5	λ	λ	λ	λ	λ	λ	λ	λ	λ
w_1	w_3	w_4	w_5	λ	w_7	λ	λ	λ	λ	λ	λ	λ
w_1	w_3	w_4	w_5	w_6	λ	λ	λ	λ	λ	λ	λ	λ
w_1	w_3	w_4	w_5	w_6	w_7	λ	λ	λ	λ	λ	λ	λ
w_1	w_3	w_4	w_6	λ	λ	λ	λ	λ	λ	λ	λ	λ
w_1	w_3	w_4	w_6	λ	w_7	λ	λ	λ	λ	λ	λ	λ
w_1	w_3	w_7	λ	λ	λ	λ	λ	λ	λ	λ	λ	λ

Fig. B.2 Solution to Problem 2.1.

of the node of T_2 and, whenever sibling nodes of T_1 are substituted or replaced with sibling nodes of T_2, the substitution preserves their relative order. This requirement is formalized by a mapping M of an ordered tree $T_1 = (V_1, E_1)$ to an ordered tree $T_2 = (V_2, E_2)$, that is, a bijection $M \subseteq W_1 \times W_2$, where $W_1 \subseteq V_1$, and $W_2 \subseteq V_2$, such that, for all nodes $v_1, v_2 \in W_1$ and $w_1, w_2 \in W_2$ with $(v_1, w_1), (v_2, w_2) \in M$, v_1 is an ancestor of v_2 if and only if w_1 is an ancestor of w_2, and v_1 is a left sibling of v_2 if and only if w_1 is a left sibling of w_2, as first shown in [14].

2.3 Since a node in a tree can have at most one incoming edge, the information attached to an edge can be attached to the target node of the edge instead, and edge labels are thus not really needed. Then, it suffices to replace the second condition in Definition 2.5 with the following condition: $\gamma(v, w) = 0$ if and only if $v \neq \lambda$, $w \neq \lambda$, and $label[v] = label[w]$. That is, only *identical* node substitutions have zero cost.

2.4 In the tree inclusion problem, deletions are allowed on leaves and nonleaves as well, and insertions are forbidden. The formulation of the tree inclusion problem is thus identical to the formulation of the general tree edit distance problem of Problem 2.2, where the elementary edit operations are deletions and substitutions.

2.5 The division of a tree into the subtree with nodes up to a certain node in preorder and the subtree with the remaining nodes, which has the advantage that subtrees are defined by just giving their first and last node in preorder, cannot be improved to produce a division of the trees into subtrees of about the same size, since one of the subtrees has to be a bottom-up subtree (rooted at a node along the path of last children) and a bottom-up subtree of about half the size of the tree does not necessarily exist. An alternative formulation consists of making the division based on a node with the smallest difference between the size of the tree and the size of the subtree rooted at the node, and replacing the range of nodes in preorder with the set of nodes in the subtree. Notice that this only guarantees a division of a tree into subtrees of about the same size if the tree is *balanced*.

2.6 For the transformation of ordered trees with n nodes, the backtracking tree has $O(2^n)$ nodes, the branch-and-bound tree has $O(n^2)$ nodes, the divide-and-conquer tree has $O(n^3)$ nodes, and the dynamic programming directed acyclic graph has $O(n^2)$ vertices. These bounds are tight for unlabeled, complete binary trees.

B.3 Tree Traversal

3.1 During a preorder traversal of the rooted tree in Fig. 3.11, nodes are visited in the following order: v_1, v_2, v_5, v_{11}, v_{12}, v_{13}, v_{14}, v_{15}, v_3, v_6, v_7, v_8, v_9, v_4, v_{10}. During a top-down traversal, nodes are visited in the following order: v_1, v_2, v_3, v_4, v_5, v_6, v_7, v_8, v_9, v_{10}, v_{11}, v_{12}, v_{13}, v_{14}, v_{15}.

3.2 During a postorder traversal of the rooted tree in Fig. 3.11, nodes are visited in the following order: v_{11}, v_{12}, v_{13}, v_{14}, v_{15}, v_5, v_2, v_6, v_7, v_8, v_9, v_3, v_{10}, v_4, v_1. During a bottom-up traversal, nodes are visited in the following order: v_6, v_7, v_8, v_9, v_{10}, v_{11}, v_{12}, v_{13}, v_{14}, v_{15}, v_3, v_4, v_5, v_2, v_1.

3.3 The evolution of the stack of nodes during execution of the iterative preorder traversal algorithm upon the tree of Fig. 3.11 is illustrated in Fig. B.3.

3.4 The evolution of the queue of nodes during execution of the top-down traversal algorithm upon the tree of Fig. 3.11 is illustrated in Fig. B.4.

3.5 The preorder traversal and the top-down traversal of the following full binary tree on $2k + 1$ nodes shown in Fig. B.5 are identical. The order in which vertices are visited during each traversal is v_1, v_2, $\ldots$, v_{2k+1}.

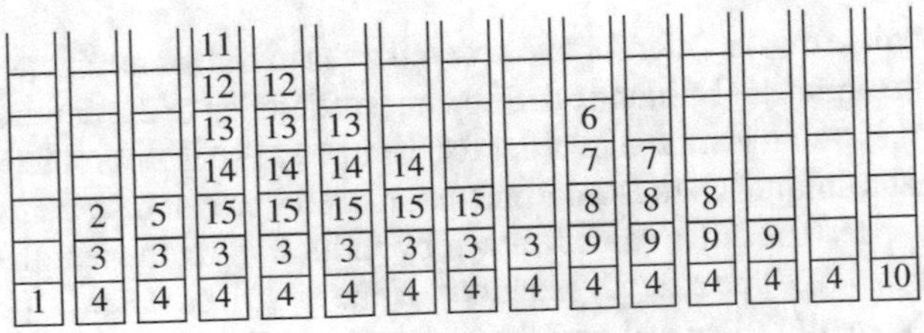

Fig. B.3 Solution to Problem 3.3.

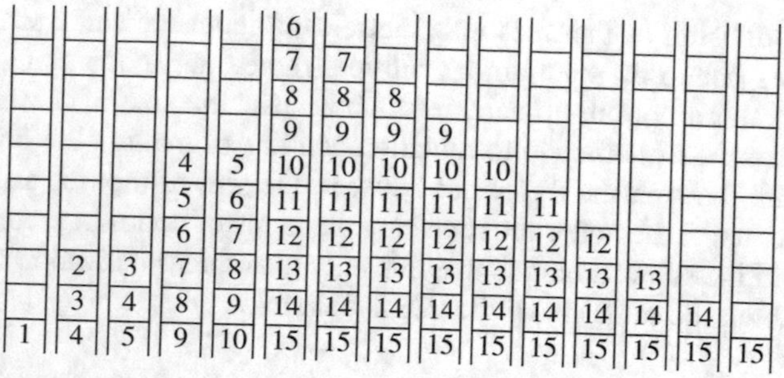

					6									
					7	7								
					8	8	8							
					9	9	9	9						
			4	5	10	10	10	10	10					
			5	6	11	11	11	11	11	11				
			6	7	12	12	12	12	12	12	12			
	2	3	7	8	13	13	13	13	13	13	13	13		
	3	4	8	9	14	14	14	14	14	14	14	14	14	
1	4	5	9	10	15	15	15	15	15	15	15	15	15	15

Fig. B.4 Solution to Problem 3.4.

Fig. B.5 Solution to Problems 3.5 and 3.6.

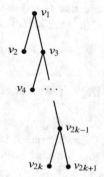

3.6 The postorder traversal and the bottom-up traversal of the full binary tree on $2k + 1$ nodes shown in Fig. B.5 are identical. The order in which vertices are visited during each traversal is $v_2, v_4, \ldots, v_{2k}, v_{2k+1}, v_{2k-1}, \ldots, v_3, v_1$.

B.4 Tree Isomorphism

4.1 The given trees T_1 and T_2 are isomorphic, and the isomorphism code of their root is [24, 1, 1, 3, 1, 1, 8, 1, 1, 5, 1, 3, 1, 1, 10, 1, 3, 1, 1, 5, 1, 3, 1, 1], as shown in Fig. B.6. The canonical ordered tree of T_2 is T_2 itself, which is thus the canonical ordered tree of T_1 as well.

4.2 It is immediate that in order for two ordered trees to be isomorphic, it is necessary that their depth sequences be identical. Showing sufficiency of the depth sequences of two ordered trees being identical in order for the trees to be isomorphic, is equivalent to showing that a unique, up to isomorphism, ordered tree can be reconstructed from its depth sequence. The latter can be shown by induction on the length of the depth sequence. Let $depth[T]$ be a depth sequence. If the length of $depth[T]$ is equal to 1, then it must be $depth[T] = [0]$ and the ordered tree T consists of a single node. Assume now that a unique, up to isomorphism, ordered tree T can be reconstructed

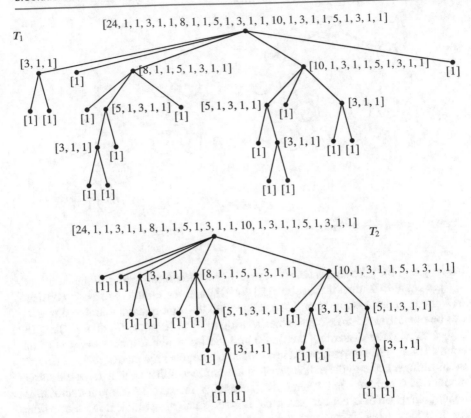

Fig. B.6 Solution to Problem 4.1.

from a depth sequence $depth[T]$ of length i, for all $1 \leqslant i \leqslant n$. After discarding the first integer, the rest of a sequence $depth[T]$ of length $n + 1$ can be partitioned in a unique way into a sequence of disjoint depth sequences, because the order in which the nodes are visited during a preorder traversal coincides with the order of corresponding integers in the depth sequence and, since the depth of the children of a node is 1 plus the depth of the node, these disjoint sequences all start at depth $0 + 1 = 1$. Now, since none of the sequences resulting from partitioning the depth sequence can have more than n integers, a unique, up to isomorphism, ordered tree can be reconstructed from each of them, after subtracting 1 from each of the integers in these sequences. Then, the ordered tree T consists of an additional node, the root of the tree, together with these reconstructed ordered trees as subtrees, in left-to-right order.

4.3 It is immediate that in order for two unordered trees to be isomorphic, it is necessary that their parenthesis strings be identical. Showing sufficiency of the parenthesis strings of two unordered trees being identical in order for the trees to be isomorphic, is equivalent to showing that a unique, up to isomorphism, unordered tree can be reconstructed from its parenthesis string. The latter can be shown by induction on

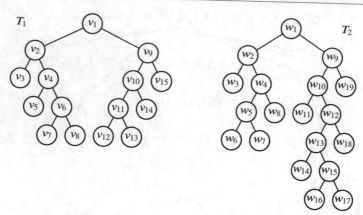

Fig. B.7 Ordered trees for Problem 4.5.

the length of the parenthesis string. Let $s[T]$ be a parenthesis string. If the length of $s[T]$ is equal to 2, then it must be $s[T] = [01]$ and the unordered tree T consists of a single node. Assume now that a unique, up to isomorphism, unordered tree T can be reconstructed from a parenthesis string $s[T]$ of length i, for all $i = 2k$ with $1 \leqslant k \leqslant n$. After discarding the first 0 and the last 1, the rest of a string $s[T]$ of length $2(n + 1)$ can be partitioned in a unique way into a sequence of disjoint parenthesis strings, because parenthesis strings are *balanced*, that is, they have the same number of 0 symbols and 1 symbols. (Arranging these strings in non-decreasing lexicographic order is not necessary, but it gives a unique, canonical representation for unordered trees.) Now, since none of the strings resulting from partitioning the parenthesis string can have more than n symbols, a unique, up to isomorphism, unordered tree can be reconstructed from each of them. Then, the unordered tree T consists of an additional node, the root of the tree, together with these reconstructed unordered trees as subtrees.

4.4 There are at least $\lceil \lg n_2 \rceil - \lceil \lg n_1 \rceil$ and at most $n_2 - n_1$ top-down subtree isomorphisms of an ordered tree on n_1 nodes into an ordered tree on n_2 nodes, with $n_1 \leqslant n_2$. The lower bound corresponds to complete binary trees, and the upper bound is for path graphs. On the other hand, there are at least one and at most $2^{\lceil \lg n_2 \rceil} / 2^{\lceil \lg n_1 \rceil}$ bottom-up subtree isomorphisms. The lower bound now corresponds to path graphs, while the upper bound is for complete binary trees.

4.5 Let the nodes be numbered according to the order in which they are visited during a postorder traversal, as shown in Fig. B.7. Mapping $\{(v_1, w_1), (v_2, w_2), (v_3, w_3), (v_4, w_4), (v_5, w_5), (v_6, w_8), (v_9, w_9), (v_{10}, w_{10}), (v_{11}, w_{11}), (v_{14}, w_{12}), (v_{15}, w_{19})\}$ is a top-down maximum common subtree isomorphism, and mapping $\{(v_4, w_{13}), (v_5, w_{14}), (v_6, w_{15}), (v_7, w_{16}), (v_8, w_{17})\}$ is a bottom-up maximum common subtree isomorphism of the ordered trees. On the other hand, $\{(v_1, w_1), (v_2, w_2), (v_3, w_3), (v_4, w_4), (v_5, w_8), (v_6, w_5), (v_7, w_6), (v_8, w_7), (v_9, w_9), (v_{10}, w_{10}), (v_{11}, w_{12}), (v_{12}, w_{13}), (v_{13}, w_{18}), (v_{14}, w_{11}), (v_{15}, w_{19})\}$ is a top-down maximum common subtree isomorphism, and $\{(v_2, w_2), (v_3, w_3), (v_4, w_4), (v_5, w_8), (v_6, w_5), (v_7, w_6), (v_8, w_7)\}$ is a

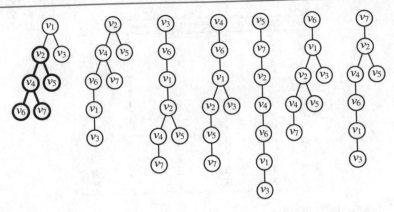

Fig. B.8 Solution to Problem 5.1.

bottom-up maximum common subtree isomorphism of the underlying unordered trees.

4.6 Any distance between sets based on set intersection yields a distance between (ordered or unordered) trees based on (top-down or bottom-up) maximum common subtree isomorphism. For example, the real function δ defined by $\delta(T_i, T_j) = 1 - |T_i \cap T_j| / \max(|T_i|, |T_j|)$ for all trees T_i and T_j, where $|T_i \cap T_j|$ is the size of a maximum common subtree isomorphism between T_i and T_j, is a distance between trees. See [5, 11, 12] for a proof that $d(X, Y) = 1 - |X \cap Y|/|X \cup Y|$ is a distance between nonempty finite sets, and see [3, 7] for alternative ways to define distances based on common structures.

B.5 Graph Traversal

5.1 The depth-first trees rooted in turn at each of the vertices of the given graph are shown in Fig. B.8. Since all vertices are reachable from vertex v_1, which is the first vertex in the representation of the graph, the depth-first forest coincides with the depth-first tree rooted at vertex v_1. The depth-first tree rooted at vertex v_2 in the depth-first forest of the graph, shown highlighted, contains only those unvisited vertices which are reachable from vertex v_2. In the depth-first forest, vertex v_1 is visited before vertex v_2, and vertex v_3 can only be reached from vertex v_1.

5.2 The first visit order and completion order for the source and target vertices of the edges in the depth-first traversal of the graph in Fig. 5.13, along with their classification as tree (T), forward (F), backward (B), or cross (C) edges, are shown in Fig. B.9.

v	w	$order[v]$	$order[w]$	$comp[v]$	$comp[w]$	class
1	2	1	2	7	5	T
1	3	1	7	7	6	T
1	4	1	3	7	3	F
2	4	2	3	5	3	T
2	5	2	6	5	4	T
3	6	7	4	6	1	C
4	6	3	4	3	1	T
4	7	3	5	3	2	T
5	7	6	5	4	2	C
6	1	4	1	1	7	B
7	2	5	2	2	5	B
7	6	5	4	2	1	C

Fig. B.9 Solution to Problem 5.2.

trail of vertices along traversed edges	order
$v_1\ v_2\ v_4\ v_1\ v_4\ v_6\ v_3\ v_1\ v_3\ v_6\ v_7\ v_5\ v_2\ v_5\ v_7\ v_4\ v_7\ v_6\ v_4\ v_2\ v_1$	1 2 5 3 7 4 6
$v_1\ v_3\ v_6\ v_7\ v_5\ v_2\ v_1\ v_2\ v_4\ v_1\ v_4\ v_6\ v_4\ v_7\ v_4\ v_2\ v_5\ v_7\ v_6\ v_3\ v_1$	1 6 2 7 5 3 4
$v_1\ v_4\ v_6\ v_3\ v_1\ v_3\ v_6\ v_7\ v_5\ v_2\ v_1\ v_2\ v_4\ v_2\ v_5\ v_7\ v_4\ v_7\ v_6\ v_4\ v_1$	1 7 4 2 6 3 5
$v_2\ v_1\ v_3\ v_6\ v_7\ v_5\ v_2\ v_5\ v_7\ v_4\ v_2\ v_4\ v_1\ v_4\ v_6\ v_4\ v_7\ v_6\ v_3\ v_1\ v_2$	2 1 3 7 6 4 5
$v_2\ v_4\ v_1\ v_2\ v_1\ v_3\ v_6\ v_7\ v_5\ v_2\ v_5\ v_7\ v_4\ v_7\ v_6\ v_4\ v_6\ v_3\ v_1\ v_4\ v_2$	3 1 4 2 7 5 6
$v_2\ v_5\ v_7\ v_4\ v_2\ v_4\ v_1\ v_2\ v_1\ v_3\ v_6\ v_7\ v_6\ v_4\ v_6\ v_3\ v_1\ v_4\ v_7\ v_5\ v_2$	5 1 6 4 2 7 3
$v_3\ v_1\ v_4\ v_6\ v_3\ v_6\ v_7\ v_5\ v_2\ v_1\ v_2\ v_4\ v_2\ v_5\ v_7\ v_4\ v_7\ v_6\ v_4\ v_1\ v_3$	2 7 1 3 6 4 5
$v_3\ v_6\ v_7\ v_5\ v_2\ v_1\ v_3\ v_1\ v_4\ v_6\ v_4\ v_7\ v_4\ v_2\ v_4\ v_1\ v_2\ v_5\ v_7\ v_6\ v_3$	6 5 1 7 4 2 3
$v_4\ v_1\ v_2\ v_4\ v_2\ v_5\ v_7\ v_4\ v_7\ v_6\ v_4\ v_6\ v_3\ v_1\ v_3\ v_6\ v_7\ v_5\ v_2\ v_1\ v_4$	2 3 7 1 4 6 5
$v_4\ v_6\ v_3\ v_1\ v_4\ v_1\ v_2\ v_4\ v_2\ v_5\ v_7\ v_4\ v_7\ v_6\ v_7\ v_5\ v_2\ v_1\ v_3\ v_6\ v_4$	4 5 3 1 6 2 7
$v_4\ v_7\ v_6\ v_4\ v_6\ v_3\ v_1\ v_4\ v_1\ v_2\ v_4\ v_2\ v_5\ v_7\ v_5\ v_2\ v_1\ v_3\ v_6\ v_7\ v_4$	5 6 4 1 7 3 2
$v_4\ v_2\ v_5\ v_7\ v_4\ v_7\ v_6\ v_4\ v_6\ v_3\ v_1\ v_4\ v_1\ v_2\ v_1\ v_3\ v_6\ v_7\ v_5\ v_2\ v_4$	7 2 6 1 3 5 4
$v_5\ v_2\ v_1\ v_3\ v_6\ v_7\ v_5\ v_7\ v_4\ v_2\ v_4\ v_1\ v_4\ v_6\ v_4\ v_7\ v_6\ v_3\ v_1\ v_2\ v_5$	3 2 4 7 1 5 6
$v_5\ v_7\ v_4\ v_2\ v_5\ v_2\ v_1\ v_3\ v_6\ v_7\ v_6\ v_4\ v_6\ v_3\ v_1\ v_4\ v_1\ v_2\ v_4\ v_7\ v_5$	5 4 6 3 1 7 2
$v_6\ v_3\ v_1\ v_4\ v_6\ v_4\ v_7\ v_6\ v_7\ v_5\ v_2\ v_1\ v_2\ v_4\ v_2\ v_5\ v_7\ v_4\ v_1\ v_3\ v_6$	3 7 2 4 6 1 5
$v_6\ v_7\ v_5\ v_2\ v_1\ v_3\ v_6\ v_3\ v_1\ v_4\ v_6\ v_4\ v_7\ v_4\ v_2\ v_4\ v_1\ v_2\ v_5\ v_7\ v_6$	5 4 6 7 3 1 2
$v_6\ v_4\ v_7\ v_6\ v_7\ v_5\ v_2\ v_1\ v_3\ v_6\ v_3\ v_1\ v_4\ v_1\ v_2\ v_4\ v_2\ v_5\ v_7\ v_4\ v_6$	6 5 7 2 4 1 3
$v_7\ v_4\ v_2\ v_5\ v_7\ v_5\ v_2\ v_1\ v_3\ v_6\ v_7\ v_6\ v_4\ v_6\ v_3\ v_1\ v_4\ v_1\ v_2\ v_4\ v_7$	5 3 6 2 4 7 1
$v_7\ v_6\ v_4\ v_7\ v_4\ v_2\ v_5\ v_7\ v_5\ v_2\ v_1\ v_3\ v_6\ v_3\ v_1\ v_4\ v_1\ v_2\ v_4\ v_6\ v_7$	6 4 7 3 5 2 1
$v_7\ v_5\ v_2\ v_1\ v_3\ v_6\ v_7\ v_6\ v_4\ v_7\ v_4\ v_2\ v_4\ v_1\ v_4\ v_6\ v_3\ v_1\ v_2\ v_5\ v_7$	4 3 5 7 2 6 1

Fig. B.10 Solution to Problem 5.3.

5.3 The closed trail of vertices along the leftmost depth-first traversal of the bidirected graph in Fig. 5.13, starting in turn at each of the edges of the graph, along with the order in which the vertices are first visited, are shown in Fig. B.10. In half of these traversals, the order in which the vertices are first visited is the reversal of the order in the other half of the traversals.

5.4 The breadth-first trees rooted in turn at each of the vertices of the given graph are shown in Fig. B.11. Since all vertices are reachable from vertex v_1, which is the first vertex in the representation of the graph, the breadth-first forest coincides with the breadth-first tree rooted at vertex v_1. The breadth-first tree rooted at vertex v_2 in the breadth-first forest of the graph, shown highlighted, contains only those unvisited

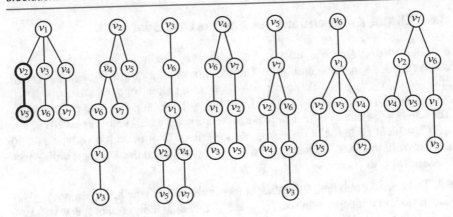

Fig. B.11 Solution to Problem 5.4.

Fig. B.12 Solution to
Problem 5.5.

v	w	depth[v]	depth[w]	class
1	2	0	1	T
1	3	0	1	T
1	4	0	1	T
2	4	1	1	C
2	5	1	2	T
3	6	1	2	T
4	6	1	2	C
4	7	1	2	T
5	7	2	2	C
6	1	2	0	B
7	2	2	1	C
7	6	2	2	C

vertices which are reachable from vertex v_2. In the breadth-first forest, vertex v_1 is visited before vertex v_2, and vertices v_4, v_6, v_7 are visited first along a different path.

5.5 The depth in the breadth-first spanning tree for the source and target vertices of the edges in the graph in Fig. 5.13, along with their classification as tree (T), backward (B), or cross (C) edges, are shown in Fig. B.12. It can be shown that $depth[w] = depth[v] + 1$ for all tree edges, $0 \leqslant depth[w] < depth[v]$ for all backward edges, and $depth[w] \leqslant depth[v] + 1$ for all cross edges (v, w). See [4, Problem 22-1].

5.6 The ordered graphs are isomorphic. The vertex mapping $M \subseteq V_1 \times V_2$ given by $\{(v_1, w_1), (v_2, w_3), (v_3, w_7), (v_4, w_5 v_5, w_6), (v_6, w_4), (v_7, w_8), (v_8, w_2)\}$ is an ordered graph isomorphism of $G_1 = (V_1, E_1)$ to $G_2 = (V_2, E_2)$.

B.6 Clique, Independent Set, and Vertex Cover

6.1 There are 43 nontrivial cliques, 9 of which are maximal, and 2 of the maximal cliques are also maximum cliques. The maximum cliques are $\{v_2, v_4, v_6, v_8, v_{10}\}$ and $\{v_3, v_5, v_7, v_9, v_{11}\}$. The other $9 - 2 = 7$ maximal cliques are $\{v_1, v_2, v_4\}$, $\{v_1, v_3, v_5\}, \{v_1, v_4, v_5\}, \{v_4, v_5, v_8, v_9\}, \{v_8, v_9, v_{12}\}, \{v_8, v_{10}, v_{12}\}$, and $\{v_9, v_{11}, v_{12}\}$. The $43 - 9 = 34$ non-maximal cliques are the 5 cliques with 4 vertices and 10 cliques with 3 vertices included in $\{v_2, v_4, v_6, v_8, v_{10}\}$, the 5 cliques with 4 vertices and 10 cliques with 3 vertices included in $\{v_3, v_5, v_7, v_9, v_{11}\}$, and the 4 cliques with 3 vertices included in $\{v_4, v_5, v_8, v_9\}$.

6.2 There are 21 maximal independent sets, only one of which, $\{v_1, v_6, v_7, v_{12}\}$, is also a maximum independent set. There are 14 maximal independent sets with 3 vertices: $\{v_1, v_6, v_9\}$, $\{v_1, v_6, v_{11}\}$, $\{v_1, v_7, v_8\}$, $\{v_1, v_7, v_{10}\}$, $\{v_1, v_8, v_{11}\}$, $\{v_1, v_9, v_{10}\}$, $\{v_1, v_{10}, v_{11}\}$, $\{v_2, v_3, v_{12}\}$, $\{v_2, v_5, v_{12}\}$, $\{v_2, v_7, v_{12}\}$, $\{v_3, v_4, v_{12}\}$, $\{v_3, v_6, v_{12}\}$, $\{v_4, v_7, v_{12}\}$, $\{v_5, v_6, v_{12}\}$, and 6 maximal independent sets with only 2 vertices: $\{v_2, v_9\}, \{v_2, v_{11}\}, \{v_3, v_8\}, \{v_3, v_{10}\}, \{v_4, v_{11}\}, \{v_5, v_{10}\}$.

6.3 The graph complement has vertices $\{v_1, v_2, v_3, v_4, v_5, v_6, v_7, v_8, v_9, v_{10}, v_{11}, v_{12}\}$ and edges $\{(v_1, v_6), (v_1, v_7), (v_1, v_8), (v_1, v_9), (v_1, v_{10}), (v_1, v_{11}), (v_1, v_{12}), (v_2, v_3), (v_2, v_5), (v_2, v_7), (v_2, v_9), (v_2, v_{11}), (v_2, v_{12}), (v_3, v_4), (v_3, v_6), (v_3, v_8), (v_3, v_{10}), (v_3, v_{12}), (v_4, v_7), (v_4, v_{11}), (v_4, v_{12}), (v_5, v_6), (v_5, v_{10}), (v_5, v_{12}), (v_6, v_7), (v_6, v_9), (v_6, v_{11}), (v_6, v_{12}), (v_7, v_8), (v_7, v_{10}), (v_7, v_{12}), (v_8, v_{11}), (v_9, v_{10}), (v_{10}, v_{11})\}$. The maximal and maximum cliques of the graph complement are the maximal and maximum independent sets given in the solution to Problem 6.2.

6.4 There are 21 minimal vertex covers, only one of which, $\{v_2, v_3, v_4, v_5, v_8, v_9, v_{10}, v_{11}\}$, is also a minimum vertex cover. There are 14 minimal vertex covers with 9 vertices: $\{v_1, v_2, v_3, v_4, v_7, v_8, v_9, v_{10}, v_{11}\}, \{v_1, v_2, v_3, v_5, v_6, v_8, v_9, v_{10}, v_{11}\}, \{v_1, v_2, v_4, v_5, v_7, v_8, v_9, v_{10}, v_{11}\}$, $\{v_1, v_2, v_5, v_6, v_7, v_8, v_9, v_{10}, v_{11}\}$, $\{v_1, v_3, v_4, v_5, v_6, v_8, v_9, v_{10}, v_{11}\}$, $\{v_1, v_3, v_4, v_6, v_7, v_8, v_9, v_{10}, v_{11}\}$, $\{v_1, v_4, v_5, v_6, v_7, v_8, v_9, v_{10}, v_{11}\}$, $\{v_2, v_3, v_4, v_5, v_6, v_7, v_8, v_9, v_{12}\}$, $\{v_2, v_3, v_4, v_5, v_6, v_7, v_8, v_{11}, v_{12}\}$, $\{v_2, v_3, v_4, v_5, v_6, v_7, v_9, v_{10}, v_{12}\}$, $\{v_2, v_3, v_4, v_5, v_6, v_8, v_9, v_{11}, v_{12}\}$, $\{v_2, v_3, v_4, v_5, v_6, v_9, v_{10}, v_{11}, v_{12}\}$, $\{v_2, v_3, v_4, v_5, v_7, v_8, v_9, v_{10}, v_{12}\}$, $\{v_2, v_3, v_4, v_5, v_7, v_8, v_{10}, v_{11}, v_{12}\}$, and 6 minimal vertex covers with 10 vertices: $\{v_1, v_2, v_3, v_4, v_6, v_7, v_8, v_9, v_{11}, v_{12}\}$, $\{v_1, v_2, v_3, v_5, v_6, v_7, v_8, v_9, v_{10}, v_{12}\}$, $\{v_1, v_2, v_4, v_5, v_6, v_7, v_8, v_9, v_{11}, v_{12}\}$, $\{v_1, v_2, v_4, v_5, v_6, v_7, v_9, v_{10}, v_{11}, v_{12}\}$, $\{v_1, v_3, v_4, v_5, v_6, v_7, v_8, v_9, v_{10}, v_{12}\}$, $\{v_1, v_3, v_4, v_5, v_6, v_7, v_8, v_{10}, v_{11}, v_{12}\}$.

6.5 The graph complement of $K_{p,q}$ is the disjoint union of K_p and K_q. Therefore, $K_{p,q}$ has two maximal independent sets, with, respectively, p and q vertices.

6.6 The vertices of a maximum clique are mutually adjacent and thus, a vertex coloring of an undirected graph G using less than $\omega(G)$ colors would result in two adjacent vertices from a maximum clique of G being assigned the same color. Therefore, $\chi(G) \geqslant \omega(G)$. On the other hand, at most $\beta(G)$ vertices can be assigned the same color and thus, the number of different colors used in a vertex coloring of an undirected G with n vertices must be at least $\lceil n/\beta(G) \rceil$, that is, $\chi(G) \geqslant \lceil n/\beta(G) \rceil$.

B.7 Graph Isomorphism

7.1 An invariant for graph isomorphism or *graph invariant* is a function ϱ defined on graphs that is invariant under graph isomorphisms [13, Sect. 3.10], that is, a function ϱ such that $G \cong G'$ implies $\varrho(G) = \varrho(G')$. Simple examples of graph invariants are the order (number of vertices) of a graph, the size (number of edges) of a graph, the minimum degree of the vertices, the maximum degree of the vertices, and the degree sequence of a graph. Any invariant for graph isomorphism is also an invariant for subgraph isomorphism, that is, $S \cong S' \subseteq G'$ implies $\varrho(S) = \varrho(S')$. Further, any invariant for graph isomorphism is also an invariant for (maximal) common subgraph isomorphism, that is, $G' \supseteq S' \cong S \cong S'' \subseteq G''$ implies $\varrho(S') = \varrho(S) = \varrho(S'')$.

7.2 The automorphism group of the wheel graph on 5 vertices W_5 is of order 8. The permutations of the vertex set shown in Fig. B.13 correspond to the graph automorphisms of W_5.

7.3 The smallest connected graph is the path graph, and the largest connected graph is the complete graph. The lower bound is 0, since the complete graph on n_1 vertices is not isomorphic to any subgraph of the path graph on n_2 vertices. The upper bound is $n_2!/(n_2 - n_1)!$, since any permutation of a subset of n_1 vertices of a set of n_2 vertices, where $n_1 \leqslant n_2$, gives a subgraph isomorphism of the path graph on n_1 vertices into the complete graph on n_2 vertices.

7.4 There are at least $n_2 - n_1$ and at most $n_1!\binom{n_2}{n_1} = n_2!/n_1!$ maximal common induced subgraph isomorphisms between two connected graphs with n_1 and n_2 vertices, where $n_1 \leqslant n_2$. The lower bound corresponds to circle graphs, and the common induced subgraph is a path graph. (Path graphs also give the same lower bound.) The upper bound is for complete graphs, and the common induced subgraph is also a complete graph, with n_1 vertices. Every subset of n_1 vertices from the n_2 vertices of the larger graph induces a maximal common induced subgraph, and every permutation of these n_1 vertices gives a maximal common induced subgraph isomorphism.

7.5 There are 264 maximal common induced subgraph isomorphisms between the complete bipartite graph $K_{3,4}$ and the wheel graph W_5. The maximal common induced subgraph is the path graph P_3 in 120 of the maximal common induced subgraph isomorphisms, and it is the cycle graph C_4 in the remaining 144 maximal common induced subgraph isomorphisms. Some of them are shown in Fig. B.14.

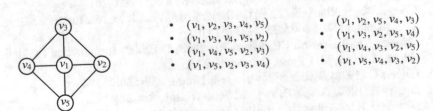

- $(v_1, v_2, v_3, v_4, v_5)$
- $(v_1, v_3, v_4, v_5, v_2)$
- $(v_1, v_4, v_5, v_2, v_3)$
- $(v_1, v_5, v_2, v_3, v_4)$

- $(v_1, v_2, v_5, v_4, v_3)$
- $(v_1, v_3, v_2, v_5, v_4)$
- $(v_1, v_4, v_3, v_2, v_5)$
- $(v_1, v_5, v_4, v_3, v_2)$

Fig. B.13 Automorphism group of the wheel graph W_5.

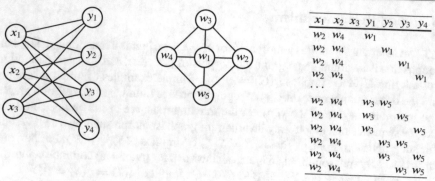

x_1	x_2	x_3	y_1	y_2	y_3	y_4
w_2	w_4		w_1			
w_2	w_4			w_1		
w_2	w_4				w_1	
w_2	w_4					w_1
...						
w_2	w_4		w_3	w_5		
w_2	w_4		w_3		w_5	
w_2	w_4		w_3			w_5
w_2	w_4			w_3	w_5	
w_2	w_4			w_3		w_5
w_2	w_4				w_3	w_5

Fig. B.14 Maximal common induced subgraph isomorphisms between the complete bipartite graph $K_{3,4}$ and the wheel graph W_5.

7.6 Let $G_1 = (V_1, E_1)$ and $G_2 = (V_2, E_2)$ be two graphs, let S_1 be a subgraph of G_1, let S_2 be a subgraph of G_2, and let M be a graph isomorphism of S_1 to S_2. Because of Theorem 7.2, it suffices to show that (S_1, S_2, M) is a maximal common subgraph isomorphism of G_1 to G_2 if and only if $(L(S_1), L(S_2), M')$ is a maximal common induced subgraph isomorphism of $L(G_1)$ to $L(G_2)$, where the **line graph** of a graph $G = (V, E)$ is the graph $L(G) = (E, \{(e, e') \in E \times E \mid e \neq e', e \cap e' \neq \emptyset\})$, that is, the **intersection graph** of the set of edges of the graph [6, Chap. 8], and $M' = \{((v, v'), (w, w')) \in E_1 \times E_2 \mid (v, w), (v', w') \in M\}$. That is, it suffices to show that $L(S_1)$ is an induced subgraph of $L(G_1)$ and $L(S_2)$ is an induced subgraph of $L(G_2)$. But it follows from the definition of line graph that a subgraph of the line graph of a graph is an induced subgraph of the graph. In fact, let G be a graph, let S be a subgraph of G, and let e, e' be two vertices of $L(S)$. If (e, e') is an edge of $L(G)$, it must be $e \cap e' \neq \emptyset$ and then, (e, e') is also an edge of $L(S)$.

References

1. Chaiken S, Kleitman DJ (1978) Matrix tree theorems. J Comb Theory, Ser A 24(3):377–381
2. Chakraborty M, Chowdhury S, Chakraborty J, Mehera R, Pal RK (2019) Algorithms for generating all possible spanning trees of a simple undirected connected graph: an extensive review. Complex Intell Syst 5(3):265–281
3. Chen S, Ma B, Zhang K (2009) On the similarity metric and the distance metric. Theor Comput Sci 24–25:2365–2376 (410)
4. Cormen TH, Leiserson CE, Rivest RL, Stein C (2009) Introduction to Algorithms, 3rd edn. MIT Press, Cambridge MA
5. Gilbert G (1972) Distance between sets. Nature 239(5368), 174
6. Harary F (1969) Graph Theory. Addison-Wesley, Reading MA
7. Horadam KJ, Nyblom MA (2014) Distances between sets based on set commonality. Discret Appl Math 167(1), 310–314

8. Kirchhoff G (1847) Ueber die Auflösung der Gleichungen, auf welche man bei der Untersuchung der linearen Verteilung galvanischer Ströme geführt wird. Annalen der Physik und Chemie 72(12):497–508
9. Kirchhoff G (1958) On the solution of the equations obtained from the investigation of the linear distribution of galvanic currents. IRE Trans Circuit Theory 5(1):4–7
10. König, D (1936) Theorie der endlichen und unendlichen Graphen, Archiv zur Mathematik, vol 6. Teubner, Leipzig, Germany
11. Levandowsky M, Winter D (1971) Distance between sets. Nature 234(5323):34–35
12. Mason JH (1972) Distance between sets. Nat Phys Sci 235:80
13. Nešetřil J, de Mendez PO (2012) Sparsity: graphs, structures, and algorithms. In: Algorithms and combinatorics, vol 28. Springer, Berlin Heidelberg
14. Tai KC (1979) The tree-to-tree correction problem. J ACM 26(3):422–433
15. Tutte WT (1948) The dissection of equilateral triangles into equilateral triangles. Math Proc Cambridge Philoso Soc 44(4):463–482

Citing Publications

1. Abboud A, Bačkurs A, Hansen TD, Williams VV, Zamir O (2018) Subtree isomorphism revisited. ACM Trans Algorithms 14(3):27:1–27:23
2. Adamoli A, Hauswirth M (2010) Trevis: A context tree visualization and analysis framework and its use for classifying performance failure reports. In: Telea A, Görg C, Reiss S, Anslow C (eds) Proceedings 5th international symppsium software visualization. IEEE, pp 73–82
3. Ahvenjärvi M (2008) *Ohjelmistokoodin Ohjelmoinnin Apuvälineet*. BSc thesis, Satakunta University of Applied Sciences, Finland
4. Akhtman J (2007) Smart Antenna-Aided Multicarrier Transceivers for Mobile Communications. PhD thesis, University of Southampton, England
5. Akhtman J, Wolfgang A, Chen S, Hanzo L (2007) An optimized-hierarchy-aided approximate log-MAP detector for MIMO systems. IEEE Trans Wirel Commun 6(5):1900–1909
6. Akutsu T, Fukagawa D, Takasu A (2006) Approximating tree edit distance through string edit distance. In: Asano T (ed) Proceedings 17rd international symposium algorithms and computation, vol 4288 of Lecture notes in computer science. Springer, Berlin, pp 90–99
7. Akutsu T, Fukagawa D, Takasu A (2010) Approximating tree edit distance through string edit distance. Algorithmica 57(2):325–348
8. Akutsu T, Tamura T, Melkman AA, Takasu A (2013) On the complexity of finding a largest common subtree of bounded degree. In: Gasieniec L, Wolter F (eds) Proceedings 19th international symposium fundamentals of computation theory, vol 8070 of Lecture notes in computer science. Springer, Berlin, pp 4–15
9. Akutsu T, Tamura T, Melkman AA, Takasu A (2015) On the complexity of finding a largest common subtree of bounded degree. Theor Comput Sci 590(1):2–16
10. Alekseev S, Kajrys A, Karoly A (2014) Graph theoretical algorithms for control flow graph comparison. In: Proceedings IASTED international conference software engineering. ACTA Press, pp 77–82
11. Almukhaizim S, Drineas P, Makris Y (2004) Roving concurrent error detection for logic circuits. In: Proceedings 13th IEEE North Atlantic test workshop, pp 65–71
12. Alvaro Muñoz F (2010) Off-line Recognition of Printed Mathematical Expressions Using Stochastic Context-Free Grammars. MSc thesis, Technical University of Valencia, Spain

© The Editor(s) (if applicable) and The Author(s), under exclusive license
to Springer Nature Switzerland AG 2021
G. Valiente, *Algorithms on Trees and Graphs*, Texts in Computer Science,
https://doi.org/10.1007/978-3-030-81885-2

13. Andrei O (2008) A Rewriting Calculus for Graphs: Applications to Biology and Autonomous Systems. PhD thesis, National Polytechnic Institute of Lorraine, France

14. Andrews M (2013) Method and/or system for performing tree matching. US Patent 7.620,632

15. Antal I (2005) Párhuzamos algoritmusok. ELTE Eötvös Kiadó, Budapest, Hungary

16. Antonelli D, Caroleo B (2006) Analysis of collaborative networks using the extremal graph theory. IFAC Proc Vol 39(3):517–522

17. Antonelli D, Caroleo B (2012) An integrated methodology for the analysis of collaboration in industry networks. J Intell Manuf 23(6):2443–2450

18. Arjomandi-Nezhad A, Fotuhi-Firuzabad M, Mazaheri H, Moeini-Aghtaie M (2018) Developing a MILP method for distribution system reconfiguration after natural disasters. In: Proceedings 2018 electrical power distribution conference. IEEE, pp 28–33

19. Arvind V, Köbler J, Rattan G, Verbitsky O (2017) Graph isomorphism, color refinement, and compactness. Comput Complex 26(3):627–685

20. Arvind V, Köbler J, Rattan G, Verbitsky O (2015) On the power of color refinement. In: Kosowski A, Walukiewicz I (eds) Proceedings 20th international symposium fundamentals of computation theory, volume 9210 of Lecture Notes in Computer Science. Springer, Berlin, pp 339–350

21. Asano T, Evans P, Uehara R, Valiente G (2006) Site consistency in phylogenetic networks with recombination. In: Iliopoulos CS, Park K, Steinhöfel K (eds) Algorithms in Bioinformatics, vol 6 of Texts in Algorithmics, Chap 2. College Publications, pp 15–26

22. Athanasopoulos D (2014) Designing and Developing Service-Oriented Software with respect to Fundamental Object-Oriented Principles. PhD thesis, University of Ioannina, Greece

23. Athanasopoulos D (2018) Digital ecclesia: Towards an online direct-democracy framework. In: Crnkovic I, Isssarny V, Dustdar S (eds) Proceedings 40th international conference software engineering: software engineering in society. ACM, pp 91–94

24. Athanasopoulos D, Zarras AV (2018) Mining abstract XML data-types. ACM Trans Web 13(1):2:1–2:37

25. Athanasopoulos D, Zarras AV, Miskos G, Issarny V, Vassiliadis P (2015) Cohesion-driven decomposition of service interfaces without access to source code. IEEE Trans Services Comput 8(4):550–562

26. Augenstein I, Padó S, Rudolph S (2012) LODifier: Generating linked data from unstructured text. In: Simperl E, Cimiano P, Polleres A, Corcho O, Presutti V (eds) Proceedings 9th extended semantic web conference, vol 7295 of Lecture notes in computer science. Springer, Berlin, pp 210–224

27. Baggenstos D (2006) Implementation and evaluation of graph isomorphism algorithms for RDF-graphs. MSc thesis, University of Zurich, Switzerland

28. Balcázar JL, Bifet A, Lozano A (2006) Intersection algorithms and a closure operator on unordered trees. In: Gärtner T, Garriga GC, Meinl T (eds) Proceedings 4th international workshop on mining and learning with graphs, pp 1–12

29. Balcázar JL, Bifet A, Lozano A (2007) Mining frequent closed unordered trees through natural representations. In: Priss U, Polovina S, Hill R (eds) proceedings international conference conceptual structures, vol 4604 of Lecture notes in computer science. Springer, Berlin, pp 347–359

30. Balcázar JL, Bifet A, Lozano A (2007) Subtree testing and closed tree mining through natural representations. In: Tjoa AM, Wagner RR (eds) Proceedings 18th international workshop on database and expert systems applications. IEEE, pp 499–503

31. Balcázar JL, Bifet A, Lozano A (2010) Mining frequent closed rooted trees. Mach Learn 78(1–2):1–33

32. Barzegar AR (2019) Computationally Efficient Methods for Solving Optimal Power Flow Problems by Exploiting Supplementary Information. PhD thesis, Nanyang Technological University, Singapore

33. Barzegar AR, Molzahn DK, Su R (2019) A method for quickly bounding the optimal objective value of an OPF problem using a semidefinite relaxation and a local solution. Electric Power Syst Res 177(1):105954:1–105954:9

34. Bezerra de Melo JC (2005) Análise de Estruturas de Proteínas. PhD thesis, Federal University of Pernambuco, Brasil

35. Bifet A (2009) Adaptive Learning and Mining for Data Streams and Frequent Patterns. PhD thesis, Technical University of Catalonia, Spain

36. Bifet A (2010) Adaptive stream mining: pattern learning and mining from evolving data streams, volume 207 of frontiers in artificial intelligence and applications. IOS Press

37. Bifet A, Gavaldà R (2008) Mining adaptively frequent closed unlabeled rooted trees in data streams. In: Li Y, Liu B, Sarawagi S (eds) Proceedings 14th ACM SIGKDD international conference knowledge discovery and data mining. ACM, pp 34–42

38. Bifet A, Gavaldà R (2011) Mining frequent closed trees in evolving data streams. Intell Data Anal 15(1):29–48

39. Binz T, Fehling C, Leymann F, Nowak A, Schumm D (2012) Formalizing the cloud through enterprise topology graphs. In: Foster I, Feig E, Yau SS, Chang R (eds) Proceedings 2012 IEEE 5th international conference cloud computing. IEEE, pp 742–749

40. Binz T, Leymann F, Nowak A, Schumm D (2012) Improving the manageability of enterprise topologies through segmentation, graph transformation, and analysis strategies. In: Chi C-H, Gašević D, van den Heuvel W-J (eds) Proceedings 2012 IEEE 16th international enterprise distributed object computing conference. IEEE, pp 61–70

41. Birkedal L, Damgaard TC, Glenstrup AJ, Milner R (2007) Matching of bigraphs. Electron Notes Theor Comput Sci 175(4):3–19

42. Biswas P, Venkataramani G (2008) Comprehensive isomorphic subtree enumeration. In: Altman E (eds) Proceedings 2008 international conference compilers, architectures and synthesis for embedded systems. ACM, pp 177–186

43. Blažek T (2020) *Interpret Petriho Sítí*. MSc thesis, Brno University of Technology, Czech Republic

44. Bloedorn E, Rothleder NJ, DeBarr D, Rosen L (2005) Relational graph analysis with real-world constraints: an application in IRS tax fraud detection. In: Mladenic D, Milic-Frayling N, Grobelnik M (eds) Proceedings 2005 AAAI workshop on link analysis. AAAI Press, pp 30–38

45. -/ Bonnin G (2006) Aide Syntaxique pour la Rraduction Automatique Franco-Allemande. MSc thesis, Montpellier 2 University, France

46. Bošnački D, Donaldson AF, Leuschel M, Massart T (2007) Efficient approximate verification of Promela models via symmetry markers. In: Namjoshi KS, Yoneda T, Higashino T, Okamura Y (eds) Proceedings 5th international symposium automated technology for verification and analysis, volume 4762 of Lecture notes in computer science. Springer, Berlin, pp 300–315

47. Brandel S, Schneider S, Perrin M, Guiard N, Rainaud J-F, Lienhard P, Bertrand Y (2005) Automatic building of structured geological models. J Comput Inf Sci Eng 5(2):138–148

48. Brimkov VE, Leach A, Mastroianni M, Wu J (2011) Guarding a set of line segments in the plane. Theor Comput Sci 412(15):1313–1324

49. Bringmann K, Gawrychowski P, Mozes S, Weimann O (2018) Tree edit distance cannot be computed in strongly subcubic time. In: Czumaj A (eds) Proceedings 2018 Annual ACM-SIAM symposium discrete algorithms. SIAM, pp 1190–1206

50. Bringmann K, Gawrychowski P, Mozes S, Weimann O (2020) Tree edit distance cannot be computed in strongly subcubic time. ACM Trans Algorithms 16(4):48:1–48:22

51. Bublík T, Virius M (2014) Scripting language for Java source code recognition. IERI Procedia 10(1):119–130

52. Bublík T, Virius M (2014) Tool for fast detection of Java code snippets. Int J Comput Inf Syst Control Eng 8(11):1973–1979

53. Bush BJ (2011) Solving The Shirokuro Puzzle Constraint Satisfaction Problem with Back-tracking: A Theoretical Foundation. MSc thesis, California State University, California CA

54. Bóna M, Sitharam M, Vince A (2011) Enumeration of viral capsid assembly pathways: Tree orbits under permutation group action. Bull Math Biol 73(4):726–753

55. Camillo MHM (2013) Avaliação de uma Metodologia para Restabelecimento de Energia baseada em Algoritmos Evolutivos Multi-Objetivos no Sistema de Distribuição de Energia da COPEL na Cidade de Londrina. MSc thesis, University of São Paulo, Brasil

56. Canthadai AM (2011) Shopping search and the semantic web. In: Proceedings 2011 IEEE consumer communications and networking conference. IEEE, pp 699–700

57. Cao J, Yao Y, Wang Y (2012) Mining change operations for workflow platform as a service. World Wide Web 18(4):1071–1092

58. Cardellini G (2018) Forest Products: Contribution to Carbon Storage and Climate Change Mitigation. PhD thesis, KU Leuven, Belgium

59. Cardellini G, Mutel CL, Vial E, Muys B (2018) Temporalis, a generic method and tool for dynamic life cycle assessment. Sci Total Environ 645(15):585–595

60. Cardona G, Llabrés M, Rosselló F, Valiente G (2008) On Nakhleh's metric for reduced phylogenetic networks. IEEE/ACM Trans Comput Biol Bioinf 6(4):629–638

61. Cardona G, Llabrés M, Rosselló F, Valiente G (2009) Metrics for phylogenetic networks I: Generalizations of the Robinson-Foulds metric. IEEE/ACM Trans Comput Biol Bioinf 6(1):46–61

62. Carlsson JG, Behroozi M, Li X (2016) Geometric partitioning and robust ad-hoc network design. Ann Oper Res 238(1–2):41–68

63. Cechová A (2010) Porovnávání Proteinů Reprezentovaných obecným Grafem. BSc thesis, Masaryk University, Czech Republic

64. Chan HY (2008) Graph-theoretic approach to the non-binary index assignment problem. PhD thesis, Hong Kong University of Science and Technology, Hong Kong

65. Chang HJ, Fischer T, Petit M, Zambelli M, Demiris Y (2016) Kinematic structure correspon-dences via hypergraph matching. In: Tuytelaars T, Li F-F, Bajcsy R (eds) Proceedings 2016 IEEE conference on computer vision and pattern recognition. IEEE, pp 4216–4225

66. Chang HJ, Fischer T, Petit M, Zambelli M, Demiris Y (2018) Learning kinematic struc-ture correspondences using multi-order similarities. IEEE Trans Pattern Anal Mach Intell 40(12):2920–2934

67. Chehreghani MH, Chehreghani MH, Lucas C, Rahgozar M, Ghadimi E (2009) Efficient rule based structural algorithms for classification of tree structured data. Intell Data Anal 13(1):165–188

68. Chehreghani MH, Rahgozar M, Lucas C, Chehreghani MH (2007) A heuristic algorithm for clustering rooted ordered trees. Intell Data Anal 11(4):355–376

69. Chen L, Bhowmick SS, Nejdl W (2009) NEAR-Miner: Mining evolution associations of web site directories for efficient maintenance of web archives. Proc VLDB Endowment 2(1):1150–1161

70. Chen S, Zhu F (2012) Traffic simulation using web information of activities location. In: Miller J, Zhang W-B, Wang Y (eds) Proceedings 2012 15th international IEEE conference intelligent transportation systems. IEEE, pp 758–763

71. Chi Y (2005) Mining Databases of Labeled Trees using Canonical Forms. PhD thesis, Uni-versity of California, Los Angeles, California CA

72. Chi Y, Nijsseny S, Muntz RR (2005) Canonical forms for labelled trees and their applications in frequent subtree mining. Knowl Inf Syst 8(2):203–234

73. Chi Y, Nijsseny S, Muntz RR, Kok JN (2005) Frequent subtree mining: An overview. Funda-menta Informaticae 66(1–2):161–198

74. Chiang H-K, Chen H-W, Yang J-R (2008) The development and application of an automatic link analysis algorithm for social networks. J Inf Manag 15(3):158–180

75. Chien Y-C, Liu M-C, Wu T-T, Lai C-H, Huang Y-M (2016) Enriching search queries to construct comprehensive concept maps for online inquiries: a case study of a food web. J Internet Technol 17(1):19–27

76. Chowdhury AS, Bhandarkar SM (2011) Graph-theoretic foundations. In: Chowdhury AS, Bhandarkar SM (eds) Computer vision-guided virtual craniofacial surgery: a graph-theoretic and statistical perspective, Advances in computer vision and pattern recognition. Springer, Berlin, pp 15–23

77. Chowdhury AS, Bhandarkar SM, Robinson RW, Yu JC (2007) Novel graph theoretic enhancements to ICP-based virtual craniofacial reconstruction. In: Fessler J, Denney T (eds) Proceedings 2007 4th IEEE international symposium biomedical imaging. IEEE, pp 1136–1139

78. Chowdhury AS, Bhandarkar SM, Robinson RW, Yu JC (2009) Virtual craniofacial reconstruction using computer vision, graph theory and geometric constraints. Pattern Recognit Lett 30(10):931–938

79. Chowdhury IJ, Nayak R (2013) A novel method for finding similarities between unordered trees using matrix data model. In: Lin X, Manolopoulos Y, Srivastava D, Huang G (eds) Proceedings 14th international conference web information systems engineering, volume 8180 of Lecture notes in computer science. Springer, Berlin, pp 421–430

80. Chowdhury IJ, Nayak R (2014) BEST: an efficient algorithm for mining frequent unordered embedded subtrees. In: Pham D-N, Park S-B (eds) Proceedings 13th pacific rim international conference artificial intelligence, volume 8862 of Lecture notes in computer science. Springer, Berlin, pp 459–471

81. Chowdhury IJ, Nayak R (2014) BOSTER: An efficient algorithm for mining frequent unordered induced subtrees. In: Benatallah B, Bestavros A, Manolopoulos Y, Vakali A, Zhang Y (eds) Proceedings 15th international conference web information systems engineering, volume 8786 of Lecture notes in computer science. Springer, Berlin, pp 146–155

82. Chowdhury IJ, Nayak R (2015) FreeS: A fast algorithm to discover frequent free subtrees using a novel canonical form. In: Wang J, Cellary W, Wang D, Wang H, Chen S-C, Li T, Zhang Y (eds) Proceedings 16th international conference web information systems engineering, volume 9418 of Lecture notes in computer science. Springer, Berlin, pp 123–137

83. Cibej U, Mihelič J (2014) Search strategies for subgraph isomorphism algorithms. In: Gupta P, Zaroliagis C (eds) Proceedings 1st international conference applied algorithms, volume 8321 of Lecture notes in computer science. Springer, Berlin, pp 77–88

84. Cibej U, Mihelič J (2015) Improvements to Ullmann's algorithm for the subgraph isomorphism problem. Int J Pattern Recognit Artif Intell 29(7):1550025:1–1550025:26

85. Clemente JC (2007) Comparative Analysis of Metabolic Pathways. PhD thesis, Japan Advanced Institute of Science and Technology, Japan

86. Clemente JC, Jansson J, Valiente G (2010) Accurate taxonomic assignment of short pyrosequencing reads. In: Altman RB, Dunker AK, Hunter L, Murray TA, Klein TE (eds) Proceedings pacific symposium on biocomputing 2010. World Scientific, pp 3–9

87. Clemente JC, Jansson J, Valiente G (2011) Flexible taxonomic assignment of ambiguous sequencing reads. BMC Bioinf 12:8

88. Correia P, Paquete L, Figueira JR (2021) Finding multi-objective supported efficient spanning trees. Comput Optim Appl 78(2):491–528

89. Czarnecki K, Wasowski A (2007) Feature diagrams and logics: there and back again. In: Kang KC, Kishi T, Muthig D (eds) Proceedings 11th international software product line conference. IEEE, pp 23–32

90. Damgaard TC (2008) Developing Bigraphical Languages. PhD thesis, IT University of Copenhagen, Denmark

91. Damgaard TC, Glenstrup AJ, Birkedal L, Milner R (2013) An inductive characterization of matching in binding bigraphs. Formal Aspects Comput 25(2):257–288

92. Damon J (2007) Tree structure for contractible regions in $\mathbb{R}^3$. Int J Comput Vis 74(2):103–116

93. de Silva E, Stumpf MPH (2005) Complex networks and simple models in biology. J R Soc Interface 2(5):419–430

94. del Razo Lopez F (2007) Recherche de Sous-Structures Arborescentes Ordonnées Fréquentes au sein de Bases de Données semi-structurées. PhD thesis, Montpellier 2 University, France

95. Demaine ED, Mozes S, Rossman B, Weimann O (2007) An optimal decomposition algorithm for tree edit distance. In: Arge L, Cachin C, Jurdziński T, Tarlecki A (eds) Proceedings 34th international colloquium automata, languages, and programming, volume 4596 of Lecture notes in computer science. Springer, Berlin, pp 146–157

96. Demaine ED, Mozes S, Rossman B, Weimann O (2009) An optimal decomposition algorithm for tree edit distance. ACM Trans Algorithms 6(1):2:1–2:19

97. Djelloul M (2009) Algorithmes de Graphes pour la Recherche de Motifs Récurrents dans les Structures Tertiaires d'ARN. PhD thesis, Paris-Sud University, France

98. Djelloul M, Denise A (2008) Automated motif extraction and classification in RNA tertiary structures. RNA 14(12):2489–2497

99. Djelloul M, Denise A (2008) A method for automated discovering of RNA tertiary motifs. In Proc. 2008 Journées Ouvertes Biologie, Informatique et Mathématiques. SFBI, pp 47–52

100. Dondelinger F (2013) A Machine Learning Approach to Reconstructing Signalling Pathways and Interaction Networks in Biology. PhD thesis, University of Edinburgh, Scotland

101. Droschinsky A (2014) Effiziente Enumerationsalgorithmen für Common Subtree Probleme. MSc thesis, Technische Universität Dortmund, Germany

102. Droschinsky A, Kriege NM, Mutzel P (2016) Faster algorithms for the maximum common subtree isomorphism problem. In: Faliszewski P, Muscholl A, Niedermeier R (eds) Proceedings 41st international symposium mathematical foundations of computer science, volume 58 of Leibniz international proceedings in informatics, pp 33:1–33:14

103. Droschinsky A, Kriege NM, Mutzel P (2018) Largest weight common subtree embeddings with distance penalties. In: Potapov I, Spirakis P, Worrell J (eds) Proceedings 43rd international symposium mathematical foundations of computer science, volume 117 of Leibniz international proceedings in informatics, pp 54:1–54:15

104. Dubois J-E (2002) Chemical complexity and molecular topology: The DARC concepts and applications. In: Rayward WB, Bowden ME (eds) Proceedings conference on the history and heritage of science information systems. American Society for Information Science and Technology, pp 149–167

105. Dubois J-E (2008) Chemical complexity and molecular topology: The DARC concepts and applications. L'Actualité Chimique 320–321(1):37–42

106. Dulucq S, Touzet H (2005) Decomposition algorithms for the tree edit distance problem. J Discret Algorithms 3(2–4):448–471

107. Dvovák O, Pergl R, Kroha P (2017) Tackling the flexibility-usability trade-off in component-based software development. In: Rocha A, Correia AM, Adeli H, Reis LP, Costanzo S (eds) Recent advances in information systems and technologies, volume 569 of advances in intelligent systems and computing. Springer, Berlin, pp 861–871

108. Dyer M, Jerrum M, Müller H, Vuškovič K (2021) Counting weighted independent sets beyond the permanent. SIAM J Discret Math. In press

109. Escardó M, Oliva P (2010) Selection functions, bar recursion and backward induction. Math Struct Comput Sci 20(2):127–168

110. Fadlallah BH (2012) Representing and Matching Multi-Object Images With Holes Using Concavity Trees. PhD thesis, American University of Beirut, Lebanon

111. Faisal A, Dondelinger F, Husmeier D, Beale CM (2010) Inferring species interaction networks from species abundance data: A comparative evaluation of various statistical and machine learning methods. Ecolo Inf 5(6):451–464

112. Fathalla R (2017) Holistic Interpretation of Visual Data based on Topology: Semantic Segmentation of Architectural Facades. PhD thesis, Aston University, England

113. Fellini S, Salizzoni P, Soulhac L, Ridolfi L (2019) Propagation of toxic substances in the urban atmosphere: A complex network perspective. Atmos Environ 198(1):291–301

114. Feuillâtre H (2016) Détermination Automatique de l'Incidence Optimale pour l'Observation des Lésions Coronaires en Imagerie Rotationnelle R-X. PhD thesis, University of Rennes 1, France

115. Fischer I, Meinl T (2006) Subgraph mining. In: Wang J (ed) Encyclopedia of Data Warehousing and Mining, Chapter 199. Idea Group, Inc., 1st edn, pp 1059–1063

116. Fischer I, Meinl T (2009) Subgraph mining. In: Wang J (ed) Encyclopedia of Data Warehousing and Mining, Chap 285. Idea Group, Inc., 2nd edn, pp 1865–1870

117. Flores-Lamas A, Fernández-Zepeda JA, Trejo-Sánchez JA (2020) A distributed algorithm for a maximal 2-packing set in Halin graphs. J Parallel Distributed Comput 142:62–76

118. Fluri B (2008) Change Distilling: Enriching Software Evolution Analysis with Fine-Grained Source Code Change Histories. PhD thesis, University of Zurich, Switzerland

119. Fluri B, Gall HC (2006) Classifying change types for qualifying change couplings. In: Kontogiannis K, Ebert J, Linos P (eds) Proceedings 14th IEEE international conference on program comprehension. IEEE, pp 35–45

120. Fluri B, Wursch M, PInzger M, Gall H (2007) Change distilling: Tree differencing for fine-grained source code change extraction. IEEE Trans Softw Eng 33(11):725–743

121. Fukagawa D, Akutsu T (2004) Fast algorithms for comparison of similar unordered trees. In: Fleischer R, Trippen G (eds) Proceedings 15th international symposium algorithms and computation, volume 3341 of Lecture notes in computer science. Springer, Berlin, pp 452–463

122. Fukagawa D, Akutsu T (2006) Fast algorithms for comparison of similar unordered trees. Int J Found Comput Sci 17(3):703–729

123. Gadde R, Marlet R, Paragios N (2016) Learning grammars for architecture-specific facade parsing. Int J Comput Vis 117(3):290–316

124. Gadkar A (2010) Time Slotted Optical Networks: Architectures and Performance Evaluation. PhD thesis, George Washington University, Washington DC

125. Gadkar A, Subramaniam S (2010) Connection scheduling in wavelength-constrained optical time-slotted networks. In: Proceedings 2010 IEEE international conference communications. IEEE, pp 1–5

126. Gadkar A, Subramaniam S (2010) Wavelength-reuse in optical time-slotted networks. Opt Switch Netw 7(4):153–164

127. Gaidon A (2012) Structured Models for Action Recognition in Real-World Videos. PhD thesis, Université de Grenoble, France

128. García Zapata JL, Rico–Gallego J (2020) Classification of maps on a finite set under permutation. Math Methods Appl Sci. In press

129. Gensane T (2009) Generation of optimal packings from optimal packings. Electron J Combinator 19(1):1–15

130. Gençer M (2020) Structural analysis with graphs. In: Applied Social Network Analysis With R: Emerging Research and Opportunities, Advances in Computer and Electrical Engineering, chapter 2. IGI Global, pp 31–47

131. Giro S, Frydman C (2008) Simulation of Petri nets with multiple instances using DEVS. In: Lamouri S, Thomas A, Artiba A, Vernadat F (eds) Proceedings 7è Conference International Modélisation et Simulation, pp 306–315

132. Giunchiglia F, McNeill F, Yatskevich M, Pane J, Besana P, Shvaiko P (2008) Approximate structure-preserving semantic matching. In: Meersman R, Tari Z (eds) Proceedings OTM 2008 confederated international conference on the move to meaningful internet systems, volume 5332 of Lecture notes in computer science. Springer, Berlin, pp 1217–1234

133. Giunchiglia F, Yatskevich M, McNeill F (2007) Structure preserving semantic matching. In: Shvaiko P, Euzenat J, Giunchiglia F, He B (eds) Proceedings 2nd international workshop on ontology matching, volume 304. CEUR Workshop Proceedings, pp 13–24

134. Glockner M (2007) Methoden zur Analyse von Rückwärtskompatibilität von Steuergeräten. PhD thesis, Chemnitz University of Technology, Germany

135. Graham MW (2008) Robust Methods for Human Airway-Tree Segmentation and Anatomical-Tree Matching. PhD thesis, The Pennsylvania State University, Pennsylvania PE

136. Grzelak D, Aßmann U (2021) A canonical string encoding for pure bigraphs. SN Comput Sci 2(4):246

137. Gurin D, Prokhorchenko A, Kravchenko M, Shapoval G (2020) Development of a method for modelling delay propagation in railway networks using epidemiological SIR models. Eastern-Euro J Enterprise Technol 6(3):6–13

138. A. Gutkin. *Towards Formal Structural Representation of Spoken Language: An Evolving Transformation System (ETS) Approach.* PhD thesis, University of Edinburgh, Scotland, 2005

139. Gutkin A, Gay DR (2005) Structural representation and matching of articulatory speech structures based on the evolving transformation system (ETS) formalism. In: Hofbaur M, Rinner B, Wotawa F (eds) Proceedings 19th international workshop on qualitative reasoning, pp 89–96

140. Hadzic F, Dillon TS, Chang E (2008) Knowledge analysis with tree patterns. In: Sprague RH (ed) Proceedings 41st annual hawaii international conference system sciences. IEEE, pp 1–10

141. Hadzic F, Tan H, Dillon TS (2007) Uni3: Efficient algorithm for mining unordered induced subtrees using TMG candidate generation. In: Proceedings 2007 IEEE symposium computational intelligence and data mining. IEEE, pp 568–575

142. Hadzic F, Tan H, Dillon TS (2008) Mining unordered distance-constrained embedded subtrees. In: Jean-Fran J-F, Berthold MR, Horváth T (eds) Proceedings 11th international conference discovery science, volume 5255 of Lecture notes in computer science. Springer, Berlin pp 272–283

143. Hadzic F, Tan H, Dillon TS (2008) U3: Mining unordered embedded subtrees using TMG candidate generation. In: Li Y, Pasi G, Zhang C, Cercone N, Cao L (eds) Proceedings 2008 IEEE/WIC/ACM international conference web intelligence and intelligent agent technology. IEEE, pp 285–292

144. Hadzic F, Tan H, Dillon TS (2010) Model guided algorithm for mining unordered embedded subtrees. Web Intell Agent Syst 8(4):413–430

145. Hadzic F, Tan H, Dillon TS (2011) Tree mining problem. In: Hadzic F, Tan H, Dillon TS (eds) Mining of data with complex structures, volume 333 of studies in computational intelligence. Springer, Berlin, pp 23–40

146. Haghighi M (2007) Graph-Theoretic Analysis of Gene Expression Networks. MSc thesis, University of Ottawa, Canada

147. Hamilton JAG (2008) Static Source Code Analysis Tools and their Application to the Detection of Plagiarism in Java Programs. MSc thesis, Goldsmiths, University of London, England

148. Han S-E (2005) On the simplicial complex stemmed from a digital graph. Honam Math J 27(1):115–129

149. Han S-E (2007) Digital fundamental group and Euler characteristic of a connected sum of digital closed surfaces. Inf Sci 177(16):3314–3326

150. Hanzo L, Akhtman Y, Wang L, Jiang M (2011) MIMO-OFDM for LTE. Coherent versus Non-coherent and Cooperative Turbo-transceivers. Wiley, Wi-Fi and WiMAX

151. Hessellund A, Wasowski A (2008) Interfaces and metainterfaces for models and metamodels. In: Czarnecki K, Ober I, Bruel J-M, Uhl A, Völter M (eds) Proceedings international conference model driven engineering languages and systems, volume 5301 of Lecture notes in computer science. Springer, Berlin, pp 401–415

152. Hokazono T, Kan T, Yamamoto Y, Hirata K (2012) An isolated-subtree inclusion for unordered trees. In: Proceedings 2012 IIAI international conference advanced applied informatics. IEEE, pp 345–350

153. Hollenstein S (2005) XQuery Similarity Joins. MSc thesis, University of Zurich, Switzerland

154. Holzer JT (2014) Optimization and Equilibrium Methods in Power Systems. PhD thesis, University of Wisconsin, Wisconsin WI

155. Hourai Y, Nishida A, Oyanagi Y (2004) Optimal broadcast scheduling on tree-structured networks. J Inf Processing 45(SIG03(ACS5)):100–108

156. Hsieh S-K (2006) Hanzi, Concept and Computation: A Preliminary Survey of Chinese Characters as a Knowledge Resource in NLP. PhD thesis, Universität Tübingen, Germany

157. Hsieh S-Y (2007) Finding maximal leaf-agreement isomorphic descendent subtrees from phylogenetic trees with different species. Theor Comput Sci 370(1–3):299–308

158. Hsieh S-Y, Huang C-W (2007) An efficient strategy for generating all descendant subtree patterns from phylogenetic trees with its implementation. Appl Math Comput 193(2):408–418

159. Huangjun Z (2012) Quantum State Estimation and Symmetric Informationally Complete POMs. PhD thesis, National University of Singapore, Singapore

160. Huth M, Fabian B (2016) Inferring business relationships in the internet backbone. Int J Netw Virtual Organ 16(4):315–345

161. Ibrahim A, Fletcher GHL (2014) Efficient processing of containment queries on nested sets. In: Guerrini G, Paton NW (eds) Proceedings 16th international conference extending database technology. ACM, pp 227–238

162. Immanuel S, Haran AP (2014) Big data streaming using adaptive machine learning and mining algorithms. Australian J Basic Appl Sci 8(13):211–217

163. Jabr RA (2012) Exploiting sparsity in SDP relaxations of the OPF problem. IEEE Trans Power Syst 27(2):1138–1139

164. Jahn K, Beerenwinkel N, Zhang L (2020) The Bourque distances for mutation trees of cancer. In: Proceedings 20th international workshop on algorithms in bioinformatics, volume 172 of Leibniz international proceedings in informatics, 14:1–14:23

165. Jahn K, Beerenwinkel N, Zhang L (2021) The Bourque distances for mutation trees of cancer. Algorithms Molecular Biol 16(1):9

166. Jamil HM (2009) An efficient logic programming approach to sub-graph isomorphic queries. In: Proceedings 18th international conference applications of declarative programming and knowledge management, pp 83–96

167. Jamil HM (2009) A novel knowledge representation framework for computing sub-graph isomorphic queries in interaction network databases. In: Chung SM, Ziavras SG (eds) Proceedings 2009 21st IEEE international conference tools with artificial intelligence. IEEE, pp 131–138

168. Jamil HM (2011) Computing subgraph isomorphic queries using structural unification and minimum graph structures. In: Chu W, Wong WE, Palakal MJ, Hung C-C (eds) Proceedings 2011 ACM symposium applied computing. ACM, pp 1053–1058

169. Jamil HM (2011) Integrating large and distributed life sciences resources for systems biology research: progress and new challenges. In: Hameurlain A, Küng J, Wagner R (eds) Transactions on large-scale data- and knowledge-centered systems III, volume 6790 of lecture notes in computer science. Springer, Berlin, pp 208–237

170. Jamil HM (2012) Design of declarative graph query languages: On the choice between value, pattern and object based representations for graphs. In: Wang XS, Gehrke J, Ooi BC, Pitoura E (eds) Proceedings 2012 IEEE 28th international conference data engineering workshops. IEEE, pp 178–185

171. Jiang C (2011) Frequent Subgraph Mining Algorithms on Weighted Graphs. PhD thesis, University of Liverpool, England

172. Jiang C, Coenen F, Zito M (2012) A survey of frequent subgraph mining algorithms. Knowl Eng Rev 28(1):75–105

173. Jiao J, Zhang L (2008) Data mining applications of process platform formation for high variety production. In: Liao TW, Triantaphyllou E (eds) Recent advances in data mining of enterprise data: algorithms and applications. World Scientific, pp 247–286

174. Jiao J, Zhang L (2009) Process platform formation for product families. In: Proceedings 33rd Design Automation Conference. ASME, pp 1049–1057
175. Jiao J, Zhang L, Pokharel S (2006) Process platform and production configuration for product families. In: Simpson TW, Siddique Z, Jiao J (eds) Product Platform and Product Family Design, Chapter 16. Springer, Berlin, pp 377–402
176. Jiao J, Zhang L, Pokharel S (2006) Process platform and production configuration for product families. In: Simpson TW, Siddique Z, Jiao JR (eds) Product platform and product family design: methods and applications. Springer, Berlin, pp 377–402
177. Jiao J, Zhang L, Pokharel S, He Z (2007) Identifying generic routings for product families based on text mining and tree matching. Decision Support Syst 43(3):866–883
178. Kan T, Higuchi S, Hirata K (2014) Segmental mapping and distance for rooted labeled ordered trees. Fundamenta Informaticae 132(4):461–483
179. Kasper W, Steffen J, Zhang Y (2008) News annotations for navigation by semantic similarity. In: Dengel AR, Berns K, Breuel TM, Bomarius F, Roth-Berghofer TR (eds) Proceedings 31st annual german conf. artificial intelligence, volume 5243 of Lecture notes in computer science. Springer, Berlin, pp 233–240
180. Kelenc A (2019) Distance-Based Invariants and Measures in Graphs. PhD thesis, University of Maribor, Slovenia
181. Kelenc A, Taranenko A (2015) On the Hausdorff distance between some families of chemical graphs. MATCH Commun Math Comput Chem 74(2):223–246
182. Khan MS (2018) Malvidence: A Cognitive Malware Characterization Framework. PhD thesis, University of Manitoba, Canada
183. Kiefer C (2008) Non-Deductive Reasoning for the Semantic Web and Software Analysis. PhD thesis, University of Zurich, Switzerland
184. Kiefer C, Bernstein A (2007) Analyzing software with iSPARQL. In: Proceedings 3rd international workshop on semantic web enabled software engineering, pp 1–15
185. Kiefer C, Bernstein A (2011) Application and evaluation of inductive reasoning methods for the semantic web and software analysis. In: Polleres A, d'Amato C, Arenas M, Handschuh S, Kroner P, Ossowski S, Patel-Schneider P (eds) Proceedings 7th international summer school reasoning web, volume 6848 of Lecture notes in computer science. Springer, Berlin, pp 460–503
186. Kiefer C, Bernstein A, Locher A (2008) Adding data mining support to SPARQL via statistical relational learning methods. In: Bechhofer S, Hauswirth M, Hoffmann J, Koubarakis M (eds) Proceedings 5th European semantic web conference, volume 5021 of Lecture notes in computer science. Springer, Berlin, pp 478–492
187. Kiefer C, Bernstein A, Tappolet J (2007) Mining software repositories with iSPAROL and a software evolution ontology. In: Proceedings 4th international workshop on mining software repositories. IEEE, pp 1–8
188. Kim I (2011) Solutions to Decision-Making Problems in Management Engineering Using Molecular Computational Algorithms and Experimentations. PhD thesis, Waseda University, Japan
189. Kou XY, Tan ST (2007) A systematic approach for integrated computer-aided design and finite element analysis of functionally-graded-material objects. Mater Des 28(10):2549–2565
190. Koščák J (2014) Analýza Záznamů Řešení Matematických Úloh. MSc thesis, Masaryk University, Czech Republic
191. Kriege N (2009) Erweiterte Substruktursuche in Moleküldatenbanken und ihre Integration in Scaffold Hunter. MSc thesis, TU Dortmund University, Germany
192. Krishnan GP (2014) Improving the Unification of Software Clones using Tree and Graph Matching Algorithms. MSc thesis, Concordia University, Canada
193. Krishnan GP, Tsantalis N (2013) Refactoring clones: An optimization problem. In: Serebrenik A, Mens T, Guéhéneuc Y-G (eds) Proceedings 2013 IEEE international conference software maintenance. IEEE, pp 360–363

194. Krishnan GP, Tsantalis N (2014) Unification and refactoring of clones. In: Demeyer S, Binkley D, Ricca F (eds) Proceedings 2014 IEEE conference software maintenance, reengineering, and reverse engineering. IEEE, pp 104–113

195. Kristensen TG (2011) Virtual Screening Algorithms. PhD thesis, Aarhus University, Denmark

196. Krpec O (2015) Rozpoznání Plagiátů Zdrojového Kódu v Jazyce PHP. BSc thesis, Brno University of Technology, Czech Republic

197. Kuboyama T (2007) Matching and Learning in Trees. PhD thesis, The University of Tokyo, Japan

198. Kuczenski B, Mutel C, Srocka M, Scanlon K, Ingwersen W (2021) Prototypes for automating product system model assembly. Int J Life Cycle Assessment

199. Labarre A (2010) Combinatorial Aspects of Genome Rearrangements and Haplotype Networks. PhD thesis, Free University of Brussels, France

200. Lachowski R, Pellenz ME, Penna MC, Jamhour E, Souza RD (2015) An efficient distributed algorithm for constructing spanning trees in wireless sensor networks. Sensors 15(1):1518–1536

201. Lahlou TA (2016) Decentralized Signal Processing Systems With Conservation Principles. PhD thesis, Massachusetts Institute of Technology, Massachusetts MA

202. Lakkaraju P, Evalyn S, Speretta M (2008) Document similarity based on concept tree distance. In: Brusilovsky P, Davis H (eds) Proceedings 19th ACM Conf. Hypertext and Hypermedia. ACM, pp 127–132

203. Lamas AF (2018) Estudio de Factibilidad para Resolver el Problema 2-Packing Máximo en Tiempo Polinomial en Grafos Outerplanares. PhD thesis, CICESE, Mexico

204. Lapointe A (2008) Issues in Performance Evaluation of Mathematical Notation Recognition Systems. MSc thesis, Queen's University, Canada

205. Lee G (2004) Verification of Graph Algorithms in Mizar. MSc thesis, University of Alberta, Canada

206. Lee H-J, Klein M (2008) Semantic process retrieval with similarity algorithms. Asia Pac J Inf Syst 18(1):79–96

207. Lee M-S, Lee Y-P, Song I-H, Yoon S-H (2011) A novel decoding scheme for MIMO signals using combined depth- and breadth-first search and tree partitioning. J Korean Inst Commun Inf Sci 36(1C):37–47

208. Lee M-S, Song I-H, Lee Y-P, Yoon S-H (2010) A novel ML decoding scheme for MIMO signals. In: Latifi S (ed) Proceedings 2010 7th international conference information technology. IEEE, pp 867–872

209. LeTourneau JJ (2009) Method and/or system for manipulating tree expressions. US Patent 7.627,591

210. LeTourneau JJ (2010) Method and/or system for tagging trees. US Patent 7,801,923

211. LeTourneau JJ (2010) Method and/or system for transforming between trees and strings. US Patent 7,681,177

212. LeTourneau JJ (2011) File location naming hierarchy. US Patent 7.882,147

213. LeTourneau JJ (2011) Manipulating sets of hierarchical data. US Patent 8,037,102

214. LeTourneau JJ (2013) Method and/or system for transforming between trees and arrays. US Patent 8,356,040

215. LeTourneau JJ (2013) Method and/or system for tree transformation. US Patent 8,615,530

216. LeTourneau JJ (2017) Method and/or system for simplifying tree expressions such as for query reduction. US Patent 9,646,107

217. LeTourneau JJ (2019) Method and/or system for transmitting and/or receiving data. US Patent 10,411,878

218. LeTourneau JJ (2020) Method and/or system for simplifying tree expressions, such as for pattern matching. US Patent 10,733,234

219. LeTourneau JJ (2020) Method and/or system for transforming between trees and strings. US Patent 10,713,274

220. Levin MS (2015) Aggregation of structured solutions. In: Modular system design and evaluation, decision engineering, Chapter 9. Springer, Berlin, pp 191–246

221. Li Q (2010) An Algorithm for Finding Optimal Descent Trees in Genealogies Conditional on the Observed Data. MSc thesis, Memorial University, Canada

222. Li Q-Q, Yang A-M (2007) Research on graphical annotation and retrieval of image semantic. In: Proceedings 6th international conference machine learning and cybernetics. IEEE, pp 1565–1569

223. Ligaarden OS (2007) Detection of Plagiarism in Computer Programming Using Abstract Syntax Trees. MSc thesis, University of Oslo, Norway

224. Lin K, Chen D, Sun C, Dromey G (2005) Maintaining constraints in collaborative graphic systems: The CoGSE approach. In: Gellersen H, Schmidt K, Beaudouin-Lafon M, Mackay W (eds) Proceedings 9th European conference computer-supported cooperative work. Springer, Berlin, pp 185–204

225. Lin K, Chen D, Sun C, Dromey RG (2004) Tree structure maintenance in a collaborative genetic software engineering system. In: Proceedings 6th international workshop on collaborative editing systems, Chicago, IEEE Computer Society Press

226. Lindow N, Baum D, Hege H-C (2018) Atomic accessibility radii for molecular dynamics analysis. In: Byska J, Krone M, Sommer B (eds) Proceedings workshop on molecular graphics and visual analysis of molecular data. The Eurographics Association, pp 9–17

227. Linse B (2010) Data Integration on the (Semantic) Web with Rules and Rich Unification. PhD thesis, Ludwig Maximilian University of Munich, Germany

228. Liu M-C, Kinshuk, Huang Y-M, Wen D (2012) Boosting semantic relations for example population in concept learning. In: Graf S, Lin F, Kinshuk, McGreal R (eds) Intelligent and adaptive learning systems: technology enhanced support for learners and teachers, Chapter 11. IGI Global, pp 165–181

229. Llabrés M, Rosselló F, Valiente G (2020) A generalized Robinson-Foulds distance for clonal trees, mutation trees, and phylogenetic trees and networks. In: Aluru S, Kalyanaraman A, Wang MD (eds) Proceedings 11th ACM international conference bioinformatics, computational biology and health informatics. New York, NY, ACM Press, pp 13:1–13:10

230. Locher A (2007) SPARQL-ML: Knowledge Discovery for the Semantic Web. MSc thesis, University of Zurich, Switzerland

231. ohrey M, Maneth S, Reh CP (2017) Compression of unordered XML trees. In: Benedikt M, Orsi G (eds) Proceedings 20th international conference database theory, volume 68 of Leibniz international proceedings in informatics, pp 18:1–18:17

232. Lozano A, Pinter RY, Rokhlenko O, Valiente G, Ziv-Ukelson M (2007) Seeded tree alignment and planar tanglegram layout. In: Giancarlo R, Hannenhalli S(ed) Proceedings 7th interntional workshop on algorithms in bioinformatics, volume 4645 of Lecture notes in computer science. Springer, Berlin, pp 98–110

233. Lozano A, Pinter RY, Rokhlenko O, Valiente G, Ziv-Ukelson M (2008) Seeded tree alignment. IEEE/ACM Trans Comput Biol Bioinf 5(4):503–513

234. Lozano A, Valiente G (2004) On the maximum common embedded subtree problem for ordered trees. In: Iliopoulos CS, Lecroq T (eds) String algorithmics, volume 2 of texts in algorithmics, Chapter 7. College Publications, pp 155–169

235. Luccio F, Pagli L, Enriquez AM, Rieumont PO (2007) Bottom-up subtree isomorphism for unordered labeled trees. Int J Pure Appl Math 38(3):325–343

236. Lux M (2009) An evaluation of metrics for retrieval of MPEG-7 semantic descriptions. In: Proceedings 2009 11th IEEE international symposium multimedia. IEEE, pp 546–551

237. Lux M (2012) How to search in MPEG-7 based semantic descriptions: an evaluation of metrics. Multimed Tools Appl 59(2):673–690

238. Lux M, Granitzer M (2005) A fast and simple path index based retrieval approach for graph based semantic descriptions. In: Proceedings 2nd international workshop on text-based information retrieval, pp 29–41

239. Lux M, zu Eissen SM, Granitzer M (2006) Graph retrieval with the suffix tree model. In: Stein B, Kao O (eds) Proceedings 3rd international workshop on text-based information retrieval, pp 30–34

240. Madhusudan T, Zhao JL, Marshall B (2004) A case-based reasoning framework for workflow model management. Data Knowl Eng 50(1):87–115

241. Magdevski P (2016) Algoritmi za Problem Izomorfizma Dreves. MSc thesis, University of Ljubljana, Slovenia

242. Marchese M, Shvaiko P, Vaccari L, Pane J (2008) An application of approximate ontology matching in eResponse. In: Fiedrich F, de Walle V (eds) Proceedings 5th international conference information systems for crisis response and management. ISCRAM, pp 294–304

243. Marchette DJ (2004) Random Graphs for Statistical Pattern Recognition. Wiley, New York

244. Martinez-Rodriguez DE, Contreras-Cruz MA, Hernandez-Belmonte U-H, Bereg S, Ayala-Ramírez V (2018) Saliency improvement through genetic programming. In: Balakrishnan P, McMahan RP (eds) Proceedings 3rd international workshop on interactive and spatial computing. ACM, pp 29–38

245. Mashia K (2019) The Role of Graph Theory in Big Data Analytics. MSc thesis, University of the Witwatersrand, Johannesburg, South Africa

246. Mazumdar S (2009) TREE-D-SEEK: A Framework for Retrieving Three-Dimensional Scenes. PhD thesis, Old Dominion University, Virginia VA

247. Melo LTC (2007) Uma Biblioteca para Desenho de Grafos Construída sob o Paradigma de Programação Genérica. MSc thesis, Federal University of Minas Gerais, Brasil

248. Metwalli SA, Gall FL, Meter RV (2020) Finding small and large k-clique instances on a quantum computer. IEEE Trans Quantum Eng 1(1):3102911

249. Mihelič J, Čbej U, Fürst L (2021) A backtracking algorithmic toolbox for solving the subgraph isomorphism problem. In: Handbook of research on methodologies and applications of supercomputing, advances in systems analysis, software engineering, and high performance computing, Chapter 14. IGI Global, pp 208–246

250. Mir A, Rosselló F, Rotger L (2013) A new balance index for phylogenetic trees. Math Biosci 241(1):125–136

251. Mlinarić D, Milašinović B, Mornar V (2020) Tree inheritance distance. IEEE Access 8(1):52489–52504

252. Mlinarić D, Mornar V, Milašinović B (2020) Generating trees for comparison. Computers 9(2):35

253. Moala JG (2005) On Being Stuck: A Preliminary Investigation into the Essence of Mathematical Problem Solving. MSc thesis, The University of Auckland, New Zealand

254. Möller A (2002) Eine virtuelle Maschine für Graphprogramme. MSc thesis, Carl von Ossietzky Universität Oldenburg, Germany

255. Molzahn DK (2013) Application of Semidefinite Optimization Techniques to Problems in Electric Power Systems. PhD thesis, University of Wisconsin-Madison, Wisconsin WI

256. Molzahn DK, Holzer JT, Lesieutre BC, DeMarco CL (2013) Implementation of a large-scale optimal power flow solver based on semidefinite programming. IEEE Trans Power Syst 28(4):3987–3998

257. Montani S, Leonardi G, Quaglini S, Baudi A (2013) Improving process model retrieval by accounting for gateway nodes: An ongoing work. In: Giordano L, Montani S, Dupré DT (eds) Proceedings workshop AI meets business processes 2013, volume 1101. CEUR Workshop Proceedings, pp 31–40

258. Montani S, Leonardi G, Quaglini S, Cavallini A, Micieli G (2014) Improving structural medical process comparison by exploiting domain knowledge and mined information. Artif Intell Med 62(1):33–45

259. Montani S, Leonardi G, Quaglini S, Cavallini A, Micieli G (2014) Knowledge-intensive medical process similarity. In: Miksch S, Riaño D, ten Teije A (eds) Proceedings 6th international workshop on knowledge representation for health care, volume 8903 of Lecture notes in computer science. Springer, Berlin, pp 1–13

260. Montani S, Leonardi G, Quaglini S, Cavallini A, Micieli G (2015) A knowledge-intensive approach to process similarity calculation. Exp Syst with Appl 42(9):4207–4215
261. Morales Alcántara LD (2019) Sistema Web para el Conteo Eficiente de Conjuntos Independientes en Estructuras Jerárquicas (Árboles). MSc thesis, Benemérita Universidad Autónoma de Puebla, Mexico
262. Morris C (2019) Learning with Graphs: Kernel and Neural Approaches. PhD thesis, TU Dortmund University, Germany
263. Mozes S (2008) Some Lower and Upper Bounds for Tree Edit Distance. MSc thesis, Brown University, Providence RI
264. Mu W, Bénaben F, Pingaud H (2015) A methodology proposal for collaborative business process elaboration using a model-driven approach. Enterprise Inf Syst 9(4):349–383
265. Müller-Hannemann M, Schirra S (eds) Algorithm engineering: bridging the gap between algorithm theory and practice, volume 5971 of Lecture notes in computer science. Springer, Berlin
266. Nan K, Yu J, Su H, Guo S, Zhang H, Xu K (2007) Towards structural web services matching based on kernel methods. Frontiers Comput Sci China 1(4):450–458
267. Nepomniaschaya A, Kokosinski Z (2004) Associative graph processor and its properties. In: Merker R, Fettweis G, Kelber J, Trinitis C (eds) Proceedings 4th international conference on parallel computing in electrical engineering. IEEE, pp 297–302
268. Nestrud MA (2011) A Graph Theoretic Approach to Food Combination Problems. PhD thesis, Cornell University, Ithaca NY
269. Nestrud MA, Ennis JM, Lawless HT (2012) A group level validation of the supercombinatorality property: Finding high-quality ingredient combinations using pairwise information. Food Quality Preference 25(1):23–28
270. Nikolaidis AI, Charalambous CA, Mancarella P (2019) A graph-based loss allocation framework for transactive energy markets in unbalanced radial distribution networks. IEEE Trans Power Syst 34(5):4109–4118
271. Oehm S (2009) Comparing Organisms on the Level of Metabolism. PhD thesis, Bielefeld University, Germany
272. Oh J, Song I, Park J, Jeong MA, Choi MS (2008) A hybrid ML decoding scheme for multiple input multiple output signals on partitioned tree. In: Sesay AB, Badawy W (eds) Proceedings 2008 IEEE 68th vehicular technology conference. IEEE, pp 1–5
273. Oliveira Filho JA (2010) Description and Specialization of Coarse-grained Reconfigurable Architectures. PhD thesis, Eberhard Karls University of Tübingen, Germany
274. Orihara H, Utsumi A (2008) Web document clustering using HTML tags. Inf Process Soc Jpn J 49(8):2910–2921
275. Page RDM (2007) Towards a taxonomically intelligent phylogenetic database. Nat Prec
276. Page RDM, Valiente G (2005) An edit script for taxonomic classifications. BMC Bioinf 6:208
277. Pande A, Gupta M, Tripathi AK (2010) Design pattern mining for GIS application using graph matching techniques. In: Proceedings 2010 3rd international conference computer science and information technology, volume 9. IEEE, pp 477–482
278. Park JI, Lee Y, Yoon S (2013) Partition-based hybrid decoding (PHD): a class of ML decoding schemes for MIMO signals based on tree partitioning and combined depth- and breadth-first search. J Appl Res Technol 11(2):213–224
279. Parthasarathy S, Tatikonda S, Ucar D (2010) A survey of graph mining techniques for biological datasets. In: Aggarwal CC, Wang H (eds) Managing and Mining Graph Data, volume 40 of Advances in Database Systems. Springer, Berlin, pp 547–580
280. Peng W, Krueger W, Grushin A, Carlos P, Manikonda V, Santos M (2009) Graph-based methods for the analysis of large-scale multiagent systems. In: Sierra C, Castelfranchi C, Decker KS, Sichman JS (eds) Proceedings 8th international conference autonomous agents and multiagent systems. ACM, pp 545–552

281. Pergl R, Tůma J (2006) OpenCASE: A tool for ontology-centred conceptual modelling. In: Bajec M, Eder J (eds) Proceedings 2012 international conference advanced information systems engineering, volume 112 of lecture notes in business information processing. Springer, Berlin, pp 511–518

282. Pettovello PM (2008) Thor: A Universal XML Index for Efficient XPath Query Processing. PhD thesis, Wayne State University, Detroit MI

283. Poszlovszki A (2010) Process Fragment Recognition and Emphasis. MSc thesis, University of Stuttgart, Germany

284. Prema P, Ramadoss B, Balasundaram SR (2013) Identification and deletion of duplicate subtrees in classification tree for test case reduction. Int J Inf Syst Change Manag 6(4):374–388

285. Raabe O, Wacker R, Oberle D, Baumann C, Funk C (2012) Recht ex machina: Formalisierung des Rechts im Internet der Dienste. Springer, Berlin

286. Radev R (2013) Representing a relational database as a directed graph and some applications. In: Georgiadis CK, Kefalas P, Stamatis D (eds) Proceedings 6th balkan conference informatics, volume 1036. CEUR Workshop Proceedings, pp 1–8

287. Ragni M, Wölfl S (2005) Branching Allen: Reasoning with intervals in branching time. In: Freksa C, Knauff M, Krieg-Brückner B, Nebel B, Barkowsky T (eds) Proceedings international conference spatial cognition 2014, volume 3343 of Lecture notes in computer science. Springer, Berlin, pp 323–343

288. Ramamurthy A (2019) A Reinforcement Learning Framework for the Automation of Engineering Decisions in Complex Systems. PhD thesis, Georgia Institute of Technology, Georgia GA

289. Rao RS (2019) A review on design pattern detection. Int J Eng Res Technol 8(11):756–762

290. Rattan G (2016) Some Geometrical and Vertex-Partitioning Techniques for Graph Isomorphism. PhD thesis, Homi Bhabha National Institute, India

291. Reis DDC (2004) Distância de Edição em Árvores: Algoritmos e Aplicações na Web. MSc thesis, Federal University of Minas Gerais, Brasil

292. Resmi NG, Soman KP (2014) Abstract syntax tree generation using modified grammar for source code plagiarism detection. Int J Comput Technol 1(6):319–326

293. Ripperda N (2004) Graphbasiertes Matching in räumlichen Datenbanken. MSc thesis, Universität Hannover, Germany

294. Romanowski CJ (2009) Graphical data mining. In: Wang J (ed) Encyclopedia of Data Warehousing and Mining, Chapter 147. Idea Group, Inc., 2nd edition, pp 950–956

295. Romanowski CJ, Nagi R (2005) On comparing bills of materials: a similarity/distance measure for unordered trees. IEEE Trans Syst Man Cybern 35(2):249–260

296. Rosselló F, Valiente G (2006) An algebraic view of the relation between largest common subtrees and smallest common supertrees. Theor Comput Sci 362(1–3):33–53

297. Ruckel N (2018) Toward Automated Resolution of Microservice Configuration. MSc thesis, Bauhaus-Universität Weimar, Germany

298. Sager T (2006) Coogle: A Code Google Eclipse Plug-in for Detecting Similar Java Classes. MSc thesis, University of Zurich, Switzerland

299. Sager T, Bernstein A, Pinzger M, Kiefer C (2006) Detecting similar Java classes using tree algorithms. In: Diehl S, Gall H, Hassan AE (eds) Proceedings international workshop on mining software repositories. ACM, pp 65–71

300. Sajdík O (2020) Interpret Petriho sítí. MSc thesis, Brno University of Technology, Czech Republic

301. Salas-Molina F, Rodriguez-Aguilar JA, Pla-Santamaria D, García-Bernabeu A (2021) On the formal foundations of cash management systems. Oper Res 21(2):1081–1095

302. Salecker E (2012) Test and Verification of Compiler Back Ends with a Cost-Benefit Analysis. PhD thesis, Berlin Institute of Technology, Germany

303. Salecker E, Glesner S (2010) Pairwise test set calculation using k-partite graphs. In: Proceedings 2010 IEEE international conerence software maintenance. IEEE, pp 1–5

304. Sarajlić A (2015) Analysing Directed Network Data. PhD thesis, Imperial College London, England

305. Sarajlić A, Pržulj N (2014) Survey of network-based approaches to research of cardiovascular diseases. BioMed Res Int 2014:527029

306. Scatena G (2007) Development of a Stochastic Simulator for Biological Systems Based on the Calculus of Looping Sequences. MSc thesis, University of Pisa, Italy

307. Schiffmann K, LeTourneau JJ, Andrews M (2009) Enumeration of trees from finite number of nodes. US Patent 7.636,727

308. Schiffmann K, LeTourneau JJ, Andrews M (2011) Manipulation and/or analysis of hierarchical data. US Patent 7,899,821

309. K. Schiffmann, J. J. LeTourneau, and M. Andrews. Enumeration of rooted partial subtrees, 2012. US Patent 8,316,059

310. Schink H (2018) Mastering Dependencies in Multi-Language Software Applications. PhD thesis, University of Magdeburg, Germany

311. Schirmer S (2011) Comparing Forests. PhD thesis, Bielefeld University, Germany

312. Schönauer S (2004) Efficient Similarity Search in Structured Data. PhD thesis, Ludwig-Maximilians-Universität München, Germany

313. Schulz H-J, Nocke T, Schumann H (2006) A framework for visual data mining of structures. In: Estivill-Castro V, Dobbie G (eds) Proceedings 29th Australasian computer science conference, volume 48 of international conference proceeding series. ACM, pp 157–166

314. Selvaag K (2013) Symbolic Differentiation of Multivariable Functions to Arbitrary Order. MSc thesis, Norwegian University of Science and Technology, Norway

315. Selzer A (2021) Lightweight Integration of Query Decomposition Techniques into SQL-based Database Systems MSc thesis, Technical University of Vienna, Austria

316. Shen G (2010) Lifting Transforms on Graphs: Theory and Applications. PhD thesis, University of Southern California, California CA

317. Shen G, Kim W-S, Ortega A, Lee J, Wey H (2010) Edge-aware intra prediction for depth-map coding. In: Siu W-C, Cham W-K, Pereira F (eds) Proceedings 2010 IEEE international conference image processing. IEEE, pp 3393–3396

318. Shen G, Ortega A (2010) Transform-based distributed data gathering. IEEE Trans Signal Process 58(7):3802–3815

319. Shrimali S, Teotia V (2016) Undirected zero-divisor graphs of commutative ring. J Chem Biol Phys Sci 6(2):498–503

320. Sidahao N, Constantinides GA, Cheung PYK (2005) A heuristic approach for multiple restricted multiplication. In: Fujii N, Ishii R, Trajković L (eds) Proceedings ieee international symposium circuits and systems. IEEE, pp 692–695

321. Skiena SS (2008) The Algorithm Design Manual, 2nd edn. Springer, Berlin

322. Skiena SS (2020) The Algorithm Design Manual, 3rd edn. Springer, Berlin

323. Skouradaki M (2017) Workload Mix Definition for Benchmarking BPMN 2.0 Workflow Management Systems. PhD thesis, University of Stuttgart, Germany

324. Skouradaki M, Andrikopoulos V, Kopp O, Leymann F (2016) RoSE: Reoccurring structures detection in BPMN 2.0 process model collections. In: Debruyne C, Panetto H, Meersman R, Dillon T, Kühn E, O'Sullivan D, Ardagna CA (eds) Proceedings OTM 2016 conference on the move to meaningful internet systems, volume 10033 of Lecture notes in computer science. Springer, Berlin, pp 263–281

325. Soares TWDLN(2009) Estruturas de Dados Eficientes para Algoritmos Evolutivos Aplicados a Projeto de Redes. PhD thesis, University of São Paulo, Brasil

326. Socorro Llanes R (2012) Optimización del Uso de Pivotes en Tareas de Búsqueda y Clasificación. PhD thesis, University of Alicante, Spain

327. Sodenkamp MA, Tavana M, Caprio DD (2016) Production, manufacturing and logistics modeling synergies in multi-criteria supplier selection and order allocation: An application to commodity trading. Euro J Oper Res 254(3):859–874

328. Soman SJ (2015) Detecting resemblances in anti-pattern ideologies using social networks. Indian J Sci Technol 8(24):1–5

329. Song H, Kodialam M, Hao F, Lakshman TV (2010) Building scalable virtual routers with trie braiding. In: Mandyam G, Westphal C (eds) Proceedings 2010 IEEE annual joint conference: INFOCOM, IEEE Computer and communications societies. IEEE, pp 1–9

330. Song H, Kodialam M, Hao F, Lakshman TV (2012) Efficient trie braiding in scalable virtual routers. IEEE/ACM Trans Netw 20(5):1489–1500

331. Sönmez O, Ünlü Y (2014) m-potent elements in order-preserving transformation semigroups and ordered trees. Commun Algebra 42(1):332–342

332. Sotiriou CP, Sketopoulos N, Nayak A, Penzes P (2019) Extraction of structural regularity for random logic netlists. In: Proceedings 2019 panhellenic conference electronics and telecommunications. IEEE, pp 1–7

333. Stocker M (2006) The Fundamentals of iSPARQL. MSc thesis, University of Zurich, Switzerland

334. Stoess M, Doherr F, Urbas L (2012) Automated network layout for the industrial communication engineering system NetGen:X. In: Jasperneite J, Sauter T, Nolte T, Willig A (eds) Proceedings 2012 9th IEEE international workshop on factory communication systems. IEEE, pp 281–290

335. Strange H (2011) Piecewise-Linear Manifold Learning. PhD thesis, Aberystwyth University, Wales

336. Strange H, Zwiggelaar R (2010) Iterative hyperplane merging: A framework for manifold learning. In: Labrosse F, Zwiggelaar R, Liu Y, Tiddeman B (eds) Proceedings British Machine Vision Conference. BMVA Press, pp 18.1–18.11

337. Strange H, Zwiggelaar R (2010) Parallel projections for manifold learning. In: Khoshgoftaar TM, Zhu X (eds) Proceedings 2010 9th international conference machine learning and applications. IEEE, pp 266–271

338. Strange H, Zwiggelaar R (2015) Piecewise-linear manifold learning: A heuristic approach to non-linear dimensionality reduction. Intell Data Anal 19(6):1213–1232

339. Stührenberg M (2012) Auszeichnungssprachen für linguistische Korpora: Theoretische Grundlagen, De-facto-Standards, Normen. PhD thesis, Bielefeld University, Germany

340. Szabó T (2011) Incrementalizing Static Analyses in Datalog. PhD thesis, Johannes Gutenberg University of Mainz, Germany

341. Takasu A (2006) An approximate multi-word matching algorithm for robust document retrieval. In: Yu PS, Tsotras V, Fox E, Liu B (eds) Proceedings 15th ACM international conference information and knowledge management. ACM, pp 34–42

342. Tang X (2013) Lifting the Abstraction Level of Compiler Transformations. PhD thesis, Texas A&M University, Texas TX

343. Tang X, Järvi J (2007) Concept-based optimization. In: Siek JG, Tip F (2007) Proceedings 2007 symposium library-centric software design. ACM, pp 97–108

344. Taoka S, Takafuji D, Watanabe T (2013) Computing-based performance analysis of approximation algorithms for the minimum weight vertex cover problem of graphs. IEICE Trans Funda Electron Commun Comput Sci E96.A(6):1331–1339

345. Taoka S, Watanabe T (2012) Performance comparison of approximation algorithms for the minimum weight vertex cover problem. In: Sunwoo MH, Kang S-M (eds) Proceedings 2012 ieee international symposium circuits and systems. IEEE, pp 632–635

346. Tappolet J (2012) Managing Temporal Graph Data While Preserving Semantics. PhD thesis, University of Zurich, Switzerland

347. Tappolet J, Kiefer C, Bernstein A (2010) Semantic web enabled software analysis. J Web Semantics 8(2–3):225–240

348. Tappolet J, Kiefer C, Bernstein A (2014) Semantic web enabled software analysis. In: Pan JZ, Zhao Y (2014) Semantic web enabled software engineering, volume 17 of Studies on the semantic web. IOS Press, pp 109–137

349. Teelen K (2011) Geometric Uncertainty Models for Correspondence Problems in Digital Image Processing. PhD thesis, Ghent University, Belgium

350. Tiarks R, Koschke R, Falke R (2011) An extended assessment of type-3 clones as detected by state-of-the-art tools. Softw Quality J 19(2):295–331

351. Tichanow D (2009) Berechnung einer Sequenz von Editieroperationen zwischen Codefragmenten eines Typ3 Klonpaares. MSc thesis, University of Bremen, Germany

352. Tinnefeld C, Wagner B, Plattner H (2012) Operating on hierarchical enterprise data in an in-memory column store. In: Proceedings 4th international conference advances in databases, knowledge, and data applications. IARIA, pp 58–63

353. Todorov K (2011) Ontology Matching by Combining Instance-Based Concept Similarity Measures with Structure. PhD thesis, University of Osnabrück, Germany

354. Todorov K, Geibel P (2008) Ontology mapping via structural and instance-based similarity measures. In: Shvaiko P, Euzenat J, Giunchiglia F, Stuckenschmidt H (eds) Proceedings 3rd international conference ontology matching, volume 431. CEUR Workshop Proceedings, pp 224–228

355. Tomáš M (2012) Systém pro Podporu Navrhování Optimální Struktury Softwarových Projektů. MSc thesis, Technical University of Liberec, Czech Republic

356. Torsello A, Hidović D, Pelillo M (2005) Polynomial-time metrics for attributed trees. IEEE Trans Pattern Anal Mach Intell 27(7):1087–1099

357. Touzet H (2005) A linear tree edit distance algorithm for similar ordered trees. In: Apostolico A, Crochemore M, Park K (eds) Proceedings 16th annals symposium combinatorial pattern matching, volume 3537 of Lecture notes in computer science. Springer, pp 334–345

358. Touzet H (2007) Comparing similar ordered trees in linear-time. J Discret Algorithms 5(4):696–705

359. Trinh D-T (2009) XML Functional Dependencies based on Tree Homomorphisms. PhD thesis, Technical University of Clausthal, Germany

360. Tsantalis N, Mazinanian D, Krishnan GP (2015) Assessing the refactorability of software clones. IEEE Trans Softw Eng 41(11):1055–1090

361. Tůma J (2013) Methods of automated model transformations in information system analysis. Procedia Technol 8(1):612–617

362. Uppu S, Hoang DB, Hintz T (2005) A novel exception handling scheme for out patient workflow in a wireless handheld hospital environment. In: Proceedings international symposium web services and applications. CSREA Press, pp 169–175

363. Vaccari L (2009) Integration of SDI Services: An evaluation of a distributed semantic matching framework. PhD thesis, University of Trento, Italy

364. Vaccari L, Shvaiko P, Marchese M (2009) A geo-service semantic integration in spatial data infrastructures. Int J Spatial Data Infrastruct Res 4(1):24–51

365. Vaccari L, Shvaiko P, Pane J, Besana P, Marchese M (2012) An evaluation of ontology matching in geo-service applications. GeoInformatica 16(1):31–66

366. Valiente G (2003) Constrained tree inclusion. In: Baeza-Yates R, Chávez E, Crochemore M (eds) Proceedinds 14th annals symposium combinatorial pattern matching, volume 2676 of Lecture notes in computer science. Springer, pp 361–371

367. Valiente G (2004) On the algorithm of Berztiss for tree pattern matching. In: Baeza-Yates R, Marroquín JL, Chávez E (eds) Proceedings 5th Mexican international conference computer science. IEEE, pp 43–49

368. Valiente G (2005) Constrained tree inclusion. J Discret Algorithms 3(2–4):431–447

369. Valiente G (2005) A fast algorithmic technique for comparing large phylogenetic trees. In: Consens M, Navarro G (eds) Proceedings 12th international conference string processing and information retrieval, volume 3772 of Lecture notes in computer science. Springer, pp 370–375

370. Valiente G (2006) Assessing clustering results with reference taxonomies. Genome Inf 17(2):131–140

371. Valiente G (2007) Efficient algorithms on trees and graphs with unique node labels. In: Kandel A, Bunke H, Last M (eds) Applied graph theory in computer vision and pattern recognition, volume 52 of Studies in computational intelligence. Springer, pp 137–149

372. Valiente G (2009) Combinatorial Pattern Matching Algorithms in Computational Biology using Perl and R. Taylor & Francis/CRC Press

373. Velichko E, Korikov C, Korobeynikov A, Grishentsev A, Fedosovsky M (2016) Information risk analysis for logistics systems. In: Galinina O, Balandin S, Koucheryavy Y (eds) Proceedings 16th international conference internet of things and smart spaces, volume 9870 of Lecture notes in computer science. Springer, pp 776–785

374. Venkataramani G, Biswas P (2013) Identification of resource sharing patterns through isomorphic subtree enumeration. US Patent 8,352.505

375. Wang J, Zhao Y, Tang Z, Xing Z (2020) Combining dynamic and static analysis for automated grading SQL statements. J Netw Intell 5(4):179–190

376. Watson C, Li FWB, Lau RWH (2011) Learning programming languages through corrective feedback and concept visualisation. In: Leung H, Popescu E, Cao Y, Lau RWH, Nejdl W (eds) Proceedings 10th international conference advances in web-based learning, volume 7048 of Lecture notes in computer science. Springer, Berlin, pp 11–20

377. Weimann O (2009) Accelerating Dynamic Programming. PhD thesis, Massachusetts Institute of Technology, Massachusetts MA

378. Welter D (2011) Investigating Gene Ontology-Based Semantic Similarity in the Context of Functional Genomics. PhD thesis, Cardiff University, Wales

379. Weskamp N (2006) Efficient Algorithms for Robust Pattern Mining on Structured Objects with Applications to Structure-Based Drug Design. PhD thesis, Philipps-University Marburg, Germany

380. Winter A, Simon C (2006) Using GXL for exchanging business process models. Inf Syst e-Bus Manag 4(3):285–307

381. Wu D, Lu J, Zhang G (2010) A hybrid recommendation approach for hierarchical items. In: Jin X, Liu Y, Li T, Ruan D (eds) Proceedings 2010 IEEE international conference intelligent systems and knowledge engineering. IEEE, pp 492–497

382. Wu D, Lu J, Zhang G (2011) Similarity measure models and algorithms for hierarchical cases. Exp Syst Appl 38(12):15049–15056

383. Wu D, Zhang G, Lu J (2013) A fuzzy tree similarity based recommendation approach for telecom products. In: Pedrycz W, Reformat MZ (eds) Proceedings 2013 joint IFSA world congress and NAFIPS annual meeting. IEEE, pp 813–818

384. Wu D, Zhang G, Lu J (2013) A fuzzy tree similarity measure and its application in telecom product recommendation. In: Lai LL, Yeung DS (eds) Proceedings 2013 IEEE international conference systems, man, and cybernetics. IEEE, pp 3483–3488

385. Wu D, Zhang G, Lu J (2014) A fuzzy tree matching-based personalised e-learning recommender system. In: Proceedings 2014 IEEE international conference fuzzy systems. IEEE, pp 1898–1904

386. Wu D, Zhang G, Lu J (2015) A fuzzy preference tree-based recommender system for personalized business-to-business e-services. IEEE Trans Fuzzy Syst 23(1):29–43

387. Wu D, Zhang G, Lu J, Halang WA (2012) A similarity measure on tree structured business data. In: Lamp J (ed) Proceedings 23rd Australasian conference information systems. ACIS, pp 1–10

388. Xu Z, Chen L, Xi F (2008) Towards rapid machine architecture design: module layout synthesis for desired machine tool configurations. In: Proceedings IDETC/CIE 2005 computers and information in engineering conference. ASME, pp 375–384

389. Z. Xu, F. Xi, L. Liu, and L. Chen. A method for design of modular reconfigurable machine tools. *Machines*, 5(1):5:1–5:16, 2017

390. Yoshino T, Hirata K (2015) Tai mapping hierarchy for rooted labeled trees relevant to common subforest. Jpn Soc Artif Intell SIG-FPAI B5(1):1–6

391. Yoshino T, Hirata K (2017) Tai mapping hierarchy for rooted labeled trees through common subforest. Theory Comput Syst 60(4):759–783

392. Yu J, Guo S, Su H, Zhang H, Xu K (2007) A kernel based structure matching for web services search. In: Williamson C, Zurko ME, Patel-Schneider P, Shenoy P (eds) Proceedings 16th international conference world wide web. ACM, pp 1249–1250

393. Zampelli S, Deville Y, Solnon C (2010) Solving subgraph isomorphism problems with constraint programming. Constraints 15(3):327–353

394. Zdárek J (2010) Two-Dimensional Pattern Matching Using Automata Approach. PhD thesis, Czech Technical University in Prague, Czech Republic

395. Zhang B (2006) Intelligent Fusion of Evidence from Multiple Sources for Text Classification. PhD thesis, Virginia Polytechnic Institute and State University, Virginia VA

396. Zhang LL (2007) Process Platform-Based Production Configuration for Mass Customization. PhD thesis, Nanyang Technological University, Singapore

397. Zhao B, Zhao Y, Ma D (2012) A constraint mechanism for dynamic evolution of service oriented systems. In: Proceedings 2012 IEEE 15th international symposium object/component/service-oriented real-time distributed computing. IEEE, pp 103–110

398. Zhao P, Yu JX (2008) Fast frequent free tree mining in graph databases. World Wide Web 11(1):71–92

399. Zhao Y, Zhao B, Liu M, Hu C, Ma D (2014) Towards a graph grammar based verification approach for runtime constrained evolution of service-oriented architectures. In: Xu B, Bai X, Li Y (eds) Proceedings 2010 5th IEEE international symposium service oriented system engineering. IEEE, pp 159–164

400. Zhu S (2013) Stochastic Tree Models and Probabilistic Modelling of Gene Trees of Given Species Networks. PhD thesis, University of Canterbury, New Zealand

Index

ed in the United States
ker & Taylor Publisher Services

Print
by Ba